S0-AKJ-354

Advances in

High Pressure Research

Volume 3

Advances in

High Pressure Research

Edited by
R. S. BRADLEY
School of Chemistry, The University, Leeds, England

Volume 3
1969

ACADEMIC PRESS INC. (LONDON) LTD
BERKELEY SQUARE HOUSE
BERKELEY SQUARE,
LONDON, W1X 6BA

U.S. Edition published by
ACADEMIC PRESS INC.
111 FIFTH AVENUE,
NEW YORK, NEW YORK 10003

Copyright © 1969 By Academic Press Inc. (London) Ltd

All Rights Reserved

NO PART OF THIS BOOK MAY BE REPRODUCED IN ANY FORM BY
PHOTOSTAT, MICROFILM, OR ANY OTHER MEANS, WITHOUT
WRITTEN PERMISSION FROM THE PUBLISHERS

Library of Congress Catalog Card Number: 65–27317

SBN 12–021203–X

QC281
A1A33
Chem

PRINTED IN GREAT BRITAIN BY
ADLARD & SON LTD
DORKING, SURREY

10/23/69

QC 281
A1A33
v.3
CHEMISTRY
LIBRARY

List of Contributors

D. BLOCH, *Laboratoire d'Electrostatique et de Physique du Métal, Grenoble, France.* (p. 41).

H. G. DRICKAMER, *Department of Chemistry and Chemical Engineering, University of Illinois, Urbana, Illinois, U.S.A.* (p. 1).

S. C. FUNG, *Materials Research Laboratory, University of Illinois, Urbana, Illinois, U.S.A.* (p. 1).

G. K. LEWIS, JR., *Department of Chemistry and Chemical Engineering, University of Illinois, Urbana, Illinois, U.S.A.* (p. 1).

N. H. MARCH, *Department of Physics, The University, Sheffield, England.* (p. 241).

P. G. MENON, *Department of Chemical Engineering, Technological University, Twente, Enschede, The Netherlands.* (p. 313).

A. S. PAVLOVIC, *West Virginia, Morgantown, West Virginia, U.S.A.* (p. 41).

G. A. SAMARA, *Sandia Laboratories, Albuquerque, New Mexico, U.S.A.* (p.155).

293

Preface

This is the last volume of the Advances in High Pressure Research which I shall be editing. Subsequent volumes will be edited by Dr. R. H. Wentorf Jr., General Electric Company, P.O. Box 8, Schenectedy, N.Y. 12301, U.S.A.

Leeds
June 1969

R. S. BRADLEY

Contents

CHAPTER 4

Electrical Conductivity and Electronic Transitions at High Pressures

N. H. MARCH

CHAPTER 5

Adsorption of Gases at High Pressures

P. G. MENON

CHAPTER 1

High Pressure Mössbauer Studies†

H. G. DRICKAMER, S. C. FUNG and G. K. LEWIS, JR.‡

Department of Chemistry and Chemical Engineering, and Materials Research Laboratory, University of Illinois, Urbana, Illinois, U.S.A.

I. INTRODUCTION

The existence of the Mössbauer Effect (that is recoilless resonant radiation) was first discovered in 1958 (Mössbauer, 1958a, b). The principles involved have been reviewed in detail in the literature (Frauenfelder, 1963; Wertheim, 1964), so they will only be outlined here.

When a radioactive atom decays by gamma ray emission, the nucleus goes from an excited state to the ground state, and the energy of the gamma ray is a measure of the energy difference between these states. Consider first the case of a free atom. The emitted gamma ray will have associated with it a certain momentum, and in order to conserve momentum the atom must recoil. However, this recoil has associated with it kinetic energy which must reduce the energy of the gamma ray; the act of absorption of a gamma ray by a nucleus involves the inverse process. Thus the emission and absorption processes are not in resonance.

If, however, one fixes the atom in a crystal, the situation may be

† This work was supported in part by the United States Atomic Energy Commission under Contract AT(11–1)–1198. ‡ Present Address: Eastern Laboratories, E. I. duPont Company, Gibbstown, New Jersey, U.S.A.

altered. The energy and momentum can be considered as decoupled; the recoil momentum is transferred to the crystal as a whole and, if this is fixed in the laboratory, to the earth. The velocity of recoil, and hence the kinetic energy, is effectively zero. The vibrational energy of the lattice is quantized; if the lowest allowable quantum of vibrational energy (the lowest phonon energy) is large compared to the recoil energy, there will be a finite probability of recoilless decay. Since this energy peak is not Döppler broadened, it is nearly monochromatic, and the lower limit to the width is set by the lifetime of the excited state and the uncertainty principle. In actual cases there is usually additional broadening due to imperfect motion and other instrumental problems, relaxation phenomena, and other factors.

A wide variety of nuclei display Mössbauer resonance, but by far the most useful from the standpoint of the study of solids is ^{57}Fe. The process by which the 14·4 keV Mössbauer gamma ray is produced is outlined in Fig. 1(a), and in this paper we shall limit ourselves to studies involving this isotope.

The Mössbauer effect is capable of giving information concerning a

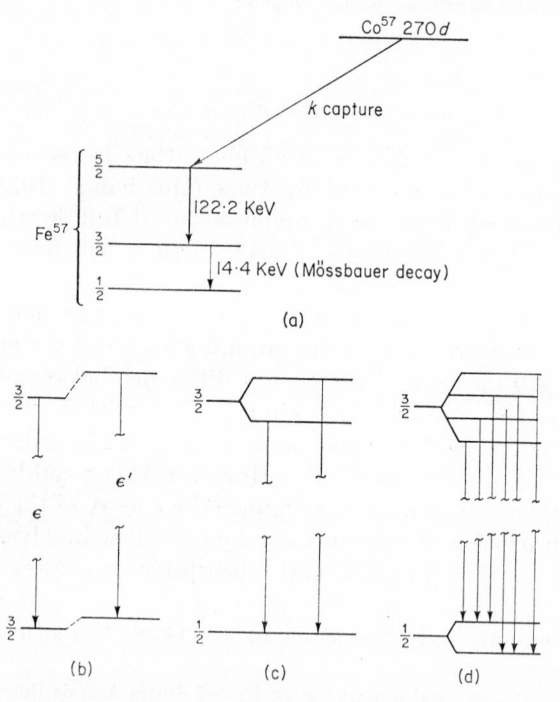

FIG. 1. Basic concepts for ^{57}Co–^{57}Fe Mössbauer system. (a) ^{57}Co decay, (b) isomer shift, (c) electric field, (d) magnetic field.

number of different aspects of the nuclear environment. First and most basic is the isomer shift. The process of decay by gamma emission involves a transition of the nucleus from an excited state to its ground state. The difference in energy between these states is slightly but measurably perturbed by the electronic environment of the nucleus. Specifically, it is affected by those electronic wave functions having non-zero amplitude at the nucleus; and these must be states of zero angular momentum (*s* states). The electronic configuration of atomic iron is $1s^2 2s^2 2p^6 3s^2 3p^6 3d^6 4s^2$. The 1*s* and 2*s* wave functions surely have significant amplitude at the nucleus, but they are largely isolated from the surrounding atoms and can tell us little about changes in the environment. The 4*s* electrons interact strongly with the neighbouring atoms, and while they are largely shielded from the nucleus, they contribute significantly to studies of isomer shift changes, particularly in metals. In ionic states of iron the 4*s* levels are usually assumed to be sparsely occupied. The 3*s* electrons do not interact to any extent with the surrounding atoms or ions, although they have their radial maximum at very nearly the same point as the 3*d* electrons which do interact with the neighbouring atoms. Changes in the 3*d* states with changing environment will then affect the degree of shielding of the 3*s* electrons from the nucleus, which will be reflected in changes in the observed difference in nuclear energy levels, that is in the isomer shift (see Fig. 1(b)). Thus, a dilute solution of ^{57}Co (^{57}Fe) in copper will not be in resonance with ^{57}Fe dissolved in chromium, or with ^{57}Fe in $FeCl_2$. If one now imparts a motion to the source with respect to the absorber, when the Döppler velocity just compensates for the difference in the isomer shifts of the two materials, resonance is obtained. A Mössbauer spectrometer is a device for measuring the velocities necessary to obtain resonance, and the differences in value of the isomer shift, $\Delta\epsilon$, in different environments is expressed in terms of the relative velocity necessary to obtain resonance. Spectrometers are widely described in the literature (Frauenfelder, 1963; Wertheim, 1964: also for high pressure, Pipkorn *et al.*, 1964; DeBrunner *et al.*, 1966), and will not be discussed here. For ^{57}Fe the energy differences (that is isomer shift differences) from material to material are of the order of a mm/sec or a fraction thereof. It is perfectly practical to measure differences smaller than 0·1 mm/sec, which corresponds to a thermal energy difference of 10^{-5}°C, or $\sim 10^{-5}$ calories ($\sim 10^{-9}$ eV) which illustrates the sensitivity of the technique.

The isomer shift is described by the equation:

$$\Delta\epsilon = \alpha[\psi_S^2(0) - \psi_A^2(0)] \tag{1}$$

where

$$\alpha = \tfrac{2}{3}\pi Ze^2[R_E^2 - R_G^2] \cong \tfrac{4}{3}\pi Ze^2[R\Delta R] \qquad (2)$$

which for iron $= 3{\cdot}52 \times 10^{10}$ $(R\Delta R)$.

Here the $\psi^2(0)$ are the amplitudes of the wave functions of the source and absorber at the nucleus, and R_G and R_G are the radii of the nucleus in the ground and excited state. From the standpoint of nuclear physics the value of ΔR is of considerable interest; from the viewpoint of solid state structure and electronic behaviour, the changes in $\psi_S^2(0) - \psi_S^2(0)$ with changing conditions are of paramount importance.

The nuclear energy levels can be perturbed by an electric field gradient at the nucleus. The ground state of spin one-half will remain unsplit, but the excited state of spin three-halves will split into two levels as indicated in Fig. 1(c). Under these conditions two transitions are possible, and one sees two peaks. The size of the splitting is then a measure of the electric field asymmetry seen by the nucleus. As will be discussed later, this effect gives important information concerning the local symmetry at the ^{57}Fe site, and the distribution and spin states of the electrons in the partially filled $3d$ shell.

A magnetic field at the iron nucleus also interacts with the nuclear levels of iron, splitting the excited level into four states and the ground level into two. When the selection rules ($\Delta m = 0, \pm 1$) are applied, one sees that a six line spectrum results as is shown in Fig. 1(d). The splittings between pairs of lines measures the magnetic field strength at the nucleus. It is thus possible to study changes in magnetic field with changing temperature, pressure, and chemical environment.

Finally, we have said that the phonon spectrum of the lattice determines the probability of recoilless decay, so that from the measured fraction of recoilless decays (proportional to the area under the Mössbauer peak) one can gain information about the lattice dynamics in the neighbourhood of the iron atom or ion. A number of rather sophisticated treatments have been given (Maradudin, 1966; Housley and Hess, 1966), but in this paper only the Debye approximation will be briefly considered where the recoilless fraction f is related to the characteristic temperature θ_D by:

$$f = \exp\left[-\frac{6Er}{k\theta_D}\left\{\frac{1}{4} + \left(\frac{T}{\theta_D}\right)^2 \int_0^{\theta_D/T} \frac{x\,dx}{e^x - 1}\right\}\right] \qquad (3)$$

where E_r is the recoil energy of the free nucleus due to decay.

Transition metal ions in crystals will be considered in this paper and it is therefore desirable to review qualitatively some of the features of ligand field and molecular orbital theory. These theories are discussed in detail in various literature (Griffith, 1964; Orgel, 1960; Ballhausen, 1962; Ballhausen and Gray, 1965).

The five $3d$ energy levels on a free transition metal ion are all degenerate. A $3d$ electron on such an ion may be in its ground state or in an excited state. The energy difference between these states can be expressed in terms of the Condon–Shortley parameters or more conveniently in terms of the Racah parameters A, B, C (see Fig. 2). These parameters can be calculated in principle, but are usually evaluated from atomic spectra; they depend on the repulsion among the $3d$ electrons on the ion.

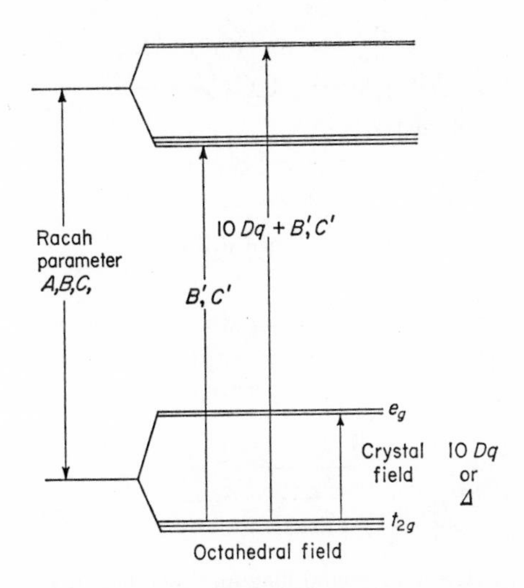

FIG. 2. Crystal field splittings and energies—octahedral symmetry.

If the ion is inserted into a crystal lattice, the surrounding ions provide a field of less than spherical symmetry. In this paper we shall discuss primarily octahedral symmetry wherein six ligands are arranged at the centres of the faces of a cube which has the iron at its centre, but we shall occasionally introduce the tetrahedral arrangement, where the ligands are at the corners of a regular tetrahedron surrounding the iron.

As seen in Fig. 2, these symmetries partially remove the degeneracy of the $3d$ levels. In octahedral symmetry the levels labelled t_{2g} (d_{xy}, d_{xz}, d_{yz}) lie below those labelled e_g (d_{z^2}, $d_{x^2-y^2}$); in tetrahedral symmetry this order is reversed.

The amount of splitting between these groups of levels depends on the intensity of the field supplied by the nearest neighbour ions (the ligands) in the crystal. For historical reasons this energy difference is usually called 10 Dq.

So far only arguments which depend on symmetry have been used. If one wishes to make even semiquantitative calculations, molecular orbitals combining realistic ligand and metal ion wave functions must be used to give orbitals which satisfy the symmetry requirements. Ballhausen and Gray have given a description of the process. A diagram such as Fig. 3 results, in which we see the combination of ligand

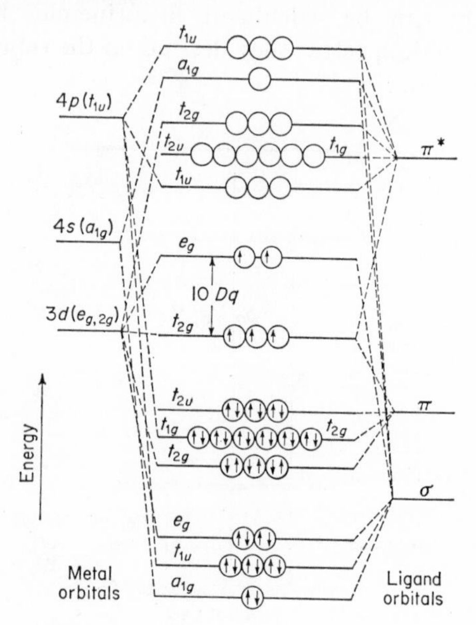

FIG. 3. Molecular orbital diagram—octahedral symmetry.

and metal ion wave functions to give bonding orbitals such as a_{1g} and t_{1u}, non-bonding orbitals such as t_{2u}, and anti-bonding orbitals such as t_{2g} and e_g. The separation between these last two is still a measure of the crystal field (10 Dq).

The $3d$ shell of iron is only partially filled. In the ideal case, the ferrous ion has six and the ferric ion five $3d$ electrons. There is more than one way to put the electrons in the levels; according to Hund's rule of atomic spectroscopy, the state of maximum spin should be lowest in energy. This case is illustrated in (a) and (b) of Fig. 4, and applies to most ionic materials. If the metal-ligand interaction is strong enough, energy may be saved by pairing spins as in (c) and (d). Potassium ferrocyanide and potassium ferricyanide are examples of this situation. For high spin ferric (a) and low spin ferrous (d) ions,

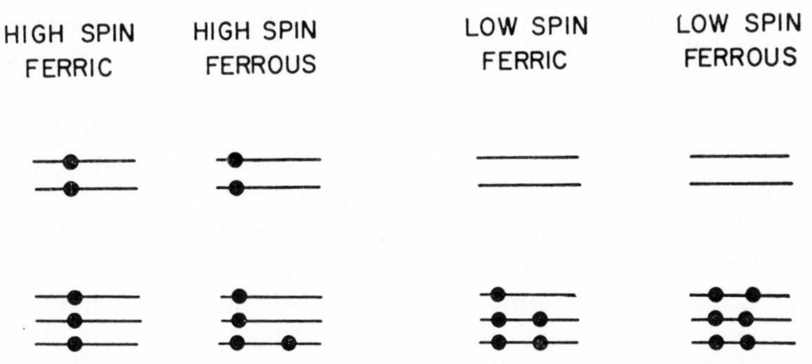

FIG. 4. Spin free and spin paired electron distributions for ferric and ferrous systems.

the 3*d* electrons present a field of spherical symmetry at the nucleus. For high spin ferrous and low spin ferric ions a strong probability of an asymmetric field exists. This is very important in understanding many aspects of quadrupole splitting.

A variety of different transitions are observed optically in crystals containing transition metal ions. There are *d*–*d* transitions which measure 10 Dq, or the Racah parameters B and C, or combinations of these two effects. These are of moderate intensity because they are allowed only due to vibrational interaction. Typically, 10 Dq is of the order of 5000–15,000 cm^{-1} (0·6–1·8 eV, 15–45 kcal) at one atmosphere. The combination peaks measuring both 10 Dq and B and C lie at somewhat higher energies, in the visible or near ultraviolet.

It is also possible to observe optical transitions which measure an electron transfer from a ligand level (t_{2u}) to a predominantly metal level (t_{2g}). This is an allowed transition and very intense, and the energy peak for such transitions typically lie at 25,000–40,000 cm^{-1} (3–5 eV), but they are broad, and the low energy tail sometimes extends through the visible and even into the infrared.

The effects of pressure on these transitions have been reviewed extensively elsewhere (Drickamer, 1963, 1965). In general, there is a marked increase in 10 Dq with pressure (10–15% in 100 kb). In the simplest order of theory, one would expect the crystal field to increase as R^{-5}, where R is the metal-ligand distance, and this is roughly what occurs. There is a measurable decrease in the Racah parameters with increasing pressure—7–11% in 100 kb. Apparently the 3*d* orbitals expand by interaction with the ligands, and this increases the average distance between electrons in the 3*d* shell, which reduces the repulsion between them.

The charge transfer peaks shift strongly to lower energy with increasing pressure, by as much as one half to 1 eV in 100 kb. This lowering of the metal energy levels vis-à-vis the ligand is apparently due primarily to the spreading of the $3d$ orbitals mentioned above. It can be shown (Vaughan, 1968) that it is possible in principle to reduce this difference by several electron volts using this mechanism. A second factor which tends to stabilize the t_{2g} orbitals is an increase in π bonding of these orbitals with excited ligand orbitals. Since the π bonding is usually considerably smaller than the σ bonding at one atmosphere, it could be expected to increase relatively more. Lewis and Drickamer (1968a) have shown that this is probably a significant factor, but not the controlling one. All of these observations will be important in the interpretation of various aspects of high pressure Mössbauer spectra for iron in ionic and covalent compounds.

To understand the Mössbauer spectrum of iron metal and of iron as a dilute solute in transition metals, a knowledge of the band structure of these materials is important. A review of even simple band structure is beyond the scope of this article, and the transition metals are far from simple, as the relevant electrons are neither completely bound nor almost free. There is, however, evidence (Slater, 1965) that all of the b.c.c. transition metals have very similar band structure, while the close packed members of the series (especially the f.c.c. metals) are very much like each other, but significantly different from the b.c.c. metals. In both cases the conduction band has a mixture of $3d$ and $4s$ character, with different emphasis in different parts of reciprocal lattice space. The relative degree of s and d character may change with pressure. We shall here assume that Stern's (1955) calculations for iron are qualitatively correct.

In the study of magnetism in transition metals, the controversy as to whether the magnetic electrons are primarily tightly bound or itinerate is a long and complex one, with outstanding proponents on each side. High pressure Mössbauer studies have so far contributed in only a minor way to the solution of this problem.

Figure 5 shows the range of isomer shifts for ^{57}Fe in different environments, relative to metallic b.c.c. iron. According to the convention used, the larger the isomer shift, the lower the electron density at the nucleus. Several facts are immediately evident.

(1) Iron as a dilute solute in a series of transition metals shows a relatively narrow range of isomer shifts although the solvent atoms have from 1–9 d electrons in their outer d shell—this would indicate that $3d$ electrons of the iron are not completely integrated into the solvent d band, although, as shall be seen later, they are closely associated with it.

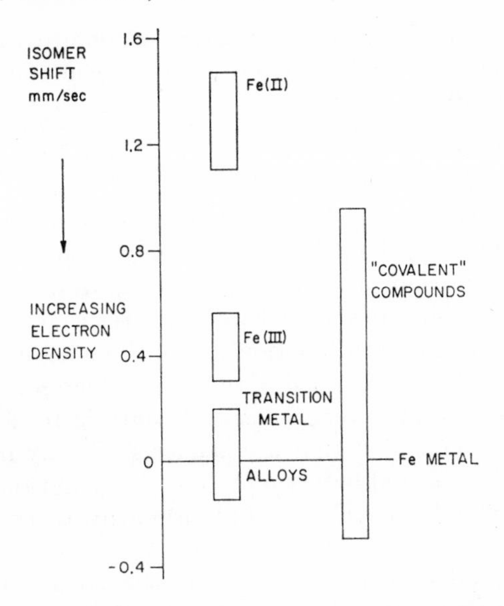

FIG. 5. Isomer shift ranges for iron and compounds.

(2) Ferrous ions typically exhibit a very low electron density. Since their outer configuration is nominally $3d^6 4s^0$, this can be attributed to the shielding of the $3s$ electrons by the $3d$.
(3) Ferric ions exhibit a significantly higher electron density than do ferrous ions. In both cases typical compounds fall in a range of electron densities, but the two ranges do not come close to overlapping. The ranges shown cover almost all the high spin compounds from the essentially completely ionic such as the fluorides to those with large covalent components such as the acetylacetonate, as long as they can be classified as ferrous or ferric. There is evidence that ferric compounds are generally more covalent than ferrous, and this would indicate some back donation from the ligands to the metal, so that the traditional $3d^5 4s^0$ ferric configuration is an oversimplification. In view of the narrow range of isomer shifts exhibited by the ferric compounds, however, the viewpoint taken here is that the major difference between the observed ferrous and ferric electron densities ($\sim 0{\cdot}9$ mm/sec) is due to the reduced shielding of the $3s$ electrons in the latter case.
(4) While the classification " covalent " is ambiguous in that all compounds exhibit some covalency, there are a number of compounds which involve a very high degree of electron sharing. Such compounds as ferrocene and potassium ferro- and ferricyanide fit this description. In addition there are compounds like $FeSe_2$, $FeTe_2$, FeP, FeP_2, FeAs

and FeSb to which it is difficult to assign a definite valence. As one might expect, this rather amorphous group of compounds has a large range of isomer shifts.

Ingalls (1967) found a linear correlation between the maximum of the square of the radial portion of the 3*d* wave function and the 3*s* electron density at the nucleus using Hartree–Fock free ion wave functions. A variational calculation, the purpose of which was to establish the effect of change of the shape of the 3*d* orbitals in going from the free ion to the metal on the 3*s* density at the nucleus, indicates that the correlation is still good for the bond wave functions, even though some of these have large electron densities in the tail of the wave function.

Thus, in trying to interpret isomer shift data in terms of covalency, it is necessary to remember that the isomer shift may not be sensitive to electron density located between the metal ion and the ligand, which is a usual criterion for covalency, but primarily to the change of 3*d* density on the ion.

As was indicated earlier in the paper, the measured difference in isomer shift between source and absorber involves a constant α. There has been a considerable controversy about the value of α, and thus of $\Delta R/R$, although it is agreed that α is negative, that is that the radius of the nucleus in the excited state is less than that in the ground state. Walker *et al.* (1961) assigned the difference in isomer shift between ferrous and ferric ion entirely to the difference in shielding of the 3*s* electrons due to the presence of one more 3*d* electron (that is they assumed configurations $3d^6 4s^0$ and $3d^5 4s^0$). Using Watson's (1959) free ion wave functions they obtained the value

$$\alpha = -0{\cdot}47\, a_0^3 \text{ mm/sec} \tag{4}$$

where a_0 is the Bohr radius.

Simanek and Stroubec (1967) however assign the difference in isomer shift between ferrous and ferric ion entirely to a difference in occupation of the 4*s* level, that is they assume that the ferrous ion, at least in compounds like the fluoride, is completely ionic, while the ferric ion has the configuration $3d^5 4s^{0\cdot 2}$. They consider overlap distortion as the only significant factor in changing $\Delta\epsilon$ with pressure, and, using published data on the oxide and fluoride, obtain a considerably smaller value of about $-0{\cdot}12$ for α.

Gol'danski (1963) and Danon (1966) also estimate a relatively small value for α. The value of α must be considered in discussing the effect of pressure on the isomer shift in metallic iron, and in iron as a dilute solute in transition metals.

II. Pressure Effects

A. ISOMER SHIFT

The factors which affect the electron density at the nucleus as a function of pressure (interatomic distance) are of two types : (a) deformation of the wave functions, and (b) transfer of electrons between orbitals. This can take the form of transfer of electrons between, say, the $3d$ and $4s$ orbitals of an iron atom or ion, of transfer of electrons between the $3d$ and $4s$ parts of the conduction band of a metal, or of transfer of electrons between metal and ligand orbitals, either bonding or non-bonding. The process by which the electron density changes with interatomic distance is a complex one, so that any description given here is, at best, an approximation and must be regarded as tentative. We shall first discuss metals and alloys for which at least a reasonably quantitative theory has been developed, and then discuss the significant factors governing the behaviour of compounds.

1. *Iron and Alloys*

The isomer shift of ^{57}Fe in b.c.c. iron has been investigated by a number of people including Pound *et al.* (1961), Nicol and Jura (1963), Pipkorn *et al.* (1964), and Moyzis and Drickamer (1968a). Figure 6 is a composite of results, largely from the last two papers. The isomer shift decreases with increasing pressure, corresponding to an increase of

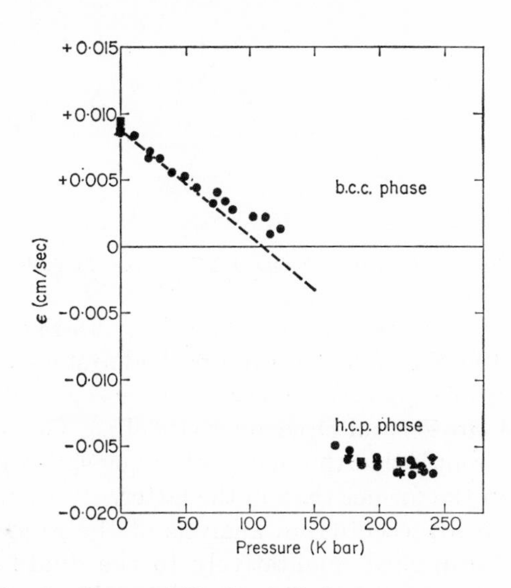

Fig. 6. Isomer shift of metallic iron *versus* pressure.

electron density at the nucleus, but the rate of increase decreases markedly at high pressure. As Moyzis has shown, $\Delta\epsilon$ is not linear in either pressure or volume. The experimental results can be expressed in the form:

$$\Delta\epsilon = -8{\cdot}1 \times 10^{-4}P + 1{\cdot}65 \times 10^{-6}P^2 \tag{5a}$$

$$= 1{\cdot}38 \frac{\Delta V}{V} + 2{\cdot}7 \left(\frac{\Delta V}{V}\right)^2 \text{ (mm/sec)} \tag{5b}$$

where P is in kb.

At about 130 kb there is a first-order phase transition in iron from the b.c.c. to the h.c.p. structure, with a large decrease in isomer shift of the order of $-0{\cdot}24$ mm/sec. The isomer shift in the high pressure h.c.p. phase decreases relatively slowly with increasing pressure or density, which, as shall be seen, is characteristic of close-packed metals.

Figures 7–9 show the isomer shift for iron as a dilute solute in a series of transition metals, plotted *versus* $\Delta V/V$ for the host. The solid line

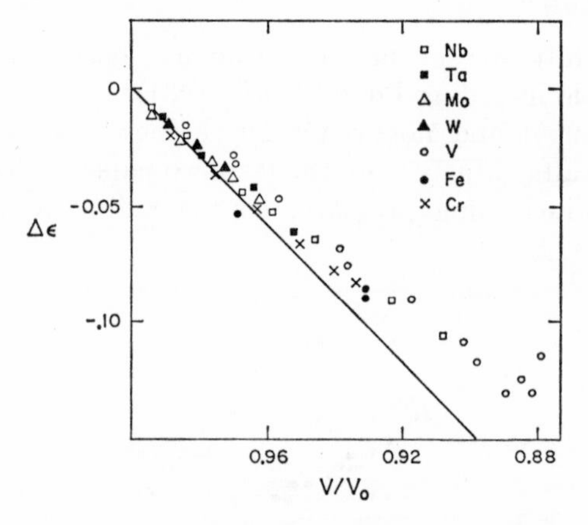

FIG. 7. Isomer shift *versus* $\Delta V/V$—b.c.c. metals.

represents the low pressure slope for b.c.c. iron. In Fig. 7 are shown the b.c.c. metals and in Figs 8 and 9 close-packed systems. These results are from the work of Pipkorn *et al.* (1964), Edge *et al.* (1965), Drickamer *et al.* (1965), and Moyzis and Drickamer (1968b). The feature which is immediately apparent is that the change of isomer shift with compression is much greater in the former than in the latter.

Ingalls (1967) has presented an analysis of the isomer shift of iron which can also be applied qualitatively to the dilute alloys (Ingalls *et al.*, 1967; Moyzis and Drickamer, 1968a, 1968b).

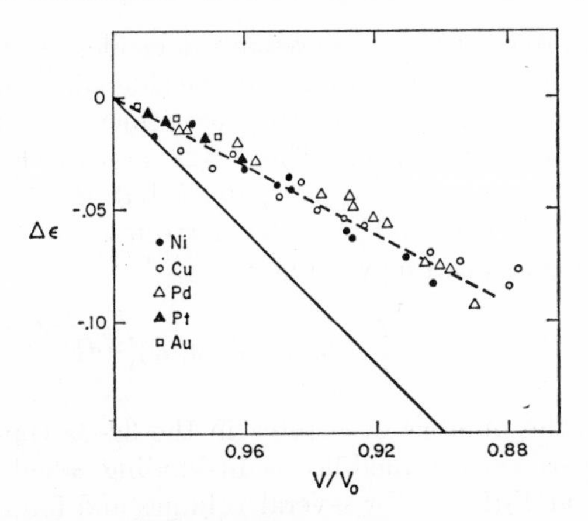

FIG. 8. Isomer shift *versus* $\Delta V/V$—f.c.c. metals.

In iron the 4*s* electrons exist in a relatively broad band, such that they can be characterized as "nearly free". This overlaps the much narrower 3*d* band. The other transition metals have a qualitatively similar structure, but within each group, b.c.c., f.c.c. and h.c.p., the similarities are much stronger than they are between groups.

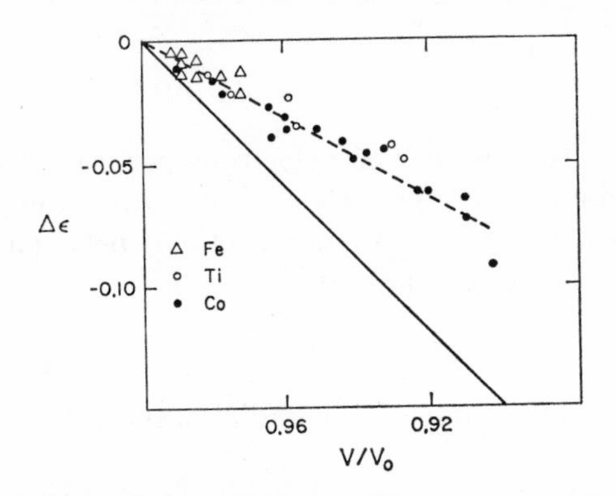

FIG. 9. Isomer shift *versus* $\Delta V/V$—h.c.p. metals.

In the first order, one can divide the effects of changing pressure (or volume) into two parts: the effect of compression of the 4*s* band $d(\Delta\epsilon)_L$, and the effect of transfer of electrons into or out of *s*-like states in the *s–d* conduction band $d(\Delta\epsilon)_B$. From the calculations of Walker *et al.* (1961) one can estimate the first effect:

$$\frac{d(\Delta\epsilon)_L}{d \ln V} = 1{\cdot}4 \text{ mm/sec.} \tag{6}$$

Since we are interested in the pressure dependence of $|\psi(0)|^2$, only the $4s$ and $3s$ electron contributions are considered. The $4s$ electrons are affected in a direct way since they are itinerant and thus can be expected to scale with volume. The $1s$, $2s$, and $3s$ electrons are not directly affected by small volume changes but the $3s$ electrons are indirectly affected by the changes in $3d$ electron wave functions.

The $4s$ contribution can be written

$$|\psi_{4s}(0)|^2 = \int_0^{E_F} N_s(E)|\psi_s(0, E)|^2 \, dE \tag{7}$$

where $N_s(E)$ is the number of s-states in the $3d$–$4s$ conduction band. Ingalls (1967) performed a modified tight-binding calculation of the $4s$ wave functions at $\Gamma_1(\mathbf{k} \equiv 0)$ for several volumes and found

$$|\psi_{\Gamma_1}(0)|^2 = 7{\cdot}1 \, a_0^{-3} \quad \text{for} \quad V = 80 \, a_0^{+3} \tag{8}$$

and

$$|\psi_{\Gamma_1}(0)|^2 = \text{const } V^{-\gamma} \tag{9}$$

where $\gamma \simeq 1{\cdot}25$. With the assumption that $|\psi_s(0, E)|^2$ equals $|\psi_{\Gamma_1}(0)|^2$, the decrease in the s-like nature of the conduction band being completely represented by the decrease in $N_s(E)$ as $\mathbf{k}$ increases, Ingalls obtained

$$|\psi_{4s}(0)|^2 = n_s | \psi_{\Gamma_1}(0)|^2 \tag{10}$$

where $n_s = 0{\cdot}53$ is the number of s-electrons per iron atom.

The $3s$ contribution is approximated by using the proportionality found by Watson (1959) and Clementi (1965) between the density of $3s$ electrons at the nucleus and $\langle nu_m^2 \rangle$

$$|\psi_{3s}(0)|^2 = \beta \langle nu_m^2 \rangle \tag{11}$$

where u_m is the maximum of the radial wave function in the wave function and

$$\langle nu_m^2 \rangle = \int_0^{E_F} N_d(E) \, u_m^2(E) \, dE \tag{12}$$

where $\beta = -5{\cdot}5 \, a_0^{-2}$. The integral in eqn (12) is performed up to the respective Fermi energy in each half of the band (spin up and spin down) using the linear relationship between $u_m^2(E)$ and E discussed by Ingalls.

Thus for the $3s$ and $4s$ contributions we have

$$|\psi(0)|^2 = n_s | \psi_{\Gamma_1}(0)|^2 + \beta \langle nu_m^2 \rangle. \tag{13}$$

Taking the volume derivative of eqn (12) gives

$$\frac{d|\psi(0)|^2}{d(\ln V)} = -n_s\gamma|\psi_{\Gamma_1}(0)|^2 + \beta\frac{\partial\langle nu_m^2\rangle}{\partial(\ln V)}$$

$$+|\psi_{\Gamma_1}(0)|^2\frac{\partial n_s}{\partial(\ln V)} + \beta\frac{\partial\langle nu_m^2\rangle}{\partial n_d}\frac{\partial n_d}{\partial(\ln V)}. \quad (14)$$

Here one sees the expression for $d(\Delta\epsilon)_L$, due to volume scaling, and for $d(\Delta\epsilon)_B$, due to $s\leftrightarrow d$ electron transfer

$$\frac{d(\Delta\epsilon)_L}{d(\ln V)} = \alpha\left[-n_s\gamma|\psi_{\Gamma_1}(0)|^2 + \beta\frac{\partial\langle nu_m^2\rangle}{\partial(\ln V)}\right] \quad (15)$$

$$\frac{d(\Delta\epsilon)_B}{d(\ln V)} = \alpha\left[|\psi_{\Gamma_1}(0)|^2\frac{\partial n_s}{\partial(\ln V)} + \beta\frac{\partial\langle nu_m^2\rangle}{\partial n_d}\frac{\partial n_d}{\partial(\ln V)}\right]. \quad (16)$$

Using the values of $|\psi_{\Gamma_1}(0)|^2$, n_s, γ, and β mentioned above and

$$\frac{\partial\langle nu_m^2\rangle}{\partial(\ln V)} = 0{\cdot}03\, a_0^{-1} \qquad \frac{\partial\langle nu_m^2\rangle}{\partial n_d} = 0{\cdot}9\, a_0^{-1} \quad (17)$$

gives

$$\frac{d(\Delta\epsilon)_L}{d(\ln V)} = -4{\cdot}86 \quad (18)$$

$$\frac{d(\Delta\epsilon)_B}{d(\ln V)} = \alpha\left[7{\cdot}1\frac{\partial n_s}{\partial(\ln V)} - 4{\cdot}95\frac{\partial n_d}{\partial(\ln V)}\right] \text{mm/sec} \quad (19)$$

where $\alpha \equiv [a_0^3 \text{ mm/sec}]$.

Equation (18) can be viewed as an alternative to eqn (6), which gave a rough value of $d(\Delta\epsilon)/d(\ln V)$ from volume scaling alone, and it will be used instead of eqn (6). To obtain agreement between these equations we would need $\alpha = -0{\cdot}35\, a_0^3$ mm/sec which is significantly different from the value of $\alpha = -0{\cdot}47\, a_0^3$ mm/sec found above.

By combining the theory with the experimental measurements and assuming the total number of electrons in the combined 4*s*–3*d* band is constant, we obtain:

$$\frac{d(\Delta\epsilon)}{d\ln V} = -4{\cdot}86\alpha + 12{\cdot}05\alpha X = 1{\cdot}32 \quad (20)$$

where
$$X = \frac{\partial n_s}{\partial\ln V} = -\frac{\partial n_d}{\partial\ln V}. \quad (21)$$

This relationship is plotted in Fig. 10.

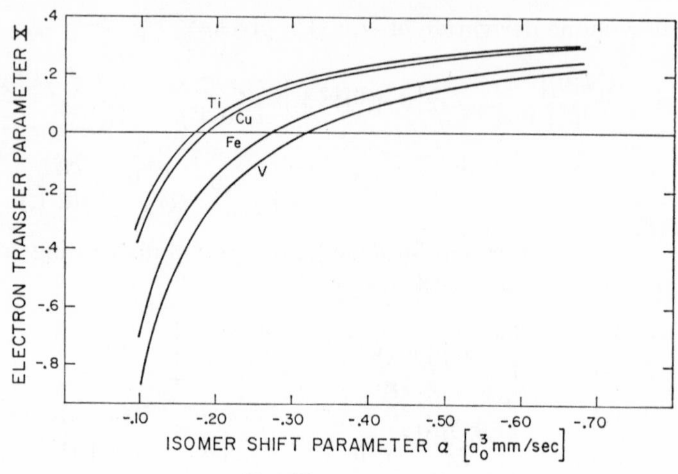

Fig. 10. *α versus X.*

While the data do not permit one to fix the value either of α or of X, we can state that they should be related to each other as shown. For values of α less than about $-0{\cdot}28\ a_0^3$ mm/sec (that is greater in magnitude than this number), X is positive which corresponds to transfer of electrons from the s to the d part of the conduction band with increasing pressure. Alternatively, the values for α suggested by Siminek and Sroubec (1967) or by Gol'danski (1963) or Danon (1966) would correspond to relatively large d to s transfer. Stern (1955) has found that as the volume decreases the d band lowers in energy with respect to the s band, making an s to d transfer energetically favourable. Unless one establishes a serious error in Stern's calculation, this is strong evidence in favour of a smaller, that is a more negative, value of α.

In Fig. 10 are also plotted values of α *versus* X for vanadium, copper, and titanium. Vanadium has the b.c.c. structure while the other two are close packed (b.c.c. and h.c.p. respectively). These are typical of the results for the classes of systems. In general, the close-packed systems exhibit a stronger tendency for s to d transfer than do the b.c.c. metals. This illustrates the basic difference in the effect of pressure on the band structure of these two classes of metals.

2. *Ionic Compounds*

Figure 11 shows isomer shifts as a function of pressure for a series of typical high spin ferrous and ferric compounds. Insofar as the classification is meaningful, these would be classified as " ionic ", and from the figure several facts are apparent. All compounds show a measurable

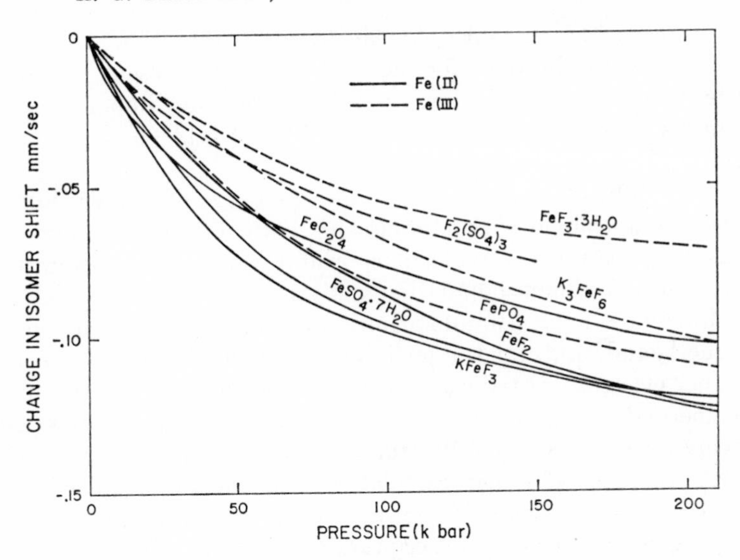

FIG. 11. Isomer shift *versus* pressure—high spin ferrous and ferric compounds.

increase in electron density at the iron nucleus with increasing pressure. On the average, the ferrous compounds show a measurably larger effect than do the ferric which is almost certainly not due to any consistent difference in compressibility. The effect in the ferrous ion is some 10–12% of the difference between normal ferrous and ferric ions in 150 kb, which represents a significant change in electronic configuration in this range.

In both cases these compounds and other relatively ionic materials such as $FeCl_3$ and $FeBr_3$ group into two quite narrow ranges. If the change with pressure were due primarily to electron transfer between ligand and metal, one would expect a much larger variation from ligand to ligand than is observed. It therefore seems reasonable to attribute the pressure effect to deformation of the metal ion wave functions. Simanek and Sroubec (1967) attribute the change with pressure for the ferrous ion entirely to compression of the "*s*" electrons, increasing the nuclear overlap. Champion *et al.* (1967) attribute the change for both ferrous and ferric ions to reduced shielding of the 3*s* electrons due to the spreading of the 3*d* orbitals discussed earlier. The ferric ions show a smaller change because there are only five 3*d* electrons in this case.

It seems most probable that neither of these factors is negligible. At present there is no apparent way to establish with certainty which is more important. The Simanek and Sroubec approach used as the sole mechanism for the pressure change gives a value of α which seems inconsistent with the data for metallic iron. At present the authors consider the change in shielding as probably the most important factor, but this is a tentative judgment, and considerable further study is needed.

It seems difficult to reconcile the narrow range of isomer shifts and the consistent change with pressure for both ferrous and ferric compounds with results of molecular orbital calculations using LCAO orbitals and a Mulliken population analysis. These indicate relatively small differences in electron distribution for ferrous and ferric ions with the same ligand, and, for the ferric ion in particular, a high covalency with the electron distribution strongly dependent on the ligand. These methods are probably particularly applicable to atoms with a relatively small number of electrons. It is also well known that one can get very satisfactory calculations of energies and of energy differences between states of a system using wave functions which are not a particularly accurate description of the true orbital. It may be that a linear combination of atomic orbitals using free ion wave functions is a rather inaccurate description of the wave function amplitude at the nucleus.

3. *Covalent Compounds*

Figure 12 exhibits the change in isomer shift with pressure for four compounds which show a high degree of electron sharing between ligand and metal. There is a very large change in isomer shift with pressure

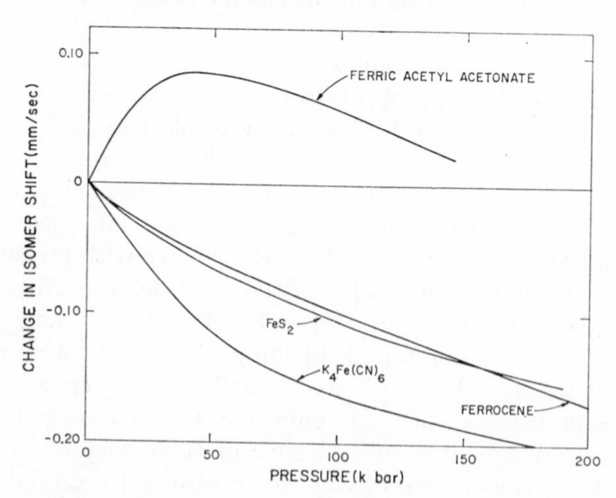

FIG. 12. Isomer shift *versus* pressure—“ covalent ” compounds.

for ferrocene, pyrites and potassium ferricyanide. This is not a compression effect, that is, it is not a result of especially high compressibility. The pyrites crystal is known to be relatively incompressible, and surely neither the ferrocene molecule nor the ferrocyanide molecular ion would show a large change in bond length with pressure. Also shown is ferric acetyl acetonate which actually exhibits a decrease in electron density with increasing pressure in the low pressure region with a reversal at higher pressures.

In these cases, in addition to changes in shielding and compression of the *s* wave functions, there must be significant change in orbital occupation with pressure. A calculation for ferrocene by Vaughan and Drickamer (1967a) indicates that at least one third of the change with pressure for that compound can be accounted for by changes in electron distribution among the orbitals. Each compound, however, requires an individual analysis so that no generalizations are possible.

B. QUADRUPOLE SPLITTING

As mentioned earlier, an electric field gradient at the nucleus can interact with nuclear states of spin equal to or greater than one, partially removing their degeneracy. The Mössbauer resonance is one of a variety of tools for investigating this effect. Here we shall discuss only briefly some high pressure results for a few compounds of iron. There are two possible sources of an electric field gradient at a transition metal nucleus. Firstly, the electrons in the partially filled $3d$ shell may exhibit less than spherical symmetry and, as mentioned earlier, this will be true, in general, for high spin ferrous or for low spin ferric compounds (see Fig. 4). Secondly, the ligands may exhibit less than cubic symmetry and thus impose an electric field gradient at the nucleus. Where the first effect is present it will dominate, since quadrupolar forces are short range and the effective radius of the $3d$ shell is much smaller than the average ligand-metal distance.

For preciseness, let us consider a high spin ferrous ion in an octahedral field. If the three t_{2g} levels are truly degenerate the orbital of the sixth electron will be spherically distributed and there will be no quadrupole splitting. As soon as the degeneracy is removed even by a very small splitting, the electron will not have an equal probability of being in the three t_{2g} states and a relatively large splitting results. Most high spin ferrous compounds exhibit a splitting of 2–3 mm/sec at room temperature. As one lowers the temperature one would expect the probability for occupation of the lowest level to increase by a Boltzmann factor and, in fact, the splitting does increase. If as the pressure increases the energy difference between states increases, one would expect the splitting to increase with pressure and ultimately to "saturate" when only the ground state had significant occupation probability. This indeed happens for many high spin ferrous systems as is illustrated in Fig. 13 for ferrous oxalate. There is a large number

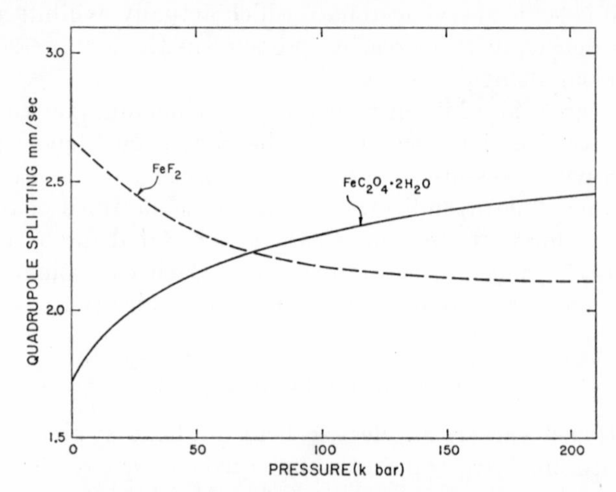

FIG. 13. Quadrupole splitting *versus* pressure—ferrous compounds.

of other examples in the literature (for example Champion *et al.*, 1967a). However, pressure may act to reduce the splitting among the t_{2g} levels and partially to equalize their occupation probability, thus reducing the quadrupole splitting as is shown for FeF_2, also in Fig. 13. For no change in relative distortion, it can be shown that the distorting field should increase as r^{-3} (or this term times a function of angle). Thus, one would expect more examples of increasing than of decreasing quadrupole splitting with increasing pressure for high spin ferrous compounds, and this is what is observed.

For high spin ferric compounds in a strictly cubic environment one would expect no quadrupole splitting. In almost all cases there appears to be some distortion, either trigonal, tetrahedral, or rhombohedral, since most high spin ferric compounds show small but measurable quadrupole splitting (0·3–0·6 mm/sec). Since the ferric quadrupole splitting responds directly to the ligand-metal distance (as r^{-3}), one would expect a large pressure effect. As typical data for K_3FeF_6 and $FeF_3 \cdot 3H_2O$ show (Fig. 14), this is indeed observed. Frequently, the increase in splitting with pressure is more rapid than r^{-3}, indicating that there is a tendency to increase the non-cubic component of the field with pressure.

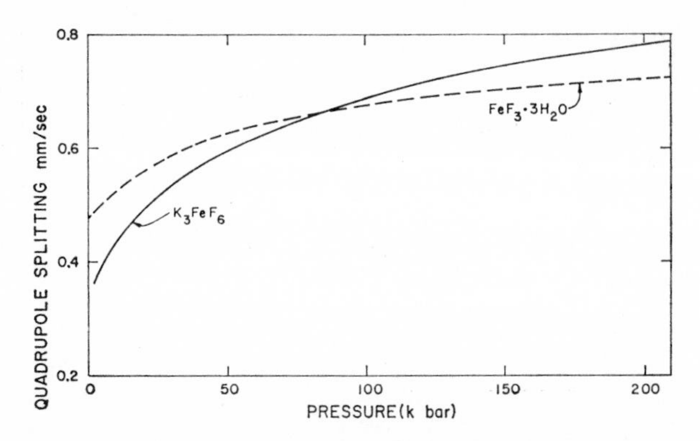

FIG. 14. Quadrupole splitting *versus* pressure—ferric compounds.

No general discussion of low spin compounds is possible, since these involve strong chemical binding and complex orbital occupation, but there are some special cases (not necessarily low spin) which deserve discussion. Firstly, ferrocene, which was mentioned briefly in the section on isomer shifts, is one of the most interesting and most thoroughly studied organometallic compounds. Iron is sandwiched between two C_5H_5 (dicyclopentadiene) rings to form a very stable molecule. The effect of pressure has been studied on both the optical and the Mössbauer spectrum of this compound (Zahner and Drickamer, 1961; Vaughan *et al.*, 1967a).

There have been a number of theoretical treatments of the electronic structure of ferrocene, but that of Dahl and Ballhausen (1961) forms the best basis for discussion of pressure effects. The application is discussed in detail by Vaughan and Drickamer (1967a). By estimating bond compressibilities from bond-force constants one can estimate orbital occupation as a function of pressure. One can show that there is a significant electron transfer from metal to ligand in the $e_{2g}(d \pm 2)$ orbitals with increasing pressure, and a reverse flow in the $e_{1g}(d \pm 1)$ orbitals. Using the treatment of Höfflinger and Voitländer (1963), one can show that this would predict a marked decrease in quadrupole splitting with increasing pressure. As shown in Fig. 15, the experimental and predicted splittings are in essentially quantitative agreement; as shown in Vaughan and Drickamer (1967a), the theory also predicts correctly the behaviour of the low energy optical transition as a function of pressure.

The compound α-Fe_2O_3 is one of the most studied of iron compounds. It has a rhombohedral crystal structure with each iron located in an octahedron of oxygen ions, although the iron is about 0·06 Å above the

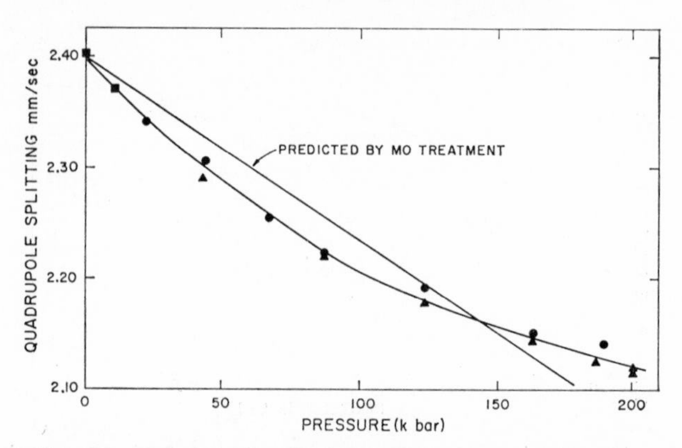

FIG. 15. Quadrupole splitting *versus* pressure—ferrocene.

centre of the octahedron. The material is antiferromagnetic below 950°K; from 950°K to about 260°K the internal magnetic field is perpendicular to the (111) body diagonal of the unit cell. Below this temperature (called the Morin temperature) it reorients through 90° and becomes parallel to the body diagonal. In such a crystal the equation for the quadrupole splitting contains a factor $(3\cos^2\theta - 1)$ which changes from -1 to $+2$ as one goes from a temperature above to one below the transition. This change in quadrupole splitting as a function of temperature has been observed by Ôno and Ito (1962).

The effect of pressure on the quadrupole splitting has been observed by Vaughan and Drickamer (1967b) using Mössbauer resonance; the low pressure region has also been studied by Worlton *et al.* (1967) using neutron diffraction. Figure 16 shows the observed splitting as a function of pressure. A sign change is obtained at about 30 kb. This has

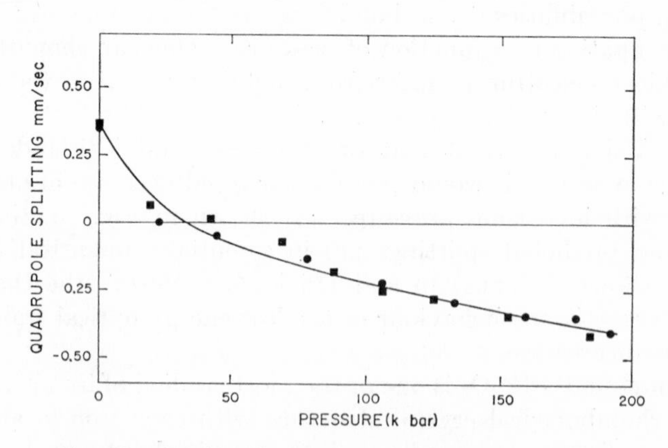

FIG. 16. Quadrupole splitting *versus* pressure—Fe_2O_3.

the qualitative features of a Morin transition, although there are several anomalous points. The transition is smeared out over a considerable range of pressure and in this transition region one does not observe the peak broadening which Ôno and Ito saw in the temperature region where two phases were present. Finally, in the high pressure region the magnitude of the splitting is not twice the atmospheric pressure value. These observations can be explained by a small movement of the iron ion in the oxygen octahedron. It is possible to show that a shift of only 0·04 Å in the position of the iron in a direction towards the centre of symmetry would be enough to cause the quadrupole splitting to go to zero. A slight shift accompanied by the Morin transition would account for the results. Worlton and Decker (1968) have shown that a more plausible argument involves the continuous change of the angle between the antiferromagnetic axis and the (111) axis of the crystal.

C. MAGNETISM

In the introduction it was pointed out that a magnetic field at the iron nucleus removed the degeneracy of both the ground state and the excited state. When one considers the selection rules, one can account for the observed six line spectrum. The changes in splitting of pairs of these lines with pressure measure the change in magnetic field. To date there have been only a limited number of Mössbauer resonance studies of magnetic fields at high pressure. We shall discuss briefly only two cases: ferromagnetism in iron, cobalt and nickel, and antiferromagnetism in cobalt oxide.

The magnetic field in iron as a function of pressure has been studied by Pound *et al.* (1961), Nicol and Jura (1963), Pipkorn *et al.* (1964), and Moyzis and Drickamer (1968a). A plot is shown in Fig. 17; the solid line in the figure represents the results of zero field n.m.r. measurements by Litster and Benedek (1963). It is interesting to see the very close agreement to 60 kb, the limit of the n.m.r. measurements, as this represents a check on both measurements but also yields additional information. The n.m.r. measurements are sensitive only to atoms at the surface of a domain, while Mössbauer resonance reflects the average field of all of the atoms in the domain. Apparently, the effects of pressure are very uniform throughout the domain. It should also be mentioned that the high pressure Mössbauer studies have shown that the high pressure (h.c.p.) phase of iron is paramagnetic at least at room temperature.

The magnetic field in nickel and in cobalt has been studied as a function of pressure, using ^{57}Fe produced by the decay of ^{57}Co as a probe by

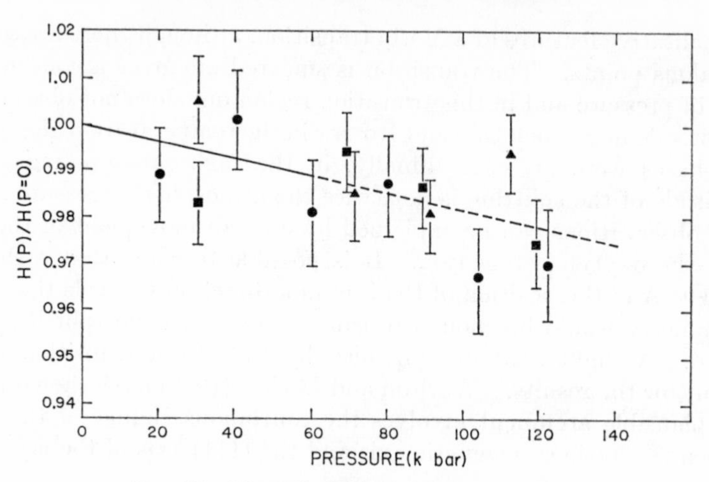

FIG. 17. Magnetic field *versus* pressure—iron.

Drickamer *et al.* (1965) and in nickel by Raimondi and Jura (1967). Zero field n.m.r. measurements in cobalt as a function of pressure have been reported by Jones and Kaminov (1960), Samara and Anderson (1963) and by Anderson (1965). In Fig. 18 are shown the smoothed values for iron in cobalt and nickel, compared with the data for pure iron discussed above. In both nickel and cobalt there is a distinct increase in magnetic field with pressure in the low pressure region, in distinct contrast to iron. In nickel there is a maximum of about 5–10 kb and in cobalt about 40–60 kb, and at higher pressures the field decreases as in iron.

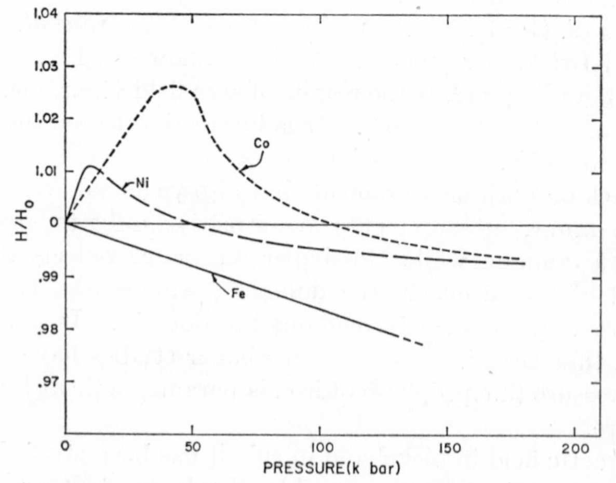

FIG. 18. Magnetic field *versus* pressure—^{57}Fe in cobalt and nickel.

Magnetism in metals and alloys is a very complex phenomenon, especially from the atomic viewpoint. There are a number of theories and no general agreement about many basic points, for example the degree to which the magnetic electrons are bound or itinerate. The theory has not advanced to the point where there is any really acceptable explanation of these pressure effects.

Cobalt oxide (CoO) at room temperature and atmospheric pressure is a cubic paramagnetic crystal with the sodium chloride (f.c.c.) structure. At 291°K it becomes antiferromagnetic, the transition being accompanied by a slight tetragonal distortion. The magnetic field has been measured as a function of temperature by Wertheim (1961) and Bearden *et al.* (1964); their results are represented on the left-hand side of Fig. 19. As one increases the pressure beyond 20 kb at 298°K, a magnetic field appears and increases with increasing pressure, as is shown on the right hand side of Fig. 19 from Coston *et al.* (1966).

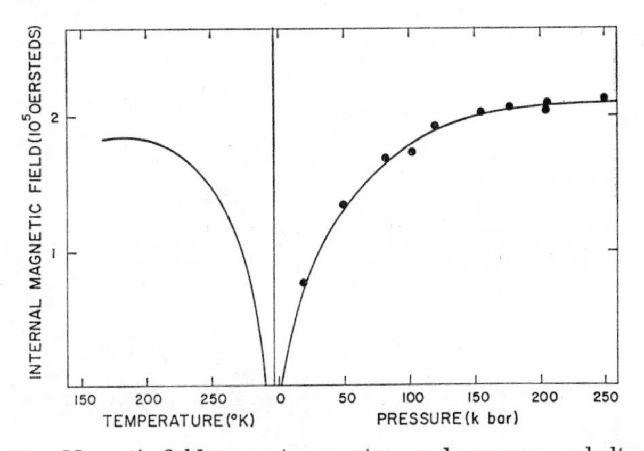

FIG. 19. Magnetic field *versus* temperature and pressure—cobalt oxide.

Since antiferromagnetism in materials like CoO is described in terms of superexchange, which depends strongly on overlap of wave functions of adjacent ions, the study of the field as a function of interatomic distance at constant temperature should be a very fruitful field of investigation.

It has been emphasized throughout this section that the initiation of ferro- or antiferromagnetism in a solid is heralded by the appearance of a six line spectrum. The observation of the appearance and disappearance of this six line spectrum as a function of pressure at constant temperature, or as a function of temperature at constant pressure, makes a very useful way of determining Curie or Néel temperatures as a function of pressure.

D. THE MEASUREMENT OF f NUMBER

While the f number, the fraction of recoilless decays, is a very basic quantity in Mössbauer theory, its quantitative experimental evaluation, even at one atmosphere, is very difficult. The main purpose of such a measurement is to study the localized vibrational structure in the solid in the neighbourhood of the decaying atom. As we shall see, in transition metals in the first order this does not differ greatly from the vibrational structure of the host.

The extensive high pressure measurements involve ^{57}Co (^{57}Fe) as a dilute solute in copper, vanadium and titanium. These have respectively, the f.c.c., and b.c.c., and h.c.p. structures, although titanium undergoes a transition near 80 kb. We shall be concerned here only with the measurement of relative f's, that is the value of f relative to its value at one atmosphere and the same temperature (294°K). Figures 20 and 21 show plots of f/f_0 *versus* $\Delta V/V_0$ for copper and vanadium

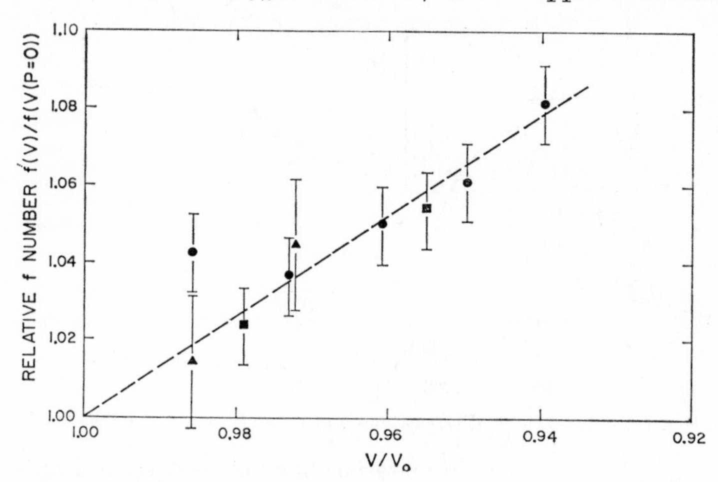

FIG. 20. f/f_0 *versus* $\Delta V/V_0$—copper.

from Moyzis and Drickamer (1968a). The titanium data are more scanty, but similar in nature. One sees that within the accuracy of the data the relationship is linear.

From eqn (3) it is convenient to write:

$$\frac{\partial \ln f}{\partial \ln V} = -\gamma Y \tag{22}$$

where $\gamma = -\dfrac{\partial \ln \theta}{\partial \ln V}$ is the Grüneisen constant,

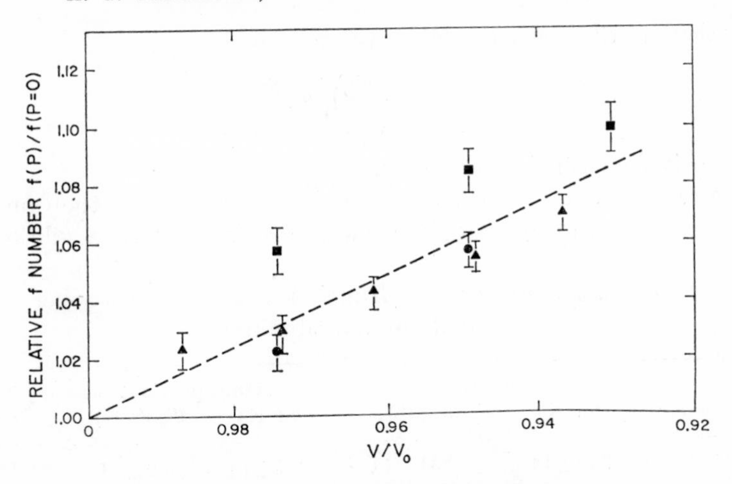

FIG. 21. f/f_0 *versus* $\Delta V/V_0$—vanadium.

and
$$Y=\frac{6E_r}{k\theta}\left[\frac{1}{4}-\frac{1}{\rho^{\theta/T}-1}+3\left(\frac{T}{\theta}\right)^2\int_0^{\theta/T}\frac{x\,dx}{\rho^x-1}\right].$$

It should be noted that the dependence on $\omega(\langle\omega^{-2}\rangle)$ is the same as for X-ray scattering but differs from the specific heat which varies as $\langle\omega^{\theta}\rangle$.

The experimental results give :

$$\left.\frac{\partial \ln f}{\partial \ln V}\right|_{\mathrm{Cu}} = -1{\cdot}283 \tag{23}$$

$$\left.\frac{\partial \ln f}{\partial \ln V}\right|_{\mathrm{V}} = -1{\cdot}380 \tag{24}$$

$$\left.\frac{\partial \ln f}{\partial \ln V}\right|_{\mathrm{Ti}} = -0{\cdot}843. \tag{25}$$

From these results it is possible to calculate a value for the characteristic temperature θ which can be corrected to apply to the host lattice by the relationship :

$$\theta_H=\left(\frac{m'}{m}\right)^{1/2}\theta_f \tag{26}$$

where $\theta_H=\theta$ of host, $\theta_f=\theta$ obtained from experiment, m, m' = mass of host and impurity atom.

Since the relationship of Figs 20 and 21 is linear, the above calculation can be assumed to apply at an average pressure of 50 kb. One can calculate the one atmosphere value from

$$\theta_0 = \left(\frac{V_p}{V_0}\right)^{\gamma} \theta_p \tag{27}$$

where γ is the Grüneisen constant.

The resultant values for θ are listed in Table I. Since the f number measures a mean value of $\langle \omega^{-2} \rangle$ where ω is the lattice vibrational

TABLE I. Mössbauer and Grüneisen constants for copper, vanadium and titanium

Element	50 kb θ_f(°K)	50 kb θ_{host}(°K)	Atmospheric θ_f(°K)	Atmospheric θ_{host}(°K)	γ
Cu	350 ± 14	331 ± 14	327 ± 14	309 ± 14	1·998
V	268 ± 10	284 ± 11	258 ± 10	274 ± 11	1·257
Ti	339 ± 35	370 ± 40	321 ± 35	350 ± 40	1·232

frequency, it is most meaningful to compare these θ's with θ obtained from X-ray diffraction measurements. Unfortunately, this can only be done for copper. Table II summarizes the available data for this metal. It can be seen that the X-ray results are in excellent agreement with the predictions from high pressure Mössbauer data.

TABLE II. θ_D from X-ray measurements for copper

Investigators	Debye parameter (°K)
Owen and Williams (1947)	314
Burie (1956)	299
Chipman and Paskin (1959)	307 (327)[a]
Flinn *et al.* (1961)	322
Graevskaya *et al.* (1965)	310

[a] Parenthetic value corrected for one and two phonon generation.

A direct comparison of these results with other data for vanadium is more difficult. The values of θ available are from low temperature specific heat measurements and are listed in Table III. It should be kept in mind that vanadium is superconducting below 5·4°K so that measurements must be made in a field 5–10 kgauss. Further, a large correction for electronic specific heat is necessary. The results in Table IV show a rather distinct trend with sample purity, with the more impure samples agreeing with our result. This is perhaps

TABLE III. θ_D from specific heat measurements for vanadium

Investigators	Sample purity (%)		γ_C (°K)
Wolcott (1955)		>99·98	380
Worley *et al.* (1955)	sample 1:	99·50	308
	sample 2:	99·80	273
Corak *et al.* (1956)		≃99·80	338
Clusius *et al.* (1960)		99·50	425[a]
Radebaugh and Keesom (1966)		>99·99	382

[a] Experiment performed in temperature range 11–23°K.

reasonable, as our samples contained ^{57}Co (^{57}Fe) impurity. It must be remembered that θ_{C_v} is a distinct function of temperature and may be 50° or more lower at room temperature.

For titanium, Wolcott (1957) and Johnson and Kothen (1953) have made C_v measurements over a long temperature range. At low temperatures they obtain $\theta = 430$°K and near room temperature a value of 360°K, in very good agreement with these results.

The results for these three metals demonstrate that, to a good approximation, γ is independent of density at least to 100 kb, which validates the Grüneisen equation of state over this range and permits the prediction of θ as a function of pressure to 100 kb at least.

TABLE IV. Constants A and B for relationship $K = AP_B$

Compound	A	B
$FeCl_3$	0·265	0·56
$FeBr_3$	0·076	0·43
$KFeCl_4$	0·092	0·50
$FePO_4$	0·078	0·46
Phosphate Glass	0·048	0·31
Ferric Acetate (418°K)	0·022	0·98
Ferric Citrate	0·112	0·35
$K_3Fe(CN)_6$	0·109	2·06

E. CONVERSION OF Fe^{III} TO Fe^{II}

As has been emphasized in the earlier sections, the Mössbauer spectra of high spin ferrous and high spin ferric iron are entirely different, both as regards isomer shift and quadrupole splitting. It is therefore easy to discern the appearance of one oxidation state in the presence of the other, and to estimate the relative amounts of the two states from computer fit areas. Although the difference in low spin states is less spectacular, the calculation is still possible. Perhaps the most remarkable of the high pressure Mössbauer observations is that ferric iron reduces to ferrous iron with increasing pressure in a wide variety of compounds, and that this is essentially a reversible process. This has been discussed in a series of articles including Champion *et al.* (1967), Champion and Drickamer (1967a, b), Lewis and Drickamer (1967a), and Fung (1968).

As an example, a series of spectra for $FeCl_3$ are shown in Figs 22(a, b). The process reverses upon release of pressure, but with considerable hysteresis. On removing the pellet and powdering the sample one almost always recovers the atmospheric spectrum. A greater or lesser degree of conversion has been observed in $FeCl_3$, $FeBr_3$, $KFeCl_4$, $FePO_4$, $Fe_2(SO_4)_3$, $Fe(NCS)_3$, $Fe(NH_3)_6Cl_3$, $K_3Fe(CN)_6$, ferric citrate, basic ferric acetate, ferric acetyl acetonate, ferric oxalate, various hydrates, hemin and hematin.

The form of the pressure dependence is essentially always the same, as is illustrated in Figs 23 and 24. If one defines an equilibrium constant $K = C_{\text{II}}/C_{\text{III}}$, then

$$K = AP^B \tag{28}$$

where A and B are independent of pressure. Typical values for these constants are shown in Table IV. It has been shown that one is observing true equilibrium and not the results of slow kinetics by the fact that successive spectra run at the same pressure are substantially identical.

The reaction is almost always endothermic, that is the conversion increases with increasing temperature, although in the case of hemin the conversion actually decreases with increasing temperature. The heat of reaction generally is in the range 0·1–0·3 eV, and usually increases with increasing temperature. For the most ionic systems the heat of reaction does not depend significantly on pressure, but in more covalent materials it may increase or decrease sharply with increasing pressure.

One is faced with two problems in discussing these results : why does the electron from ligand to metal take place, and why does the reaction have the observed pressure dependence, that is why does one not obtain discontinuous ferric to ferrous conversion at some pressure.

As discussed in the introduction, it is possible to excite an electron optically from the non-bonding ligand levels to the lowest antibonding metal levels ($t_{2u} \rightarrow t_{2g}$ in octahedral symmetry). This optical charge

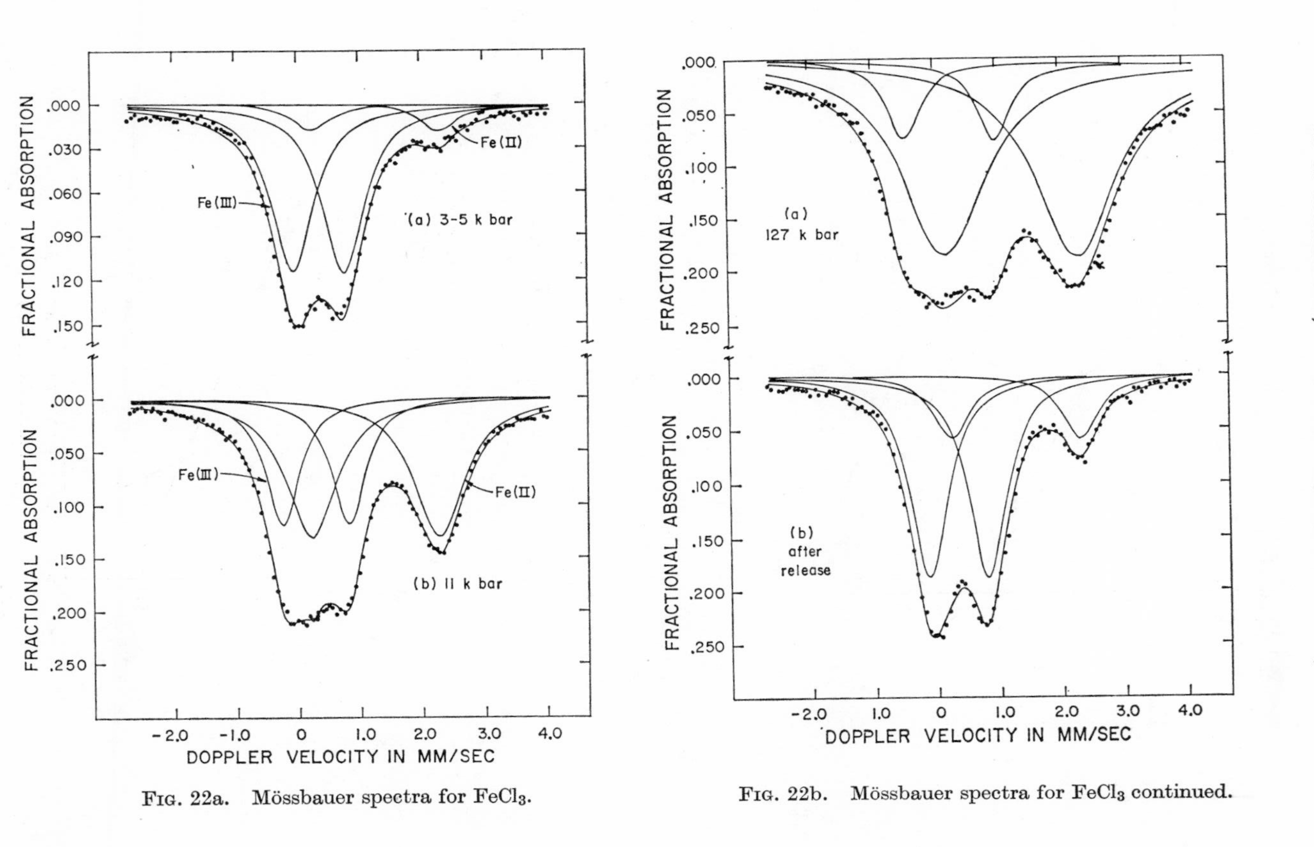

FIG. 22a. Mössbauer spectra for $FeCl_3$.

FIG. 22b. Mössbauer spectra for $FeCl_3$ continued.

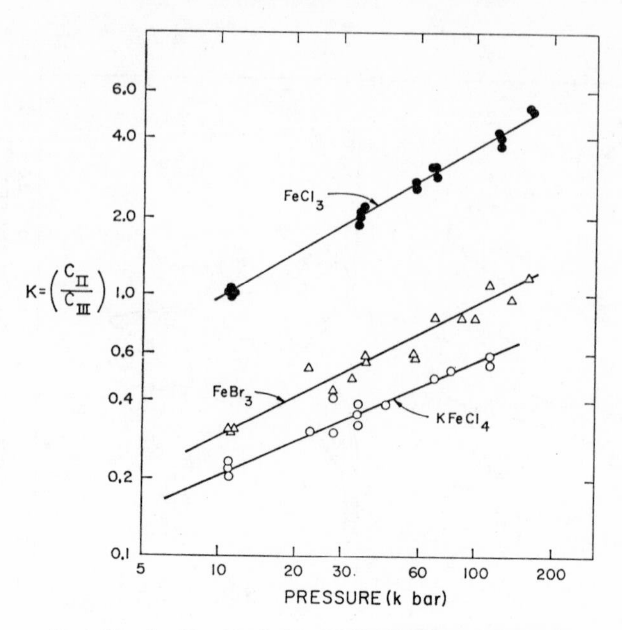

FIG. 23. ln K *versus* ln P—$FeCl_3$, $FeBr_3$ and $KFeCl_4$.

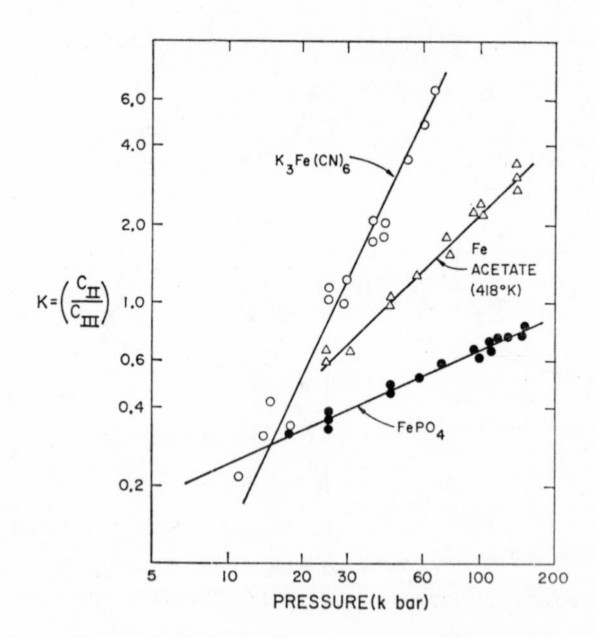

FIG. 24. ln K *versus* ln P—$FePO_4$, $K_3Fe(CN)_6$ and basic ferric acetate.

transfer peak normally has an energy of 3–6 eV, with a long low energy tail which may extend through the visible and even into the infrared part of the spectrum. As indicated earlier, a red shift (shift to lower energy) of one-half to 1 eV in 100 kb is observed experimentally, and can be explained theoretically; however this is of course not sufficient to move the optical peak to zero energy. The high pressure transition observed by the Mössbauer studies is, however, a thermal transition which takes place sufficiently slowly so that the atomic coordinates can assume their new equilibrium values, whereas an optical transition must take place vertically on a configuration diagram, according to the Franck–Condon principle. The situation is illustrated in a typical diagram in Fig. 25. The horizontal configuration coordinate is typically some vibrational mode of the system (metal plus ligands) which

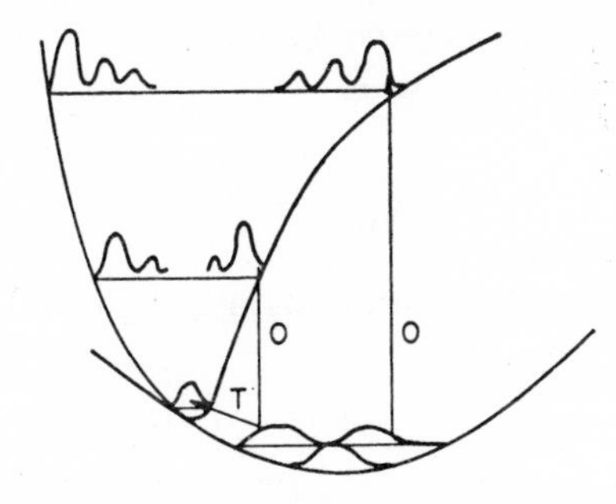

FIG. 25. Schematic configuration coordinate diagram. O = optical transition, T = thermal transition.

permits the rearrangement. The observed tail on the charge transfer peak is consistent with a very steep slope of the side of the excited-state potential well above the centre of the ground-state well, and accounts for the small thermal energy necessary for the transition. The diagram also explains the temperature coefficient of the equilibrium constant, as there is a thermal distribution for transfer of electrons from the ground to the excited electronic level, and in addition a distribution of electrons among the vibrational levels of the ground state with increasing temperature.

There remains the problem of the pressure dependence of the conversion. Qualitatively, one can see why the conversion does not go to completion at a given pressure as follows. When electron transfer takes place, one creates a ferrous ion in a ferric site plus a free radical or radical ion. (The excitation may also be smeared out over the nearest neighbour ligands.) There is thus a transfer of charge, a change of volume and a distortion. This affects the neighbouring complexes, distorting the potential wells in such a way as to make electron transfer less favourable. With increasing pressure the excited state is lowered further and one obtains more conversion, but with further increase in local strain. The sluggishness and hysteresis with which the process reverses on release of pressure can be associated with the stored up strain.

The situation can be described thermodynamically:

$$K = \exp(-\Delta\bar{G}/RT) \tag{29}$$

$$\frac{\partial \ln K}{\partial \ln P} = -\frac{P\Delta\bar{V}}{RT} = \frac{P(\bar{V}^{\mathrm{III}} - \bar{V}^{\mathrm{II}})}{RT} \tag{30}$$

where $\bar{V}^{\mathrm{III}}$ and $\bar{V}^{\mathrm{II}}$ refer to the partial molar volumes of the ferric and ferrous sites plus their associated ligands.

From our empirical observation

$$\frac{\partial \ln K}{\partial \ln P} = B \tag{31}$$

where B is a constant. One can rearrange eqn (29), remembering that $K = C_{\mathrm{II}}/1 - C_{\mathrm{II}}$. Then

$$\frac{\partial \ln C_{\mathrm{II}}}{\partial \ln P} = \frac{P(\bar{V}^{\mathrm{III}} - \bar{V}^{\mathrm{II}})}{RT}(C_{\mathrm{III}}). \tag{32}$$

Thus, the fractional increase in conversion with fractional increase in pressure is proportional to C_{III}, the number of sites available for conversion. The proportionality coefficient is the work necessary to form a ferrous site from a ferric site, measured in thermal units (that is in units of RT). It is entirely reasonable that in the first order the dependence would be first order in these variables, and that to the first approximation the coefficient is a constant. Experimentally, it appears that higher order terms are negligible.

F. HIGH PRESSURE MÖSSBAUER STUDIES ON GLASS

An interesting application of high pressure Mössbauer resonance is the investigation of the structure of glass. Lewis and Drickamer (1968) and Kurkjian and Sigety (1964) have made extensive studies, correlating the atmospheric Mössbauer spectrum of ferric ions with the site symmetry. Tischer and Drickamer (1962) have studied the effect of pressure on the optical spectra of a number of transition metal ions in various glasses. The work discussed here involves Mössbauer studies of Fe^{III} and Fe^{II} ions in a phosphate and in a silicate glass.

The phosphate glass consists of chains of PO_4^- tetrahedra with little crosslinking, in contrast to the silicate glass discussed below. Both the ferric and the ferrous ions are in octahedral sites in the phosphate glass. High pressure Mössbauer resonance studies have also been made on $FePO_4$ and $Fe_3(PO_4)_2$ in the crystalline state (Champion *et al.*, 1967a). The Fe^{III} ion in the glassy medium behaved in much the same way as it did in the crystal. This is illustrated in the change in isomer shift and quadrupole splitting for the Fe^{III} ion in Figs 26 and 27.

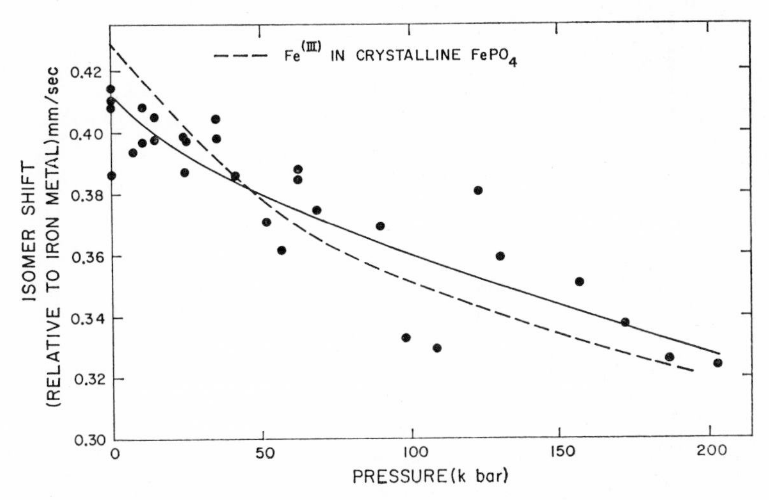

FIG. 26. Isomer shift *versus* pressure—Fe^{III} in phosphate glass.

The ferric ion reduced to the ferrous state with pressure in the glass as it did in the crystal but with different coefficients A and B. These are listed in Table IV in the previous section. The behaviour of the Fe^{II} ion in the phosphate glass was also quite similar to that in the crystal.

The silicate glass, in addition to chains and rings, includes three dimensional networks. Kurkjian and Sigety (1964) have shown that the ferric ion is in a tetrahedral site in silicate glass, while Tischer and Drickamer (1962) indicate that the ferrous ion is in a loose octahedral site. Weyl (1951) shows that the ferric ion is both a network former and a network modifier in silicate glass, that is it can replace a silicon or be in an interstice. Ferrous ion is always a network modifier.

The isomer shift for Fe^{III} ion is shown as a function of pressure in Fig. 28. One observes an increase with increasing pressure in the low

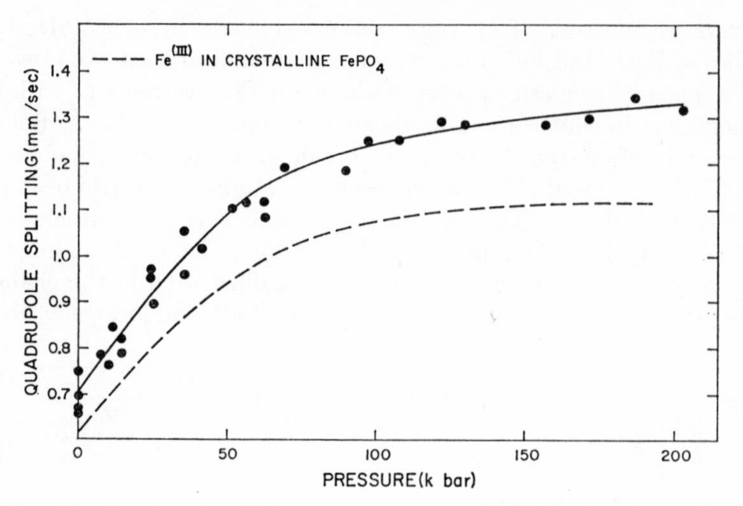

FIG. 27. Quadrupole splitting *versus* pressure—Fe^{III} in phosphate glass.

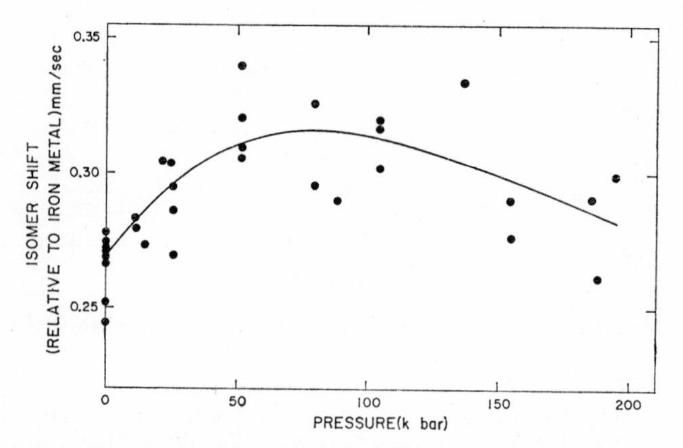

FIG. 28. Isomer shift *versus* pressure—Fe^{III} in silicate glass.

pressure region, a maximum at about 60 kb, and a decrease at higher pressures. The rather low initial value is consistent with a high degree of covalency. Apparently, in the low pressure region changes in orbital occupation dominate, while at high pressure the deformation of the wave functions controls and gives an increase in electron density at the nucleus with increasing pressure. The quadrupole splitting increases very rapidly with pressure, especially at lower pressures, indicating that substantial changes in site symmetry and/or electronic structure occur in the lower pressure region. No tendency for Fe^{III} to reduce to Fe^{II} was observed.

The ferrous ion, in a loose octahedral site, shows virtually no change in isomer shift with pressure. The quadrupole splitting behaves in an unusual manner as shown in Fig. 29; at low pressure there is a small but distinct drop, then a rise. Obviously there is a combination of site compression, change of symmetry, and electron redistribution at these sites. A run was also made with radioactive ^{57}Co introduced as cobaltous ion in silicate glass, and it, of course, does decay to ferrous ion. As can be seen in Fig. 29, the quadrupole splitting behaved

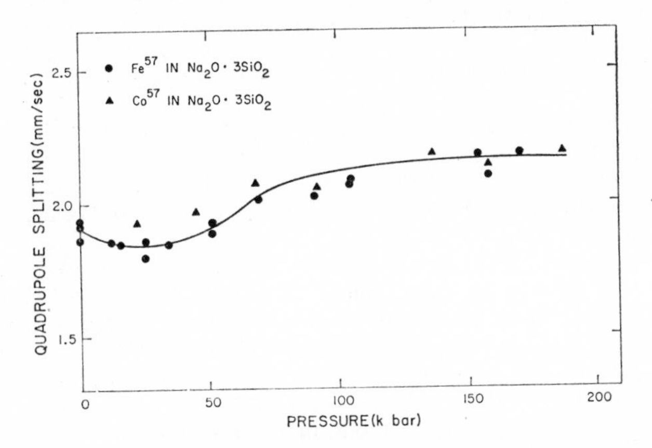

Fig. 29. Quadrupole splitting *versus* pressure—Fe^{II} in silicate glass.

identically to the ordinary ferrous ion, while the isomer shift parallels the behaviour of ordinary ferrous ion but at considerably higher electron density. The only apparent difference was that the cobalt formed a very dilute solution, while the iron was present in about 5% concentration.

In summary, it can be said that both Fe^{III} and Fe^{II} sites in phosphate glass behave very much like those in crystalline phosphates. In silicate glass the Fe^{III} sites exist in tetrahedral sites with considerable covalency and undergo significant change in orbital occupation in the low pressure region. At the ferrous sites, which are of open octahedral symmetry, there is considerable distortion with increasing pressure.

Acknowledgement

It is a pleasure to acknowledge the support of the United States Atomic Energy Commission and the collaboration of many students and Research Associates in this work.

References

Anderson, D. H. (1965). *Bull. Am. phys. Soc.* **10,** 75.

Ballhausen, C. J. (1962). "Introduction to Ligand Field Theory." McGraw-Hill, New York.

Ballhausen, C. J. and Gray, H. B. (1965). "Molecular Orbital Theory." Benjamin, New York.

Bearden, A. J., Mattern, P. L. and Hart, T. R. (1964). *Rev. mod. Phys.* **36,** 370.

Burie, B. (1956). *Acta. Cryst.* **9,** 617.

Champion, A. R. and Drickamer, H. G. (1967a). *J. chem. Phys.* **47,** 2591.

Champion, A. R. and Drickamer, H. G. (1967b). *Proc. natn Acad. Sci. U.S.A.* **58,** 876.

Champion, A. R., Vaughan, R. W. and Drickamer, H. G. (1967). *J. chem. Phys.* **47,** 2583.

Chipman, D. R. and Paskin, A. (1959). *J. appl. Phys.* **30,** 1992.

Clementi, E. (1965). *IBM Jl Res. Dev. Suppl.* **9,** 2.

Clusius, K., Franzosini, P. and Piesbergen, V. (1960). *Z. Naturf.* **15a,** 728.

Corak, W. S., Goodman, B. B., Satterthwaite, C. B. and Wexler, A. (1956). *Phys. Rev.* **102,** 656.

Coston, C. J., Ingalls, R. L. and Drickamer, H. G. (1966). *Phys. Rev.* **145,** 876.

Dahl, J. P. and Ballhausen, C. J. (1961). *K. danske Vidensk. Selsk. Skr.* **33,** 39.

Danon, J. (1966). *In* "Application of the Mössbauer Effect in Chemistry and Solid State Physics." International Atomic Energy Agency, Vienna.

DeBrunner, P., Vaughan, R. W., Champion, A. R., Cohen, J., Moyzis, J. A. and Drickamer, H. G. (1966). *Rev. sci. Inst.* **37,** 1310.

Drickamer, H. G. (1963). *In* "Solids under Pressure." ed by W. Paul and D. Warschauer. McGraw-Hill, New York.

Drickamer, H. G. (1965). *In* "Solid State Physics." Vol. 17, ed. by F. Seitz and D. Turnbull. Academic Press, New York.

Drickamer, H. G., Ingalls, R. L. and Coston, C. J. (1965). *In* "Physics of Solids at High Pressures." ed. by C. T. Tomazuka and R. M. Emrick. Academic Press, New York.

Edge, C. K., Ingalls, R. L., Debrunner, P. G., Drickamer, H. G. and Frauenfelder, H. (1965). *Phys. Rev.* **138A,** 729.

Flinn, P. A., McManus, G. M. and Rayne, J. A. (1961). *Phys. Rev.* **123,** 809.

Frauenfelder, H. (1963). "The Mössbauer Effect." Benjamin, New York.

Fung, S. C. (1968). Private communication.

Gol'danski, V. I. (1963). *In* "Proceedings of the Dubna Conference on the Mössbauer Effect." Consultation Bureau Enterprises, New York.

Graevskaya, Y. I., Iveronova, V. I. and Tarasova, V. P. (1965). *Soviet Phys. Solid St.* **7,** 1083.

Griffith, J. S. (1964). "The Theory of Transition Metal Ions." Cambridge University Press, Cambridge.

Höfflinger, V. B. and Voitländer, J. (1963). *Z. Naturf.* **18a,** 1065, 1074.

Housley, R. M. and Hess, F. (1966). *Phys. Rev.* **146,** 517.

Ingalls, R. L. (1967). *Phys. Rev.* **155,** 157.

Ingalls, R. L., dePasquali, G. and Drickamer, H. G. (1967). *Phys. Rev.* **155,** 165.

Johnson, N. L. and Kothen, C. W. (1953). *J. Am. chem. Soc.* **75,** 3101.

Jones, R. V. and Kaminov, I. P. (1960). *Bull. Am. phys. Soc.* **5,** 175.

Kurkjian, C. R. and Sigety, E. A. (1964). Proceedings VII Congress on Glasses.

Lewis, G. K., Jr. and Drickamer, H. G. (1968a). *Proc. natn. Acad. Sci* **61,** 414.
Lewis, G. K., Jr. and Frickamer, H. G. (1968b). *J. chem. Phys.* **49,** 3785.
Litster, J. D. and Benedek, G. B. (1963). *J. appl. Phys.* **34,** 688.
Maradudin, A. A. (1966). *In* " Solid State Physics," Vol. 18, ed. by F. Seitz and D. Turnbull. Academic Press, New York.
Mössbauer, R. (1958a). *Z. Phys.* **151,** 124.
Mössbauer, R. (1958b). *Naturwissenschaften* **45,** 538.
Moyzis, J. A. and Drickamer, H G (1968a) *Phys Rev* **171,** 389
Moyzis, J A and Drickamer, H. G. (1968b). *Phys. Rev.* **172,** 655.
Nicol, M. and Jura, G. (1963). *Science N.Y.* **141,** 1035.
Ôno, K. and Ito, A. (1962). *J. physiol. Soc. Japan* **17,** 1012.
Orgel, L. E. (1960). " An Introduction to Transition Metal Theory." J. Wiley and Sons, New York.
Owen, E. A. and Williams, R. W. (1947). *Proc. R. Soc.* **A188,** 509.
Pipkorn, D., Edge, C. K., Debrunner, P., dePasquali, G., Drickamer, H. G. and Frauenfelder, H. (1964). *Phys. Rev.* **135,** A1604.
Pound, R. V., Benedek, G. B. and Drever, R. (1961). *Phys. Rev. Lett.* **7,** 405.
Radebaugh, R. and Keesom, P. H. (1966). *Phys. Rev.* **149,** 209.
Raimondi, D. L. and Jura, G. (1967). *J. appl. Phys.* **38,** 2133.
Samara, G. and Anderson, D. H. (1963). *J. appl. Phys.,* **35,** 3043.
Simanek, E. and Stroubec, Z. (1967). *Phys. Rev.* **163,** 275.
Slater, J. C. (1965). " Quantum Theory of Molecules and Solids ", Vol. 2. McGraw-Hill, New York.
Stern, F. (1955). Ph.D. Thesis Princeton University (unpublished).
Tischer, R. E. and Drickamer, H. G. (1962). *J. chem. Phys.* **37,** 1554.
Vaughan, R. W. and Drickamer, H. G. (1967a). *J. chem. Phys.* **47,** 468.
Vaughan, R. W. and Drickamer, H. G. (1967b). *J. chem. Phys.* **47,** 1530.
Vaughan, R. W. (1968). Private communication.
Walker, L. R., Wertheim, G. K. and Jaccarino, V. (1961). *Phys. Rev. Lett.* **6,** 98.
Watson, R. E. (1959). Report 12, Solid State and Molecular Theory Group. Massachusetts Institute of Technology (unpublished).
Wertheim, G. K. (1961). *Phys. Rev.* **124,** 764.
Wertheim, G. K. (1964). " Mössbauer Effect: Principles and Applications." Academic Press, New York.
Weyl, W. A. (1951). " Colored Glasses." Society of Glass Technology, Sheffield.
Wolcott, N. M. (1955). *In* " Conference de Physique des Basses Temperatures." Inst. International du Froid, Paris.
Wolcott, N. M. (1957). *Phil. Mag.* **2,** 1246.
Worley, R. D., Zemansky, M. W. and Boorse, H. A. (1955). *Phys. Rev.* **99,** 447.
Worlton, T. G. and Decker, D. L. (1968). *Phys. Rev.* **171,** 596.
Worlton, T. G., Bennion, R. B. and Brugger, R. M. (1967). *Phys. Rev. Lett.* **A24,** 653.
Zahner, J. C. and Drickamer, H. G. (1961). *J. chem. Phys.* **35,** 375.

CHAPTER 2

Magnetically Ordered Materials at High Pressures

D. Bloch and A. S. Pavlovic

Laboratoire d'Electrostatique et de Physique du Métal, Grenoble, France

West Virginia University, Morgantown, West Virginia, U.S.A.

I. Introduction

Studies of the effects of high pressures on the magnetic properties of solids are playing an increasingly important role in the understanding of macroscopic phenomena associated with these magnetic solids. This derives from the fact that the interactions between atoms are a function of the distance between these atoms, so that any factor which varies this distance must alter the interactions and hence, alter the macroscopic properties of the solid. Previous attempts to effect such variations were made by dilution of the solid with foreign atoms or by varying the temperature and achieving the same result by thermal dilatation. However, the former method is not very satisfactory for, in addition to changing the interatomic distance, the electronic and/or crystallographic structures of the solid also change, which complicates any analysis of the results. In the latter case, the variation in the interatomic distance is much too small to be meaningful. Thus, the application of pressure constitutes the only simple and direct technique for varying the interatomic distance with a minimum of complexity.

The results of high pressure investigations have accumulated a large store of information pertaining to the variation of the properties of magnetic solids. The various properties investigated include the magnetization, various phase transition temperatures, compressibility, thermal dilatation, magnetostriction and magnetocrystalline anisotropy. These have led to a more complete knowledge and a better understanding of (a) the chemistry of bonding, (b) the crystallographic and magnetic structure of solids, (c) magnetic phenomena and (d) the thermodynamics of solids. From these has arisen a clearer picture of the fundamental interactions between atoms in solids. Finally, the pressure coefficients of these quantities are necessary in order to convert experimental isobaric results to theoretical isovolumic quantities and *vice versa*.

The purpose of this article is to review the investigations of magnetic materials under high pressures performed since the last review by Kouvel (1963). However, since then, the amount of work has increased at an explosive rate. Up to 1962, most measurements were made of one or two properties within small temperature and pressure ranges. They were confined to a handful of metals, alloys, insulators and intermetallic compounds. In the intervening period the activities have been expanded to include (a) pressures up to hundreds of kilobars, (b) temperatures from 4·2°K up to 1000°K, (c) a greater number of elements, alloys and compounds and (d) a greater variety of physical properties.

Due to the enormous amount of published work it is necessary to restrict the coverage of this article in order to keep it within manageable bounds. Consequently it will only be concerned with work which (a) is performed under hydrostatic or pseudo-hydrostatic pressures, (b) is concerned with magnetically ordered materials and (c) measures the isothermal macroscopic properties of solids.

Section II is concerned with a brief résumé of the experimental techniques utilized in making the studies described in this paper, as well as some new experimental developments. The reader who is more interested in conventional high-pressure techniques should consult the following books or review articles: Swenson (1960), Paul and Warschauer (1963), Bundy *et al.* (1961), and Volumes 1 and 2 of this series. Contained in Section III are the main ideas of thermodynamics, molecular field theory and indirect exchange theory employed to interpret the experimental results appearing in Sections IV and V. The pertinent mathematical relations to which constant reference is made are also given here. The magnetic materials, whose properties are discussed, are divided into two categories: non-metals (Section IV) and metals (Section V). This division is a convenient one for classifying the interactions and/or the crystallographic structures of the materials. As often happens, however, materials exist whose properties cannot clearly be assigned to one or the other of these categories, and in these cases the authors have made an arbitrary decision and placed them in one or the other.

II. Experimental Methods

A. Historical Introduction

Among the first studies concerning the influence of pressure on the magnetic properties of ferromagnetic substances were those of Weissmuth in 1882 and of Tomlinson in 1887 relative to the effects of pressure on the magnetization. No variation was observed because the pressures employed, of the order of 10 b, were much too small. In 1898, Nagaoka and Honda compared the effect of pressure on the magnetization to that of the magnetic field on the volume. A Cailletet machine permitted them to submit samples of iron and nickel to pressures up to 300 b. In 1905, Frisbie studied the magnetic permeability of iron and nickel at pressures up to 1000 b. These first efforts were inconclusive. The work of Bridgman (1958) and the interest developed by geophysicists and magneticians by the variable " pressure " contributed a great deal to the development of experimental techniques. In 1925, Yeh made a well-defined study, at room temperature, of the variation of the permeability of iron, nickel and cobalt under pressure up to 13 kb. Later, Adams and Green (1931) tried to determine the variation of the Curie points of iron, nickel and magnetite at different pressures to 3·7 kb. Steinberger (1933) observed that the alloy 70% Fe–30% Ni becomes non-magnetic at room temperature under a pressure of 12 kb. It is difficult to attribute the origin of this transformation to the variation of the magnetic moment or to that of the exchange interaction in the absence of magnetization measurements at various temperatures. Ebert and Kussman (1937, 1939) measured the effect of pressure on the saturation magnetization of several alloys of iron and nickel. The actual method of measurement either of the magnetic ordering points or of the magnetization are derived for the most part from those introduced by Adams and Green or by Ebert and Kussman. The pressures were, in general, obtained by the compression of a fluid, either liquid or gas (at low or high temperatures), and were limited in the average installation to values of the order of 10 kb. Higher pressures, up to 100 kb, have been employed in certain cases to determine the effect of pressure on the magnetic ordering points. The precision obtained is not so good because the method used to obtain the pressure results in a pseudo-hydrostatic pressure; however, these experiments frequently present supplementary information.

B. MAGNETIZATION MEASUREMENTS

The effect of pressure on the magnetization of ferro- or ferrimagnetic substances is generally weak, and it is therefore necessary to use very high pressures and very sensitive detection methods. Frequently a differential method is employed. In the method introduced by Ebert and Kussman (1937) two samples of the same substance and of the same shape and size are employed; one sample is at atmospheric pressure (*B*) (Fig. 1) on the axis of the high pressure tube which contains the other sample under high pressure (*A*). The extraction of the tube, along the axis of the pole-pieces (*C*) of an electromagnet substitutes one sample for another at the centre of a measuring coil (*D*). The deflection of a calibrated ballistic galvanometer connected to the measuring coil gives the value of the variation of the magnetization due to pressure. In this experiment the electromagnet (*E*) and the measuring coils (*D*) are placed outside the high pressure vessel (*F*) containing the sample under pressure. This vessel must be fabricated of a strong non-magnetic material. The material generally used is a beryllium copper alloy. When the magnetic field solenoid and the measuring coils are placed inside the high pressure vessel (Gugan, 1958), the magnetic field applied is reduced because the limited space available limits the size of the solenoid. The utilization of pulsed high fields produced by the discharge of condensers through the solenoid (de Blois, 1962) eliminates this difficulty, but the measurement of physical quantities is less precise. It is of obvious advantage

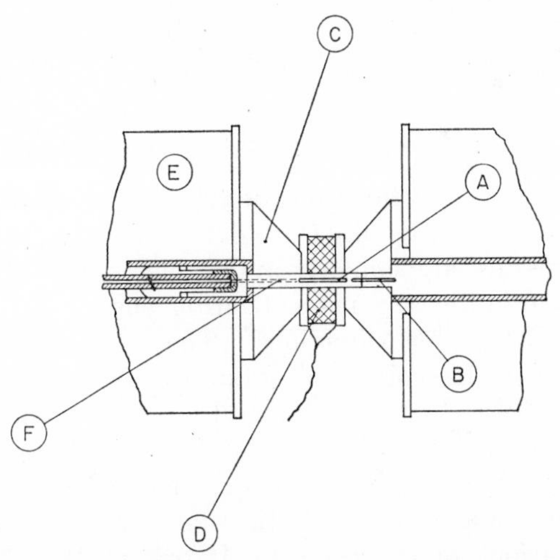

FIG. 1. An apparatus for the differential measurement of the effect of pressure on the magnetization (from Ebert and Kussman, 1937). (*A*) sample under pressure, (*B*) compensation sample at atmospheric pressure, (*C*) pole pieces, (*D*) measuring coils, (*E*) excitation coils of the electromagnet, (*F*) tube of non-magnetic berrylium-copper alloy.

to utilize here a differential method to measure the effects due to pressure (Kouvel and Wilson, 1961).

C. DETERMINATION OF MAGNETIC STRUCTURES

The study of the diffraction of neutrons by the nucleus and electrons of atoms constitutes a fruitful method for the determination of the magnetic and crystallographic structures of solids. The material used in the construction of the high pressure vessel for this purpose, as well as the pressure medium, must have small neutron absorption and scattering properties. Aluminium base alloys are generally used for the high pressure vessels and the pressure transmitting fluid is usually carbon disulfide CS_2 (Smith *et al.*, 1966). An apparatus allowing neutron diffraction studies up to 35 kb in a solid pressure medium has been described and utilized by Brugger *et al.* (1967) and Worlton *et al.* (1967).

D. MEASUREMENT OF THE MAGNETOCRYSTALLINE ANISOTROPY CONSTANTS

The free energy of a magnetic crystal is a function of the direction of its magnetic moment with respect to the crystallographic axes. One can develop this energy as a series of terms which are functions of α_1,

α_2, α_3, the direction cosines of the magnetization with respect to the crystallographic axes. For crystals of cubic symmetry, this expression is

$$E_a = K_0 + K_1(\alpha_1^2\alpha_2^2 + \alpha_2^2\alpha_3^2 + \alpha_3^2\alpha_1^2) + K_2\alpha_1^2\alpha_2^2\alpha_3^2 \tag{1}$$

where K_0, K_1 and K_2 are functions of pressure, temperature and magnetic field. The methods used to determine the constants K_1 at atmospheric pressure are similarly applicable at high pressures.

One method requires the magnetization curves of a crystal along different crystallographic directions. The magnetic field in these directions must be large enough to obtain the saturation magnetization σ_s. This method was the first to be used in high pressure experiments. Thus, in 1958, Gugan and Rowlands succeeded in obtaining the variation of the anisotropy constant K_1 for some polycrystalline samples. In this case, the shape of the magnetization curve was an average of all possible orientations of the crystal. If one neglects the anisotropy constants of order greater than one, the following expression can be written.

$$\frac{1}{K_1}\frac{\partial K_1}{\partial p} = \frac{\sigma_s^{-1}(\partial\sigma_s/\partial p)_H\sigma + H(\partial\sigma/\partial H)_p - (\partial\sigma/\partial p)_H}{H(\partial\sigma/\partial H)_p}. \tag{2}$$

The use of monocrystals however allows results which are more directly exploitable to be obtained (Kouvel, 1963). The analysis of experimental results to obtain anisotropy constants up to second order has also been done by Kouvel and Hartelius (1964a, b) for a hexagonal crystal.

One can also obtain the anisotropy constants by means of the techniques of magnetic resonance. The effect of anisotropy is equivalent to that of an applied magnetic field (Kittel, 1948). Resonance is produced in general in the range of centimetre waves. In the case of metals, considerable difficulties are encountered with such electromagnetic radiation, and as a result its use in the high pressure region has been limited to the non-metallic magnetic materials (Kaminow and Jones, 1961).

The anisotropy can be studied by measuring the torque, which is exerted on a crystal placed in a uniform magnetic field, with a torsion balance. When the magnetization vector is in a position of equilibrium, the torque due to the anisotropy energy is internally compensated by the torque exerted by the applied magnetic field. This latter torque is known from the external mechanical torque necessary to maintain the sample in equilibrium. This method is very precise because the mechanical torque measurements can attain a higher precision than that obtainable with magnetization measurements.

The entire torque balance can be placed inside the high pressure chamber. The position of equilibrium is determined either by optical means through a transparent window in the chamber wall (Veerman and Rathenau, 1965) or by an electrical signal transmitted through insulated leads passing through the chamber wall (Birss and Hegarty, 1967). Another method employed consists of isolating the high pressure gas or liquid in the high pressure vessel containing the monocrystal to be studied. This vessel is suspended from a torsion wire in a uniform magnetic field and the torque on the vessel is then measured in the usual manner.

E. MEASUREMENT OF THE MAGNETIC ORDERING TEMPERATURES

One of the correct ways of determining the magnetic ordering temperature consists of measuring the magnetization as a function of applied magnetic field at various temperatures close to the ordering temperature (Ebert and Kussman, 1939; Kaneko, 1960; Kouvel and Hartelius, 1964a).

The techniques of nuclear magnetic resonance (Benedek and Kushida, 1960; Heller and Benedek, 1962) can also be used to determine ordering points rigorously. However these measurements are difficult inside high pressure chambers so that one more frequently employs one of the other methods which, although they do not define the ordering point rigorously, give with good precision the variation of the ordering temperatures with pressure. These methods are the measurement of the thermal variation of magnetic susceptibility, electrical resistivity, thermal dilatation, magneto-resistance, and so on. The first attempt in this direction was taken by Adams and Green in 1931. The sample was the yoke of a small transformer. The secondary voltage is proportional to the permeability of the sample. Since this first attempt this technique has been employed at pressures up to the order of 100 kb (McWhan and Stevens, 1965) (Fig. 2).

One can also (Bloch and Pauthenet, 1962) place a cylindrical or spherical sample at the centre of two coils which have a mutual inductance between them; this mutual inductance is proportional to the magnetic susceptibility of the sample. These methods agree when the curves of the susceptibility at various pressures appear as a simple translation of one from the other or, better still, when the transition temperatures are characterized on the curves of the susceptibility as a discontinuity of the kind observed for a magnetic transition of the first order. The methods using the thermal variation of the resistivity or magneto-resistance present the same advantages and disadvantages.

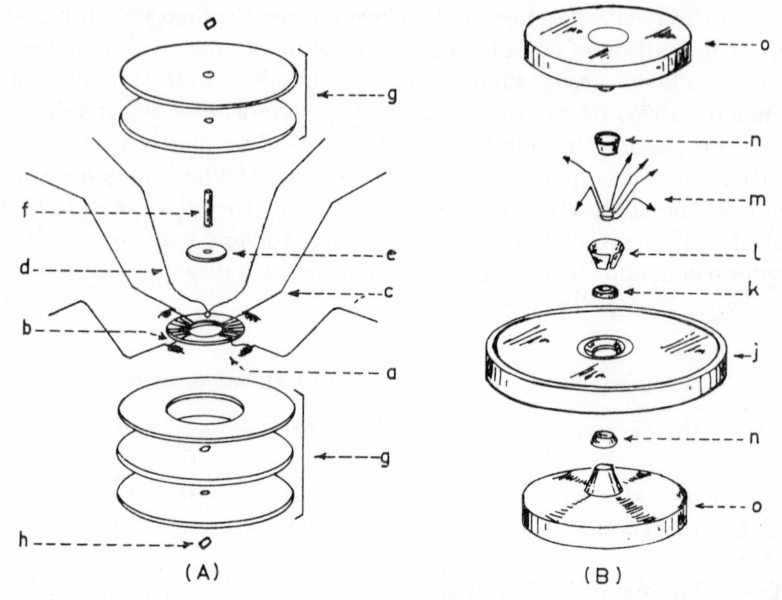

FIG. 2. Adams and Green (1931) method for the determination of magnetic ordering temperatures at high pressures extended for pressures up to 85 kb (from McWhan and Stevens, 1965). (*A*) Sample cell assembly: (a) sample, (b) No. 44 Cu wire primary and secondary coils, (c) 0·005-in Mo leads, (d) Chromel-Alumed thermocouple, (e, g) AgCl disks, (f) 0·010-in Bi wire, (h) Pt contacts. (*B*) Girdle die assembly: (j) girdle-tungsten carbide supported by maraging steel jacket, (k, n) pyrophyllite gaskets, (l) 0·001 in mica, (m) sample, cell, (o) anvils-tungsten carbide with maraging steel jacket.

In a number of cases the ordering temperatures of antiferromagnets can be located with precision from thermal dilatation studies, where the coefficient of dilatation exhibits a prominent and characteristic peak (Bloch, 1966a).

A special but important problem to consider in the studies of the variation of the ordering temperature with pressure is that of the precise determination of that temperature. The simplest solution consists of placing the junction of a thermocouple or a previously calibrated resistance just outside the high pressure chamber containing the sample. This solution is satisfactory as long as there are no thermal gradients, that is when the temperature can be stabilized. More often, the most precise measurements result from the direct measurement of the temperature by means of a thermocouple in direct contact with the sample. The pressure has, however, an effect on the electromotive force of the thermocouple (Birch, 1939, 1952; Bundy, 1961; Hanneman and Strong, 1965, 1966; Bloch and Chaissé, 1967). In all the cases noted, the thermoelectric e.m.f. due to the effect of pressure in the vicinity of room temperature is proportional to the applied pressure and to the temperature difference between the thermocouple junction and the lead exits through the wall of the pressure vessel. For this reason the corrections are often difficult to make and are in some cases a function of the experimental apparatus employed.

The other problems which are encountered such as those relative to the measurement of pressure, to pressure production, to the design calculations of the pressure chamber and to the electrical lead-ins are found in the measurement of other physical quantities under pressure as well. They are discussed in numerous articles and in particular those cited in the introduction.

III. Theoretical Relations

A. Introduction

In this section we present the many and varied mathematical relations which have been used to describe and correlate the experimental results on the effects of high-pressures on the magnetic properties of solids. These have been obtained by the application of thermodynamic methods and ideas to either theoretically or experimentally derived expressions. No pretence is made here to give a detailed discussion or derivation, but simply to give the expressions required in the discussions with sufficient explanation to make them understandable. The interested reader can go to the original literature referred to in the text.

One of the striking characteristics of magnetic materials is the large variety of magnetic transitions which do arise. These transitions are classified according to their thermodynamic order and to the types of change in the magnetic structure that take place; and usually the transition is identified by the temperature at which it occurs at atmospheric pressure. This is designated by a capital letter (T, θ, *etc.*) having appropriate subscripts or superscripts to identify the type of magnetic transition. Unfortunately, the literature is neither uniform nor consistent on this point. For the sake of consistency in this paper all transition temperatures will be designated by the symbol $\theta_{\mathrm{A}}^{\mathrm{B}}$ where A and B represent the magnetic phases below and above the transition temperature, respectively. The kinds of magnetic phases will be designated by: P = paramagnetic, F = ferromagnetic, Fi = ferrimagnetic and AF = antiferromagnetic. Thus the low temperature transition of dysprosium is designated as $\theta_{\mathrm{F}}^{\mathrm{AF}}$. The Néel temperature will continue to be designated by $\theta_N(=\theta^P_{AF})$ and also the Curie temperature $\theta_c(=\theta^P_F)$. Occasionally, more complicated situations arise. These will be explained when they appear in the text.

There is also some inconsistency in the usage of symbols for the linear and volume coefficients of expansion and the compressibility. These will be designated in the text as follows: α = the linear coefficient of expansion, β = the volume coefficient of expansion, and K = the compressibility.

B. THERMODYNAMIC RELATIONS

Let us consider a unit mass of a solid at a temperature T, under a pressure p and in a magnetic field H, such that it has a volume V and a total magnetic moment m. The Gibbs function describing the thermodynamic state of the material is

$$G = U - TS + pV - Hm \tag{3}$$

where U is the internal energy of the system and S is the entropy. If T, p and H are the independent variables, one obtains the following first derivatives of G,

$$\left(\frac{\partial G}{\partial T}\right)_{p,\,H} = -S, \qquad \left(\frac{\partial G}{\partial p}\right)_{T,\,H} = V, \qquad \left(\frac{\partial G}{\partial H}\right)_{p,\,T} = -m. \tag{4}$$

By definition a first-order transition is one in which the first derivatives of the Gibbs function exhibit a discontinuity at a critical value of the temperature, pressure or magnetic field. Thus, either one, some, or all of the derivatives of eqns (4) have a discontinuity. Since the Gibbs function G_{A} just below the transition $\theta^{\mathrm{B}}_{\mathrm{A}}$ must be equal to the Gibbs function G_{B} just above the transition, it can be shown that at constant field,

$$\frac{\partial \theta^{\mathrm{B}}_{\mathrm{A}}}{\partial p} = \frac{V_{\mathrm{B}} - V_{\mathrm{A}}}{S_{\mathrm{B}} - S_{\mathrm{A}}} = \frac{V_{\mathrm{B}} - V_{\mathrm{A}}}{L}\,\theta^{\mathrm{B}}_{\mathrm{A}} \tag{5}$$

where L is the latent heat of transformation. This is the Clapeyron equation. A second-order transition, however, is one in which the second derivatives of the Gibbs function are discontinuous. Therefore the specific heat at constant pressure C_p, the isothermal compressibility K, and the coefficient of thermal expansion α will be discontinuous. Again, because the Gibbs functions on either side of the transition must be equal, the variation of the transition temperature with pressure at constant H is

$$\frac{\partial \theta^{\mathrm{B}}_{\mathrm{A}}}{\partial p} = V\theta^{\mathrm{B}}_{\mathrm{A}}\,\frac{\beta_{\mathrm{A}} - \beta_{\mathrm{B}}}{C_{p,\,\mathrm{A}} - C_{p,\,\mathrm{B}}} = \frac{K_{\mathrm{A}} - K_{\mathrm{B}}}{\beta_{\mathrm{A}} - \beta_{\mathrm{B}}} \tag{6}$$

which is sometimes called the Ehrenfest relation. Occasionally difficulties arise from the use of eqn (6) because the discontinuities of the specific heat, the coefficient of expansion and the derivative of the magnetization are not finite and, hence, not easily measured. Such a transition is said to be a λ-type second-order transition. It can be handled by the use of

$$\frac{\partial\theta}{\partial p} = V\theta_{\mathrm{A}}^{\mathrm{B}} \frac{(\beta_{\mathrm{A}'}-\beta_{\mathrm{B}'})}{C_{p,\,\mathrm{A}'}-C_{p,\,\mathrm{B}'}} = 3V\theta_{\mathrm{A}}^{\mathrm{B}} \frac{(\alpha_{\mathrm{A}'}-\alpha_{\mathrm{B}'})}{C_{p,\,\mathrm{A}'}-C_{p,\,\mathrm{B}'}} \tag{7}$$

where the differences in β, α and C_p are the differences between these quantities occurring at two very close temperatures, $T_{\mathrm{A}'}$ and $T_{\mathrm{B}'}$, on the non-discontinuous side of the transition.

If the preceding material undergoes a small reversible process the change in the Gibbs function becomes

$$\mathrm{d}G = -S\mathrm{d}T + V\mathrm{d}p - m\mathrm{d}H$$

and, since this must be an exact differential, one immediately obtains the relation

$$\left(\frac{\partial m}{\partial p}\right)_{H,\,T} = -\left(\frac{\partial V}{\partial H}\right)_{p,\,T} \tag{8a}$$

or

$$D\left(\frac{\partial\sigma}{\partial p}\right)_{H,\,T} = -\frac{1}{V}\left(\frac{\partial V}{\partial H}\right)_{p,\,T} = -\left(\frac{\partial\omega}{\partial H}\right)_{p,\,T} \tag{8b}$$

where D is the density of the material, σ is the specific magnetization and $\left(\frac{\partial\omega}{\partial H}\right)_{p,\,T}$ the forced volume magnetostriction.

Frequently the assumption (Kouvel and Wilson, 1961) is made that the specific spontaneous magnetization of a ferromagnetic material at temperature T can be written as

$$\sigma_s = \sigma_0 \mathrm{f}\,(T/\theta_c)$$

where σ_0 is the specific spontaneous magnetization at $T = 0$°K and it is assumed that the function $\mathrm{f}\,(T/\theta_c)$ varies with the pressure by virtue only of its volume dependence of θ_c. From this relation it can be shown that

$$\frac{1}{\sigma_s}\frac{\partial\sigma_s}{\partial p} = \frac{\dfrac{1}{\sigma_0}\left(\dfrac{\partial\sigma_0}{\partial p}\right) - \dfrac{T}{\sigma_s}\left(\dfrac{\partial\sigma_s}{\partial T}\right)_p \dfrac{1}{\theta_c}\left(\dfrac{\partial\theta_c}{\partial p}\right)}{1 + 3\dfrac{\alpha T}{K}\dfrac{1}{\theta_c}\left(\dfrac{\partial\theta_c}{\partial p}\right)}. \tag{9}$$

Often $3\frac{\alpha T}{K}\frac{1}{\theta_c}\left(\frac{\partial\theta_c}{\partial p}\right) \ll 1$ so that eqn (9) can be abbreviated to

$$\frac{1}{\sigma_s}\left(\frac{\partial\sigma_s}{\partial p}\right) = \frac{1}{\sigma_0}\left(\frac{\partial\sigma_0}{\partial p}\right) - \frac{T}{\sigma_s}\left(\frac{\partial\sigma_s}{\partial T}\right)\frac{1}{\theta_c}\left(\frac{\partial\theta_c}{\partial p}\right). \tag{10}$$

This expression can be used to compute $(1/\sigma_0)(\partial\sigma_0/\partial p)$ from experimental values of the other quantities. Sometimes it has been assumed (Kornetzki, 1935, 1943) that $(1/\sigma_0)(\partial\sigma_0/\partial p)=0$ which simplifies eqn (10) and makes it possible to compute $\partial\theta_c/\partial p$.

In ferromagnetic materials there exists a volume anomaly ΔV_0 at absolute zero. This is defined as the difference between the true volume and the volume which would exist in the absence of magneto-elastic interactions. The latter volume can be obtained by extrapolation (with the aid of a Debye-type law) of the empirical volume-temperature relation from the high temperature (non-magnetic) phase. the magnetic energy W_m leads, for an isotropic substance, to (Bloch, 1966b)

$$\frac{\Delta V_0}{V_0}=-K\left(\frac{\partial W_m}{\partial V}\right)_0. \tag{11}$$

C. MOLECULAR FIELD MODEL

The generalized molecular field theory (Anderson, 1963; Smart, 1964, 1966; Elliott, 1965) has been surprisingly successful in describing the magnetic properties of many simple ferro-, ferri- and antiferromagnetic spin arrangements in solids as well as in some with more complicated spin arrangements. This description is formulated on the basis of a simple Heisenberg exchange interaction between an ion and its neighbours. In general, it is only necessary to consider the interactions of an ion with its nearest neighbours and next nearest neighbours so that the Hamiltonian of the system can be written as

$$H=-2A_1\sum_{(nn)}\mathbf{S}_i.\mathbf{S}_j-2A_2\sum_{(nnn)}\mathbf{S}_i.\mathbf{S}_j. \tag{12}$$

Parallel or antiparallel alignment of $\mathbf{S}_i$ and $\mathbf{S}_j$ is favoured according to whether the exchange integrals A_1 and A_2 are positive or negative and their relative magnitudes. If one considers these interactions on a given atom as the interaction between the magnetic moment of a central atom with the magnetic field (molecular field) created by the ions surrounding it, then one can calculate the magnetization of the solid and the temperature below which magnetic ordering will occur. The expression for the magnetic ordering temperature of various types of order and crystallographic structures have been developed by Smart (1966). These can be summarized by

$$\theta=\frac{2S(S+1)}{3k}\left[\sum_i(\pm)A_1(i)+\sum_j(\pm)A_2(j)\right] \tag{13}$$

where S is the total spin quantum number of the atom, and k is Boltzmann's constant. The sums $\sum_i (\pm) A_1(i)$ and $\sum_j (\pm) A_2(j)$ are of the exchange interactions of the nearest- and next nearest-neighbour atoms, respectively, where $(\pm)$ is associated with parallel and antiparallel configurations.

Consider, for example, a solid having a face-centred cubic (f.c.c.) lattice with twelve ferromagnetically coupled nearest-neighbour atoms and six ferromagnetically coupled next nearest-neighbour atoms. The Curie temperature of the system will be

$$\theta_c = \frac{2S(S+1)}{3k}(12A_1 + 6A_2). \tag{14}$$

In several magnetic materials A_2 is negative and much larger than A_1. This situation leads to an antiferromagnetic arrangement with the Néel temperature

$$\theta_N = -4S(S+1)\frac{A_2}{k}. \tag{15}$$

One would have also obtained the same expression if, among the twelve nearest neighbours, six were ferromagnetically coupled with the central atom and six were antiferromagnetically coupled with the central atom.

In this case, for magnetic atoms in well-defined S-states, the effect of pressure influences only the value of the exchange interaction A_2 and we have

$$\left(\frac{\partial \log \theta_N}{\partial \log V}\right)_{\theta_N} = \left(\frac{\partial \log |A_2|}{\partial \log V}\right) \tag{16}$$

where

$$\left(\frac{\partial \log \theta_N}{\partial \log V}\right)_N = \frac{-(1/K)(\partial \log \theta_N/\partial p)}{1-(3\alpha/K)(\partial \theta_N/\partial p)}. \tag{17}$$

The term $(3\alpha/K)(\partial\theta_N/\partial p)$ introduces the fact that, during the experiment, the variation of the volume arises not only from the pressure variation dp, but also from the temperature variation dθ_N. In this expression, K and α are the compressibility and the linear expansion coefficient of the non-magnetic lattice, respectively. The magnetic energy of this lattice can be approximated by

$$W_{m,0} = -\frac{3}{2}R\theta_{N,0} = -6RS^2\left|\frac{A_2}{k}\right|_0 \tag{18}$$

(R being the gas constant) and its volume dependence is given by

$$\left(\frac{\partial \log W_m}{\partial \log V}\right)_0 = \left(\frac{\partial \log |A_2|}{\partial \log V}\right)_0 = \frac{V_0(\Delta V/V_0)}{K_0 W_{m,0}} \tag{19}$$

when eqn (11) is applied. From these relations, one can obtain an expression for the exchange interaction A_2 in terms of experimentally accessible quantities (Bloch and Georges, 1968).

$$\left|\frac{A_2}{k}\right| = -\frac{V_0\theta_N(\Delta V/V_0)}{6RS^2(\partial\theta_N/\partial p)}. \tag{20}$$

Assuming the compressibility and the relative variation of exchange interaction with volume to be temperature independent, then, in a similar manner the expressions for other situations and structures can be derived.

This same approach has been applied to ferrimagnetic materials with two magnetic sub-lattices (Néel, 1948; Bloch, *et al.*, 1966) notably the ferrites. The magnetic ordering temperature is given by the expression

$$\theta_{\mathrm{Fi}} = \frac{C}{2}\left\{\lambda n_{aa}^{\theta_{\mathrm{Fi}}^p} + \mu n_{dd}^{\theta_{\mathrm{Fi}}^p} + \sqrt{\left[\left(\lambda n_{aa}^{\theta_{\mathrm{Fi}}^p} - \mu n_{dd}^{\theta_{\mathrm{Fi}}^p}\right)^2 + 4\lambda\mu\left(n_{ad}^{\theta_{\mathrm{Fi}}^p}\right)^2\right]}\right\} \tag{21}$$

in which λ and μ designate the proportion of iron ions on a and d sites, respectively. $n_{aa}^{\theta_{\mathrm{Fi}}^p}$ and $n_{dd}^{\theta_{\mathrm{Fi}}^p}$ are the molecular field coefficients describing interactions within each sub-lattice and $n_{ad}^{\theta_{\mathrm{Fi}}^p}$ between the two sub-lattices, and C is the Curie constant of the Fe^{3+} ion. Often $n_{ad}^{\theta_{\mathrm{Fi}}^p}$ is much larger than $n_{aa}^{\theta_{\mathrm{Fi}}^p}$ or $n_{dd}^{\theta_{\mathrm{Fi}}^p}$. Neglecting their variations, the volume dependence of θ_{Fi} can be related to that of the molecular field coefficient n_{ad} by the relation

$$\left(\frac{\partial \log \theta_{\mathrm{Fi}}^p}{\partial \log V}\right)_{\theta_{\mathrm{Fi}}^p} = \left(\frac{\partial \log n_{ad}}{\partial \log V}\right)_{\theta_{\mathrm{Fi}}^p}. \tag{22}$$

This relation would be applicable, too, if n_{dd}, n_{aa} and n_{ad} had the same relative volume variation.

When applied to a ferromagnetic material the molecular field theory presents the following relation for the Curie temperature

$$\theta_c = \frac{(J+1)\, nM\sigma_0^2}{3JR} \tag{23}$$

where J is the total angular momentum quantum number, M the atomic mass of the material, σ_0 the specific magnetization at absolute zero, R the gas constant, and n the molecular field coefficient. It is often considered that the volume variation of the molecular field coefficient is the interesting quantity, since this coefficient is related to the exchange integral. For a constant J, one has:

$$\left(\frac{\partial \log n}{\partial \log V}\right)_T = \left(\frac{\partial \log \theta_c}{\partial \log V}\right)_T - 2\left(\frac{\partial \log \sigma_0}{\partial \log V}\right)_T. \tag{24}$$

It is necessary in this case to consider both the variation of the Curie temperature and the specific magnetization at absolute zero with volume.

The molecular field theory presented above has also been invoked to describe the magnetic behaviour of materials which have an anti-ferromagnetic state with a helicoidal spin structure (Enz, 1961; Herpin, 1962; Nagamiya, 1962). These materials have a spin structure in which the magnetic moments in planes perpendicular to some crystallographic direction are ferromagnetically aligned in the plane, with the alignment direction rotated through some angle ω with each successive plane along the crystallographic direction.

Experimentally it has been found that the magnetic susceptibility behaviour of these materials is like that in collinear antiferromagnetic materials except that at a certain threshold magnetic field, H_t, this helicoidal spin structure collapses into a ferromagnetic fan structure. If one considers N successive planes of identical atoms with spin S and magnetic moment μ, the exchange energy can be written as

$$E_{\text{ex}} = -NS^2A(\omega) \tag{25}$$

where ω is the angle between the directions of the moments in two adjacent planes and $A(\omega)$ is the sum

$$A(\omega) = A_0 + 2A_1 \cos \omega + 2A_2 \cos 2\omega. \tag{26}$$

A_0 is the interaction constant between the spins of the atoms in the same plane, while A_1 and A_2 are the interaction constants between spins in a plane and the spins in the first and second neighbouring planes to it. Rocher (1962) has shown that it is not necessary to take into account the interaction of planes further away than the second neighbouring plane. Considering only the exchange interaction to be important, the equilibrium angle ω is given by $\cos \omega = -A_1/4A_2$; thus the helicoidal structure occurs when $|A_1/4A_2| < 1$.

Nagamiya (1962) has established the following relations between the experimental quantities H_t and χ_{AF} with the interaction constants:

$$\mu H_t = S^2 [A(\omega) - A(0)] \tag{27}$$

$$\chi_{AF} = \frac{\mu^2}{2S^2 [2A(\omega) - A(2\omega) - A(0)]}. \tag{28}$$

This model has been extended by Lee (1964), Landry (1966) and Bartholin and Bloch (1968) to a one dimensional model in which one considers only the change in the exchange energy with the distance between the ferromagnetic planes. Specifically, in some rare earth metals, the ferromagnetic planes are normal to the c-axis. Bartholin and Bloch (1968) have developed

$$\frac{\mathrm{d}(A_1/k)}{\mathrm{d}\log c} \cos\omega_\theta + \frac{\mathrm{d}(A_2/k)}{\mathrm{d}\log c} \cos 2\omega_\theta = \frac{3J\theta}{4S^2(J+1)} \frac{\mathrm{d}\log\theta}{\mathrm{d}\log c} \tag{29}$$

where θ is the highest ordering temperature; ω_θ, the helical pitch angle at θ, is equal to $qc/2$ (q is the wave vector of the helix and c the c-axis parameter).

In the preceding developments, it has always been assumed that the exchange integrals were independent of the temperature. However, if thermal expansion is considered, and if one supposes the thermal expansion coefficient β to be temperature independent, one is led to a temperature dependent exchange integral

$$A_{ij} = A_{ij}^0 \left(1 + a\frac{\Delta V}{V_0}\right) = A_{ij}^0 (1 + a\beta T) \tag{30}$$

where $$a = \left(\frac{\partial \log A_{ij}}{\partial \log V}\right)_T.$$

When θ_p, the paramagnetic Curie temperature, is proportional to A_{ij} (Néel, 1951) one obtains a temperature dependent paramagnetic Curie temperature θ_p as follows

$$\chi = \frac{C}{T - \theta_p} = \frac{C}{T - \theta_p^0(1 + a\beta T)} = \frac{C'}{T - \theta_p'} \tag{31}$$

where $$C' = C(1 + a\beta\theta_p^0) \simeq C(1 + a\beta\theta_p') \tag{32}$$

and $$\theta_p' = \theta_p^0(1 + a\beta\theta_p^0). \tag{33}$$

C', β and θ_p' are experimentally obtainable quantities.

D. INDIRECT EXCHANGE MODEL

There are good theoretical and experimental reasons to indicate that direct exchange interactions are not the dominant ones in the rare earth metals. Contrarily the most recent view (De Gennes, 1958, 1962; Rocher, 1962) is that an indirect exchange interaction between atoms occurring *via* the conduction electrons as intermediaries is the dominant interaction. These ideas result in an indirect coupling A_{mn}. If one assumes a spherical Fermi surface between two ions with magnetic moments $\boldsymbol{\mu}_m$ and $\boldsymbol{\mu}_n$

$$A_{mn} = \frac{9\pi}{4} Z_i^2 \frac{\Gamma^2 (g-1)^2}{V^2 E_F} F(2k_F R_{mn}) \, \boldsymbol{\mu}_m \cdot \boldsymbol{\mu}_n \tag{34}$$

where Z_i is the ionic charge, Γ the coupling constant between the spin of a $4f$ electron and a conduction electron, E_F the Fermi energy, k_F the wave vector of the conduction electron having E_F, V the atomic volume, g the Landé factor of the ions, R_{mn} the distance between the ions m and n, and $F(\rho)$ is the oscillatory Ruderman–Kittel function. The magnetic ordering temperature θ is given by

$$\theta = \frac{3\pi Z_i^2}{4k} \frac{\Gamma^2}{V^2 E_F} (g-1)^2 J(J+1) \sum F(2k_F R_{mn}).$$

The magnetization ΔM of the polarized conduction electrons is given by

$$\Delta M = (g-1) J \frac{3 Z_i \Gamma}{4 E_F} \mu_B$$

and the magnetic resistivity ρ_m due to the disordered spins by

$$\rho_m = \frac{3\pi}{8} \frac{m^*}{he^2} (g-1)^2 \frac{\Gamma^2}{V E_F} J(J+1)$$

(m^* is the effective electron mass).

The sum appearing in the expression for θ has been shown (Liu, 1961) to be independent of the volume. Therefore, pressure derivatives of these three quantities can easily be written in terms of the volume variations of Γ and m:

$$-\frac{1}{K}\left(\frac{1}{\theta}\frac{\partial \theta}{\partial p}\right)_{H,T} = -\frac{4}{3} + 2\left(\frac{\partial \log \Gamma}{\partial \log V}\right)_{H,T} + \left(\frac{\partial \log m^*}{\partial \log V}\right)_{H,T} \tag{35}$$

$$-\frac{1}{K}\left(\frac{1}{\rho_m}\frac{\partial \rho_m}{\partial p}\right)_{H,T} = -\frac{1}{3} + 2\left(\frac{\partial \log \Gamma}{\partial \log V}\right)_{H,T} + 2\left(\frac{\partial \log m^*}{\partial \log V}\right)_{H,T} \tag{36}$$

$$-\frac{1}{K}\left(\frac{1}{\Delta M}\frac{\partial \Delta M}{\partial p}\right)_{H,T} = \frac{2}{3} + \left(\frac{\partial \log \Gamma}{\partial \log V}\right)_{H,T} + \left(\frac{\partial \log m^*}{\partial \log V}\right)_{H,T} \tag{37}$$

IV. Non-metallic Compounds

A. Introduction

The distinction between metals and non-metals is useful in magnetism for often one can avoid a certain number of difficulties which appear in the case of metals by describing the exchange interaction in non-metals with the Heisenberg exchange Hamiltonian:

$$H = -2A_1 \sum_{(nn)} \mathbf{S}_i . \mathbf{S}_j - 2A_2 \sum_{(nnn)} \mathbf{S}_i . \mathbf{S}_j \dots \tag{12}$$

where the A_1 are the exchange integrals between first neighbour magnetic ions, and A_2 the exchange integrals between second neighbour magnetic ions. The experiments discussed in this chapter show that in the case of non-metals, the exchange interactions decrease rapidly when the distance between magnetic atoms increases, so that we will only consider the exchange interactions between first neighbours (MnF_2, EuO . . .) or the exchange interactions between first and second neighbours (MnO, FeO, CoO, α-MnS). Also discussed in this chapter are some compounds such as GdN, GdAs . . . which although having metallic conduction appear to be capable of being discussed with the Hamiltonian expression (12).

The mathematical determination of the eigen-values of eqn (12) is very complex and generally one must resort to various approximations. Thus in the approximation of the Weiss molecular field (Weiss, 1907) generalized by Néel (1932), one introduces the action of the spins $\mathbf{S}_j$ on the spin $\mathbf{S}_i$ in the form of an effective magnetic field proportional to the magnetization of j atoms neighbouring the atom i. One can then determine the energy levels and thence the principal thermodynamic properties such as the thermal variation of the magnetization and magnetic specific heat. From a comparison between theory and experiment, one can deduce the value of A_1, and A_2. However, the values thus obtained depend in a certain way on the approximations utilized in the treatment of the Hamiltonian eqn (12) (see for example, Smart, 1966). Conversely, the results of magnetic experiments under pressure are utilized to determine the variations of the exchange interactions with the interatomic distance. In certain cases, however, the same experimental results, associated with other experimental results, allow the determination of the values of the exchange interactions themselves.

A certain number of magnetic phenomena such as the antisymmetric exchange, biquadratic exchange, anisotropic exchange and the crystalline field, do not appear directly, however, in this kind of treatment. One must therefore be aware in the different cases, of the eventual importance of these effects.

B. NACL TYPE COMPOUNDS WITH TRANSITION ELEMENTS

1. *Introduction*

The concept of antiferromagnetism, introduced by Néel in 1932, has received striking confirmation by the results of neutron diffraction (Shull and Smart, 1949; Shull *et al.*, 1951). The principal experimental results and theoretical works have been reviewed by Nagamiya *et al.* (1955) and Anderson (1963). The monoxides and monosulphides of the transition elements—MnO, α–MnS, FeO and CoO—have been intensively studied because of their simple NaCl-type crystallographic structures above their magnetic ordering temperatures. Each magnetic atom has twelve first-magnetic neighbours and six second-magnetic neighbours. In the magnetically ordered phase six first neighbours have their spin directions parallel to that of the central magnetic atom whereas the six others are antiparallel (Fig. 3). In order to facilitate

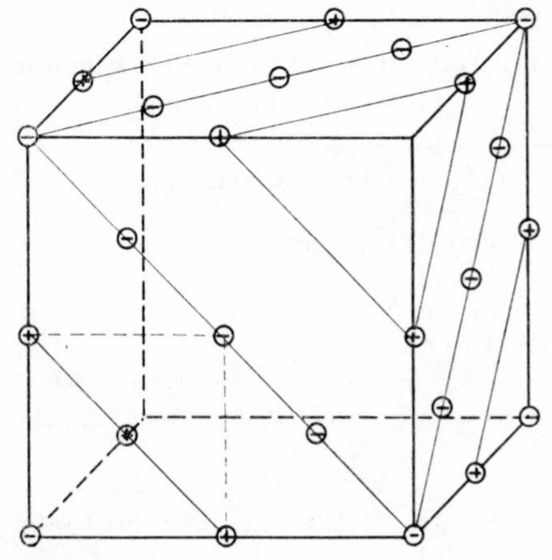

Fig. 3. The ordering of magnetic moments in MnO, FeO, CoO and α-MnS.

the discussion, some of the characteristic physical constants of these compounds have been given in Table I, and in Table II the values of the exchange interactions between first neighbours (A_1) and between second neighbours (A_2) obtained within the frame work of the molecular field theory. Equations (15) to (20) can be applied if one neglects the lattice distortion which can appear, in certain cases, below the magnetic ordering temperature.

2. *Manganese Oxide*

The Néel temperature, θ_N, of manganese oxide (MnO) defined as the temperature at which the coefficient of dilatation has a maximum,

TABLE I. Some physical data for MnO, α-MnS, FeO and CoO

	Néel (θ_N) temperature (°K) (from specific heat measurement)	Néel (θ_N) temperature (°K) (from dilatation measurement)	Atomic volume (V_x) (cm^3) (room temperature)	Initial compressibility $(K) \times 10^6$ (b^{-1}) (room temperature)
MnO	117·8[a]	116[d]	13·23[h]	0·70[h]
		117·9[e]		
α–MnS	140[b]	149·5[f]	21·46[h]	1·25[h]
FeO	188·5[a]	186[d]	12·01[h]	0·65[h]
CoO	287·3[c]	292[d]	11·63[h]	0·52[h]
		287·25[g]		

[a] From Todd and Bonnickson (1951), [b] from Anderson (1931), [c] from King (1957), [d] from Foex (1948), [e] from Bartholin *et al.* (1967), [f] from Georges (1969), [g] from Bloch (1966a), [h] from Clendenen and Drickamer (1966).

TABLE II. Exchange interaction coefficients from molecular-field theory

	Direction of the magnetic moments	$-A_1/k$ (°K)[d]	$-A_2/k$ (°K)[d]
MnO	Parallel to (111) plane[a]	7·2	3·5
α–MnS	Parallel to (111) plane[b]	4·4	4·5
FeO	Perpendicular to (111) plane[a]	7·8	8·2
CoO	Difficult to assert[a, c]	6·9	21·6

[a] From Roth (1958), [b] from Corliss *et al.* (1956), [c] from Van Laar (1965), [d] from Smart (1964).

increases with the pressure at the rate of $0{\cdot}30 \pm 0{\cdot}02$°K/kb (Bartholin *et al.*, 1967) (Fig. 4). This value is in good agreement with 0·3°K/kb deduced (Janusz, 1960) from the expression (7) when the experimental results of specific heat (Millar, 1928) and the coefficient of dilatation (Föex, 1948) are used. By assuming the compressibility of the lattice at the Néel temperature equal to the compressibility measured at room temperature (Table I) and neglecting the dilatation of the lattice (eqn 17) one has (from eqn 16)

$$\left(\frac{\partial \log |A_2|}{\partial \log V}\right)_{\theta_N} = -3{\cdot}6.$$

The volume anomaly at absolute zero $\Delta V_0/V_0$ is about -3×10^{-3} (Föex, 1948) or $-(3{\cdot}8 \pm 0{\cdot}4) \times 10^{-3}$ (Bloch *et al.*, 1968).

Molecular field theory (Table II) gives $A_2/K = -3{\cdot}5$°K while the

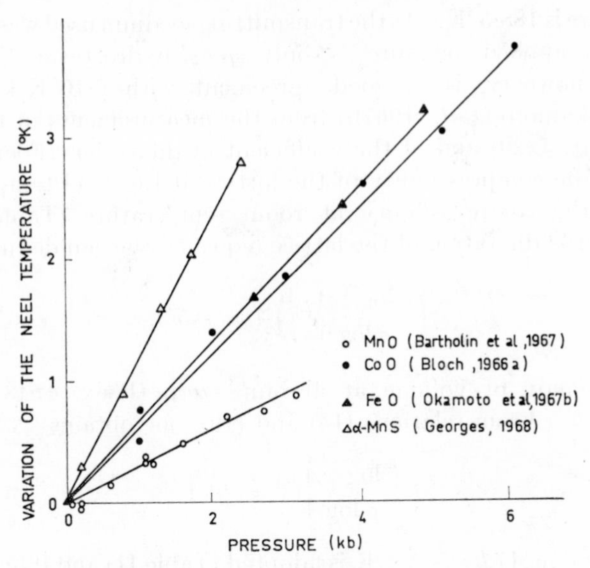

FIG. 4. Variation with pressure of the Néel temperature of MnO, α-MnS, FeO and CoO.

random phase approximation (Lines and Jones, 1965) $A_2/k = -5{\cdot}5$°K. From the eqns (18) and (19) one deduces

$$-3{\cdot}3 < \frac{\partial \log |A_2|}{\partial \log V} < 5{\cdot}2$$

depending on the value of $|A_2|$ used.

Magnetic resonance experiments of pairs of Mn^{2+} ions in MgO (Coles *et al.*, 1960) lead, for second neighbour manganese ions, to $A_2/k = -14$°K. The lattice parameter of MgO being 4·22 Å and that of MnO 4·45 Å, one notes that the value $A_2/k = -14$°K can be obtained from the value $A_2/k = 6{\cdot}4$°K (see Table III), if one assumes that the exchange interactions between second neighbours varies with a power of -15 of the interatomic distance, that is with the power of -5 of the volume. An important part of the difference between the exchange interactions A_2 between Mn^{2+} in MnO and in MgO can thus be interpreted by the modification of the interatomic distances.

3. *Iron Oxide*

According to Okamoto *et al.* (1967b), the Néel temperature of iron oxide (FeO) determined as the temperature at which the magnetic susceptibility is a maximum, increases linearly with the pressure at the rate of 0·65°K/kb (Fig. 4). At atmospheric pressure the Néel temperature is 188·5°K. As the transmitting medium used was petroleum ether, the applied pressure is only *quasi*-hydrostatic. The result obtained, however, is in good agreement with 0·70°K/kb deduced (eqn 6) (Okamoto *et al.*, 1967b) from the measurements of the specific heat (Millar, 1929) and of the coefficient of dilatation (Föex, 1948).

Taking the compressibility of the lattice at the Néel temperature as equal to the compressibility at room temperature (Table I) while neglecting the dilatation of the lattice (eqn 17), one can deduce (eqn 16)

$$\left(\frac{\partial \log |A_2|}{\partial \log V}\right)_{\theta_N} = -5{\cdot}3.$$

The anomaly of volume at absolute zero (Föex, 1948) is about $-3{\cdot}6\times 10^{-3}$. From relations (18) and (19), one obtains

$$\frac{\partial \log |A_2|}{\partial \log V} = -4{\cdot}1$$

when the value $A_2/k = -8{\cdot}2$°K is adopted (Table II) and if one supposes that the magnetic moment of the Fe^{2+} ion is independent of the temperature in the temperature region considered.

4. *Cobalt Oxide*

The Néel temperature of cobalt oxide (CoO) determined by the temperature of the maximum in the coefficient of dilatation curve, increases with the pressure at the rate of $(0{\cdot}60 \pm 0{\cdot}03)$°K/kb (Bloch, 1966a) (Fig. 4). This value is in good agreement with that of 0·64°K/kb (Okamoto *et al.*, 1967b) determined from magnetic susceptibility measurements under pressure. One can also compare these to the values 0·8°K/kb and 0·7°K/kb deduced from the thermodynamic expressions (6) (Okamoto *et al.*, 1967b) and (7) (Janusz, 1960) from the specific heat measurements of Assayag and Bizette (1954) and the coefficient of dilatation measurements of Föex (1948). The studies by the Mössbauer effect of ^{57}Co in CoO at high pressure (Coston *et al.*, 1966) confirm the increase of the Néel temperature with pressure.

If one uses the room temperature compressibility (Table I) for that at the Néel temperature and neglects the dilatation of the lattice (eqn 17), one obtains

$$\left(\frac{\partial \log |A_2|}{\partial \log V}\right)_{\theta_N} = -4{\cdot}1.$$

The volume anomaly $(\Delta V_0/V_0)$ at absolute zero (Föex, 1948) is about

-3×10^{-3} while from the expressions (8) and (19) one arrives at

$$\frac{\partial \log |A_2|}{\partial \log V} = -2{\cdot}8$$

when using $A_2/k = -21{\cdot}6°K$ (Table II).

5. *Manganese Sulphide*

The Néel temperature of α-manganese sulphide (MnS) ascertained from the maximum in the coefficient of dilatation increases with pressure at the rate of $(1{\cdot}20 \pm 0{\cdot}02)°K/kb$ (Georges, 1969) (Fig. 4). If one assumes the lattice compressibility at the Néel point to be approximately equal to that at room temperature with the lattice dilatation being neglected (eqn 17), one obtains

$$\left(\frac{\partial \log |A_2|}{\partial \log V}\right)_{\theta_N} = -6{\cdot}4.$$

The magnetic volume anomaly at absolute zero (Georges, 1969) is $\Delta V_0/V_0 = -(4{\cdot}75 \pm 0{\cdot}5)\times10^{-3}$, from which one finds

$$\frac{\partial \log |A_2|}{\partial \log V} = -5{\cdot}8$$

by using expressions (18) and (19) and the value $A_2/k = -4{\cdot}5°K$ (Table II).

6. *The Evaluation of Exchange Interactions*

The expression of the Néel temperature θ_N as a function of the exchange interaction A_2 depends on the model and approximations employed. If one considers that the temperature θ_N is proportional to A_2, the expression (16) can be applied again. The expression (18) itself, to a first approximation, is model independent, so that one can use the expression (20).

Taking the values of V_0 and θ_N from Table I and those of $\Delta V_0/V_0$ and $\partial\theta_N/\partial p$, one arrives at the values of A_2/k in Table III.

7. *Manganese Ditelluride*

Manganese ditelluride ($MnTe_2$) is a semiconductor with a pyrite (FeS_2) structure. This structure is very close to a NaCl structure, the anions being replaced by the $(\chi_2)^{2-}$ groups whose axes point in the direction of the body diagonals. $MnTe_2$ is antiferromagnetic and, at atmospheric pressure, the Néel temperature is 87°K (Sawaoka and Miyahara, 1965). In the molecular field approximation (Smart, 1964)

TABLE III. Evaluation of exchange interactions from measurements of the Néel temperature at high pressure

	S	$\frac{\Delta V_0}{V_0} \times 10^3$	$\frac{\partial \theta_N}{\partial p} \times 10^3$ (°K/b)	A_2/k (°K)
MnO	5/2	-3[a]		-5
		$-(3 \cdot 8 \pm 0 \cdot 4)$[b]	$(0 \cdot 30 \pm 0 \cdot 02)$[d]	$-(6 \cdot 4 \pm 1 \cdot 1)$
α-MnS	5/2	$-(4 \cdot 75 \pm 0 \cdot 5)$[c]	$(1 \cdot 20 \pm 0 \cdot 02)$[c]	$-(4 \cdot 1 \pm 0 \cdot 5)$
FeO	2	$-3 \cdot 6$[a]	$0 \cdot 65$[e]	$-6 \cdot 2$
CoO	3/2	-3[a]	$(0 \cdot 60 \pm 0 \cdot 03)$[f]	-15

[a] From Föex (1948), [b] from Bloch *et al.* (1968), [c] from Georges (1969), [d] from Bartholin *et al.* (1967), [e] from Okamoto *et al.* (1967b), [f] from Bloch (1966a).

$A_1/k = -7{\cdot}4$°K and $A_2/k = -1{\cdot}2$°K. The values are not compatible with the antiferromagnetic arrangement observed, which is of the first order. Each atom of manganese has eight first neighbours with their moments antiparallel with it and four first neighbours with their moments parallel, while the moments of the six second neighbours are antiparallel (Hastings *et al.*, 1959). The direction of the magnetic moments is parallel to the ferromagnetic (001) plane. It is therefore probable that the values of the exchange interaction given by the molecular field theory will not be very good. The Néel temperature whose value is given by:

$$\theta_N = \frac{2S(S+1)}{3k}(-4A_1 + 6A_2) \tag{38}$$

increases linearly with pressure (Fig. 5) at the rate of 0·17°K/kb up to 100 kb (Sawaoka *et al.*, 1966). The experimental Curie constant $C = 4{\cdot}84$ (Hastings *et al.*, 1959) is larger than the theoretical Curie constant $C = 4{\cdot}40$ in agreement (eqn 32) with the sign observed for the variation of θ_N with pressure.

C. NaCl TYPE COMPOUNDS WITH RARE EARTH ELEMENTS

1. *Introduction*

The oxide and chalcogenides of europium (EuO, EuS, EuSe and EuTe) are semi-conductors. Their crystallographic structure is of the NaCl type, with the europium ion in a divalent state and an $S = 7/2$ electronic configuration. Gadolinium nitride and arsenide, GdN and GdAs, are metallic. However, a molecular field model is able to interpret their

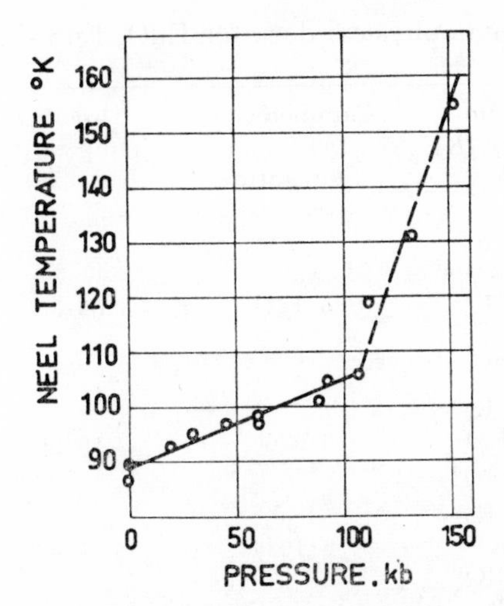

FIG. 5. Pressure dependence of the Néel temperature of $MnTe_2$. Solid circle shows data at atmospheric pressure after being compressed above 120 kb (from Sawaoka *et al.*, 1966).

magnetic behaviour. The gadolinium ion is trivalent with an S electronic state and the crystallographic structure is of the NaCl type. Some of these compounds are ferromagnetic (EuO, EuS, GdN) and some are antiferromagnetic (EuSe, EuTe, GdAs). By means of the molecular field approximation, their magnetic ordering temperatures are given by expressions (14) or (15).

2. *Europium Oxide*

The Curie temperature of europium oxide (Table IV) increases with the pressure at the rate of about $(0{\cdot}4 \pm 0{\cdot}1)$°K/kb. The exchange interactions between first neighbours (A_1/k) have an amplitude much larger than that of the exchange interactions between second neighbours (A_2/k) (Table V). One can assume that the relative variations with pressure are comparable so that one can neglect the variations o A_2/k compared to those of A_1/k. The compressibility (Table IV) being of the order of 1×10^{-3}/kb one obtains

$$\left(\frac{\partial \log |A_2|}{\partial \log V}\right)_{\theta_c} = -5{\cdot}5.$$

The origin of the exchange interactions in these compounds is still a question (de Graaf and Xavier, 1965; Xavier, 1967; Smit, 1966).
The magnetic volume anomaly at absolute zero (Argyle *et al.*, 1967)

TABLE IV. Some physical data for EuO, EuS, EuSe and EuTe

	Initial compressibility (K) $\times 10^6$ (b^{-1})	Parameter (a) (Å) (room temperature)	Ordering temperature θ_c (°K)	$\frac{\partial \theta}{\partial p} \times 10^3$ (°K/b)
EuO	4·3 (296°K)[a] 2·6 (82°K)[a] 2·0 (4·2°K)[a] 1 (RT)[b] 0·94 (RT)[c]	5·144[g]	69·35[h]	0·5[a] (0·4 ± 0·1)[b] (0·37 ± 0·1)[c]
EuS	4·8 (300°K)[d] 3·3 (80°K)[d] 2·5 (4·2°K)[d]	5·970[g]	16·2[i]	0·28[e] 0·20[k]
EuSe	4·0 (300°K)[k] 3·3 (80°K)[k] 2·5 (4·2°K)[k] ~1[e] (RT)	6·194[g]	4·58[j]	~1[e] 0·16[k]
EuTe	~3[f] (RT)	6·597[g]	9·64[j]	

[a] From Stevenson and Robinson (1965), [b] from Sokolova *et al.* (1966), [c] from McWhan *et al.* (1966a), [d] from Stevenson (1966), [e] from Schwob and Vögt (1967), [f] from Rooymans (1965), [g] from Busch *et al.* (1966b), [h] from Boyd (1966), [i] from Moruzzi and Teaney (1963), [j] from Busch *et al.* (1964), [k] from Srivastava and Stevenson (1968).

is $(\Delta V_0/V_0) = -2{\cdot}4 \times 10^{-3}$. One can obtain A_1 from the expression

$$|A_1/k| = -\frac{V_0\theta_f(\Delta V_0/V_0)}{12RS^2(\partial\theta_c/\partial p)} \tag{39}$$

analogous to expression (20).

Inserting the experimental values above gives $|A_1|/k = 0{\cdot}7$°K, in good agreement with 0·65°K computed with the molecular field approximation and 0·75°K (Boyd, 1966) from a spin wave analysis of the n.m.r. frequencies at low temperatures.

TABLE V. Exchange interaction coefficients from molecular-field theory (Busch *et al.*, 1966b)

	A_1/k (°K)	$-A_2/k$ (°K)
EuO	0·65	0·06
EuS	0·20	0·08
EuSe	0·13	0·11
EuTe	0·03	0·17

3. *Europium Sulphide*

The Curie temperature of europium sulphide increases with pressure at the rate of $(0{\cdot}28 \pm 0{\cdot}01)$°K/kb according to Schwob and Vögt (1967) and 0·20°K/kb according to Srivastava and Stevenson (1968). The exchange interactions between first neighbours are preponderant (Table V). According to Stevenson (1966) the compressibility will be about 3×10^{-3}/kb at the Curie temperature of EuS, so that one gets $(\partial \log |A_1|/\partial \log V) \simeq -5$, a value of the same order as that obtained for europium oxide.

4. *Europium Selenide*

The exchange interactions between first and second neighbours of EuSe are of the same order of magnitude. One notes (Fig. 6) a meta-

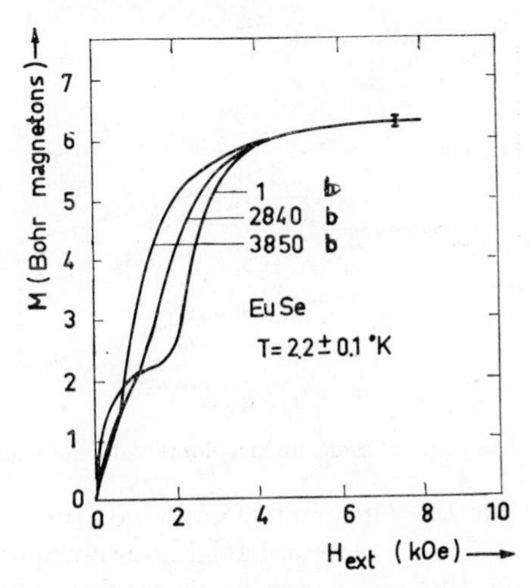

Fig. 6. Magnetization of EuSe *versus* external magnetic field at three different hydrostatic pressures (from Busch *et al.*, 1966a).

magnetic transition between the antiferromagnetic state at low fields and the ferromagnetic state at high fields. These two states are separated by a complex intermediate structure, which is suppressed by the application of a pressure of about 4 kb. Studies at higher fields than 600 Oe show that the ferromagnetic interactions A_1 augment with the applied pressure. Below 600 Oe the antiferromagnetic coupling appears favoured by the pressure, which indicates that the amplitude A_2 increases when the interatomic distances decrease. According to Srivastava and Stevenson (1968), the Néel temperature increases with pressure at the rate of 0·16°K/kb, which corresponds to a relative variation more important than in the case of EuS or EuO.

5. *Europium Telluride*

Europium telluride EuTe is antiferromagnetic, with the principal exchange interactions being negative ones between second neighbours. The magnetic volume anomaly at absolute zero is negative (Rodbell *et al.*, 1965) which indicates an increase in the amplitude of the exchange interactions when the interatomic distances decrease. If one plots (Fig. 7) the values of the exchange interactions as a function of the

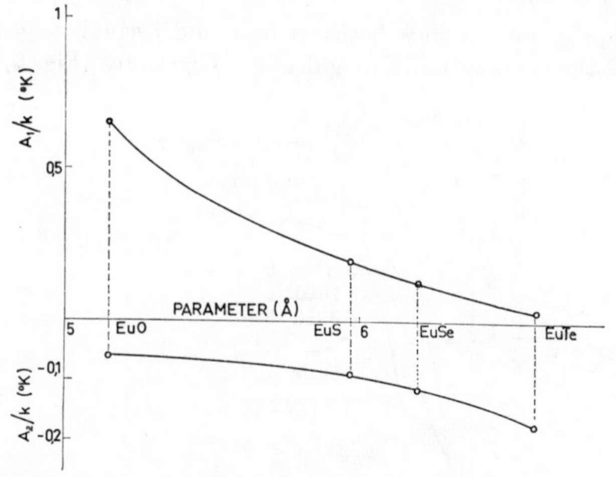

FIG. 7. Exchange interactions in europium oxide and chalcogenides.

lattice parameter for the europium compounds, one notices that the exchange interaction between second neighbour europium ions increases in magnitude when the parameter increases by substitution of one cation for another. The results obtained for EuSe and EuTe tend to show that this increase occurs essentially from chemical substitution and not from the modification of distances. However, for EuO and EuS, it indicates that the exchange interactions between first neighbours varies qualitatively in the same way whether one changes the lattice dimension by chemical substitution or by the application of pressure.

6. *Gd^{3+} Compounds*

Gadolinium nitride (GdN) is ferromagnetic whereas gadolinium arsenide (GdAs) is antiferromagnetic (Table IV). The magnetic properties of GdN have been studied rather completely under pressure

(McWhan, 1966). Its Curie temperature is of the order of 69°K at atmospheric pressure (Rebouillat and Veyssié, 1964) and increases with pressure at the rate of (0·08 ± 0·04)°K/kb. Since the initial compressibility is about $(0{\cdot}52 \pm 0{\cdot}1) \times 10^{-3}$/kb at room temperature (McWhan, 1966), the variation $(\partial \log \theta_c / \partial \log V)$ corresponds to $-2{\cdot}2 \pm 1{\cdot}5$. GdAs presents at low temperatures, below its Néel temperature, a rhombohedral distortion of its lattice similar to that observed in the case of manganese oxide (MnO). At 5°K the angle at the apex is equal to (90·034 ± 0·007)° and one deduces (Jones and Morosin, 1967) that the exchange interaction between first neighbours, responsible for this distortion, increases when their distance decreases.

D. GARNET FERRITES

1. *Introduction*

The ferrites with a garnet structure, having the chemical formula $Fe_5M_3O_{12}$, form an important class of magnetic compounds. The ferrimagnetic model of Néel (1954) allows a simple interpretation, with good precision, of their principal magnetic properties (Pauthenet, 1958). Their crystallographic structure is cubic, and isomorphic with grossularite, which is a natural garnet with the formula $Ca_3Al_2Si_3O_{12}$ (Bertaut and Forrat, 1956a). In this compound, the calcium occupies the sites 24*c*, the aluminium the sites 16*a*, and the silicon the sites 24*d*. The ferrites of the rare earths with a garnet structure, which have been much studied, result from the forementioned structure by substitution of the rare earth or yttrium ions M^{3+} at sites 24*c*, and iron ions Fe^{3+} at the sites 16*a* and 24*d*. The unit cell contains eight chemical formulae, sixteen Fe^{3+} ions occupying the 16*a* positions, twenty-four Fe^{3-} ions the positions 24*d* and twenty-four M^{3+} the positions 24*c*. The magnetic ions are separated by ninety-six O^{2-} ions. Also it is probable that the principal magnetic interactions may be superexchange interactions with

TABLE VI. Exchange coefficients and crystallographic parameters for GdN and GdAs

	GdN	GdAs
A_1/k (°K)	0·56[a]	0·10[a] −0·08[b]
$-A_2/k$ (°K)	0·14[a]	0·40[a, b]
Lattice parameter (Å) (room temperature)	4·986	5·862

[a] From Busch *et al.* (1966b), [b] from Jones and Morosin (1967).

the oxygen ions as intermediaries. Generally, it is considered that the principal magnetic interactions are negative interactions between Fe^{3+} situated on the *a* and *d* sublattices with the ferrimagnetic assemblage of Fe^{3+} ions coupled antiferromagnetically to the M^{3+} ions of the *c* sublattice.

2. *Effect of Pressure on the Curie Temperature of the Rare Earth Iron Garnets*

It has been found that the Curie temperature (θ_{Fi}^{P}) is principally defined by the interactions between iron ions and, indeed, is practically independent of the nature of the rare earth or yttrium ion in the lattice (Fig. 8). Thus, in the vicinity of the Curie point, one is dealing with a ferrimagnetic substance decomposable into two sublattices, *a* and *d*, of

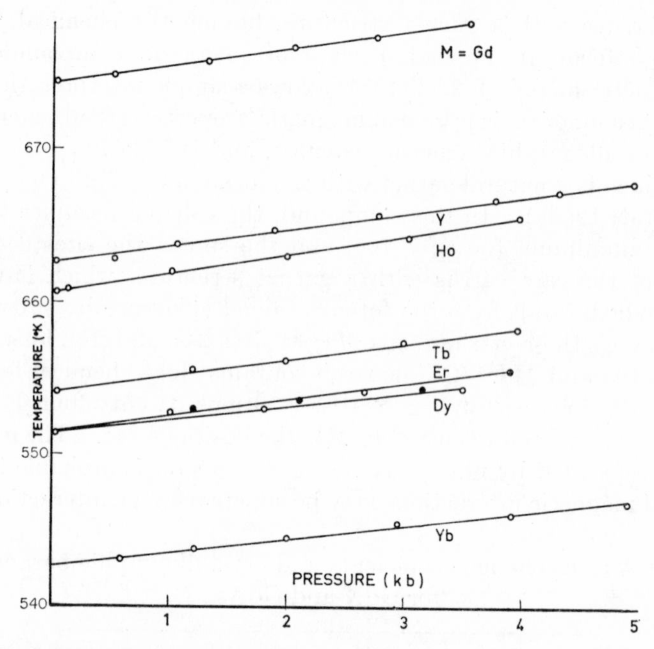

FIG. 8. Variation with pressure of the Curie temperature of rare earth iron garnets formula $5Fe_2O_3$–$3M_2O_3$ (from Bloch *et al.*, 1967).

ferric ions. The application of pressure produces a variation of the superexchange interactions n_{aa} and n_{dd} within each of the two sublattices as well as of the superexchange interactions n_{ad} between the two sublattices *a* and *d*. However, the absolute saturation moments of the ferric ions which are in *S* orbital states, are not modified. The ferrimagnetic Curie temperature θ^P_{Fi} can be expressed by the relation (21) and the variation with volume of n_{ad} by expression (22).

θ^P_{Fi} increases linearly with pressure (Fig. 8) (Table VII) with a rate close to 1·2°K/kb. At room temperature, the compressibility of the yttrium ion garnet is of the order of $6{\cdot}8\times10^{-3}$/kb (Kaminow and Jones, 1960). This value will be adopted in the following for the different rare earth iron garnets. The values of $(\partial \log \theta^P_{Fi}/\partial \log V)_{\theta^P_{Fi}}$ thus determined are of the same order of magnitude and close to the value $-10/3$ (Bloch, 1966b). This variation is opposite to that which is expected of the variation of the Curie point of the different iron garnets as a function of their parameter (Fig. 9). The superexchange interac-

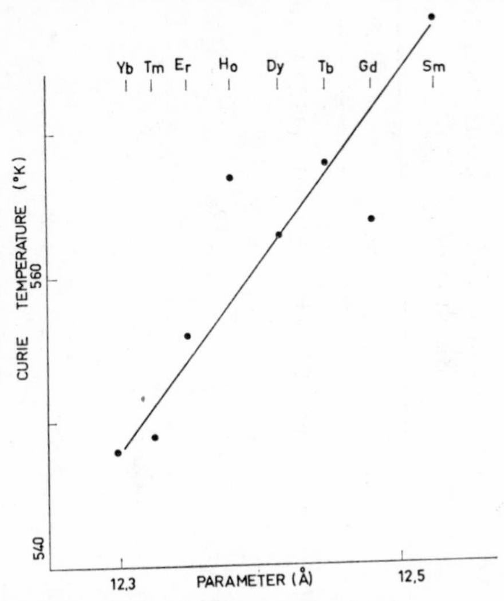

FIG. 9. Variation of the Curie temperatures of a function of the crystallographic parameter a for various rare earth iron garnets. Values of a from Bertaut and Forrat (1956). Values of θ^P_{Fi} from Pauthenet (1958).

tions are functions not only of the iron–oxygen distance but also of the angle of the interaction Fe^{3+}–O^{-2}–Fe^{3+}. One can assume, to a first approximation, that pressure modifies the interatomic distance without modifying the interaction angles. However, the substitution of one rare earth ion for another modifies (Euler and Bruce, 1965) simultaneously the distance and the angles of interaction. The crystal parameter a decreases regularly with the ionic radius of the rare earth in accordance with the lanthanide contraction rule; from the samarium iron garnet to the lutecium iron garnet; the interaction angle decreases simultaneously from $(127{\cdot}1\pm0{\cdot}3)^\circ$ for the samarium iron garnet to $(123{\cdot}9\pm0{\cdot}5)^\circ$ for the lutecium iron garnet.

TABLE VII. Relative variations with volume of the Curie and compensation temperatures of rare earth iron garnets (from Bloch *et al.*, 1967)

M^{3+}	Y^{3+}	Gd^{3+}	Tb^{3+}	Dy^{3+}	Ho^{3+}	Er^{3+}	Yb^{3+}
$\frac{\partial \theta^P_{Fi}}{\partial p} \times 10^3$ (°K/b)	$(1\cdot25 \pm 0\cdot05)$	$(1\cdot28 \pm 0\cdot05)$	$(1\cdot23 \pm 0\cdot05)$	$(1\cdot15 \pm 0\cdot05)$	$(1\cdot28 \pm 0\cdot05)$	$(1\cdot22 \pm 0\cdot05)$	$(1\cdot08 \pm 0\cdot05)$
$\left(\frac{\partial \log \theta^P_{Fi}}{\partial \log V}\right)_{\theta^P_{Fi}}$	$-3\cdot4_5$	$-3\cdot4_8$	$-3\cdot4_5$	$-3\cdot2_2$	$-3\cdot5_5$	$-3\cdot4_4$	$-3\cdot0_8$
$\frac{\partial \theta_t}{\partial p} \times 10^3$ (°K/b)	$(0\cdot95 \pm 0\cdot07)$	$(0\cdot77 \pm 0\cdot05)$	$(0\cdot40 \pm 0\cdot1)$	$(0\cdot38 \pm 0\cdot2)$			
$\left(\frac{\partial \log \theta_t}{\partial \log V}\right)_{\theta_t}$	$-4\cdot8$	$-4\cdot5$	$-2\cdot7$	$-4\cdot2$			

3. *Effect of Pressure on the Compensation Temperatures of the Rare Earth Iron Garnets*

The ferrimagnetic configuration of Fe^{3+} ions magnetizes the M^{3+} ions on the sublattice c in a direction opposite to the resultant magnetization. The iron garnets with M = Gd, Tb, Dy, Ho, Er and Tm possess a compensation point at which the total magnetization is zero. The average molecular field coefficient n representative of the interactions between the iron ions and M^{3+} ions is much more important than the molecular field coefficient n_{cc} between the M^{3+} ions themselves (Pauthenet, 1958). So to a first approximation, valid for temperatures above 100°K, the magnetization σ_c of the rare earth ions on the sublattice c in the absence of an applied field can be written as

$$\sigma_c = \frac{C_c}{T} n(\mu' \sigma_d - \lambda' \sigma_a) \tag{40}$$

where C_c is the Curie constant of the M^{3+} ions, σ_d and σ_a the magnetization of the Fe^{3+} ions on the sublattices a and d, and λ', μ', γ' are the relative proportions of the magnetic ions on the sites a, d, and c, respectively. At the compensation temperature θ_t, we have

$$\theta_t = n\gamma' C_c. \tag{41}$$

The Curie constant of the Gd^{3+} ions is independent of the volume of the sample, its orbital quantum number is zero. The orbital angular momentum quenching temperature of the ions Tb^{3+}, Dy^{3+} and Ho^{3+} is smaller than their compensation temperature so that one can write

$$\left(\frac{\partial \log \theta_t}{\partial \log V}\right)_{\theta_t} = \left(\frac{\partial \log n}{\partial \log V}\right)_{\theta_t}. \tag{42}$$

The compensation temperature of the iron garnets of gadolinium, terbium, dysprosium and holmium increase linearly with pressure (Table VII). The corresponding variations $(\partial \log \theta_t / \partial \log V)_{\theta_t}$ are close to those obtained for the variation of the Curie point $(\partial \log \theta_{Fi}^P / \partial \log V)_{\theta_{Fi}^P}$.

4. *Other Results*

Kaminow and Jones (1960) have studied the ferrimagnetic resonance under pressure of the iron garnets of yttrium, ytterbium and erbium. At room temperature, the variation with pressure of the magnetic moment of the yttrium ferrite garnet increases by $(0{\cdot}26 \pm 0{\cdot}09) \times 10^{-3}$/kb, which corresponds to a variation of the Curie temperature with pressure of 0·3°K/kb. This value can be compared to 0·69 ± 0·02°K/kb, deduced from n.m.r. experiments at pressures up to 10 kb between 196 and 346°K (Litster and Benedek, 1966) and to $1{\cdot}25 \pm 0{\cdot}05$°K/kb deduced from the direct measurement of the pressure effect on the ordering temperature described in the previous paragraph. The anisotropy constant K_1 increases rapidly with pressure (Table VIII). One notes also in the case of the erbium iron garnet a strong increase with pressure of the magnetization of the erbium ions. This can be attributed (Kaminow and Jones, 1960) to n having the variation $(\partial \log n/\partial \log V) = 7{\cdot}0$. Two other ferrites with garnet structures have likewise been studied under pressure (Foiles and Tomizuka, 1965); the ferrite of gadolinium with one atom of iron replaced by aluminium $Gd_3Fe_4AlO_{12}$, and the mixed ferrite $Gd_{1{\cdot}5}Y_{1{\cdot}5}Fe_{4{\cdot}5}Al_{0{\cdot}5}O_{12}$. Their ferrimagnetic Curie temperatures are 426 and 493°K respectively. They increase with pressure at the rate of $0{\cdot}58 \pm 0{\cdot}05$°K/kb and $1{\cdot}00 \pm 0{\cdot}05$°K/kb which corresponds to a variation $(\partial \log \theta^P_{Fi}/\partial \log V)_{\theta^P_{Fi}}$ of -2 and -3. It should be noted that the magnetic volume anomaly of yttrium iron garnet is negative (Geller and Gilleo, 1957) in agreement with the positive sign of the Curie temperature variation with pressure. Similarly, because of the thermal dilatation, the experimental Curie constant has a higher value than that theoretically anticipated, which corresponds likewise to a diminution of the exchange interaction when the volume increases (Table IX).

E. SPINEL FERRITES

The spinel ferrites have the chemical formula MFe_2O_4 where M is a divalent ion and Fe is a trivalent iron ion. Their crystallographic structure is cubic with a unit cell containing eight chemical formulae, that is sixteen Fe^{3+}, thirty-two O^{-2} ions and eight M^{2+} ions. The stacking of oxygen ion layers defines two types of interstitial sites, the A sites at a centre of an oxygen ion tetrahedron and the B sites at the centre of an oxygen ion octahedron. In a unit cell there exist sixty-four A sites of which only eight are occupied and thirty-two B sites of which sixteen are occupied. One can classify the divalent ions according to their preference for A sites (Mn^{2+}, Zn^{2+}) or for B sites (Co^{2+}, Fe^{2+}, Ni^{2+}). In the normal spinel structure, the eight divalent ions are in the A sites and the sixteen trivalent iron ions are in the B sites. If the eight divalent ions go into B sites while eight Fe^{3+} ions go into A sites, one has then an inverse spinel structure. The exchange interactions are all negative, the preponderant exchange interactions being the interactions between magnetic ions situated on the A and B sublattices.

TABLE VIII. Variation of the anisotropy constants of cubic compounds with pressure

	Fe_3O_4	$MnFe_2O_4$	$Mn_{0\cdot8}Zn_{0\cdot2}Fe_2O_4$	$5Fe_2O_3.3Y_2O_3$	$5Fe_2O_3.3Yb_2O_3$
$\frac{1}{K_1}\frac{\partial K_1}{\partial p}\times 10^6$/b	$-13\cdot5\pm0\cdot5$[ab] (RT)	$7\cdot0\pm0\cdot5$[b] (RT) $11\cdot0\pm0\cdot5$[b] (77°K)	$6\cdot0\pm0\cdot5$[b] (RT) $11\cdot8\pm0\cdot05$[b] (77°K)	$7\cdot2\pm0\cdot15$[c] (RT)	15 ± 1[c] (RT)
	$5Fe_2O_3.3Er_2O_3$	$MgFe_2O_4$	$NiFe_2O_4$	$Ni_{0\cdot95}Co_{0\cdot05}Fe_2O_4$	$Ni_{0\cdot9}Co_{0\cdot1}Fe_2O_4$
$\frac{1}{K_1}\frac{\partial K_1}{\partial p}\times 10^6$/b	$(7\cdot4\pm0\cdot5)$[c] (RT)	$3\cdot5$ to $4\cdot1$[c] (RT) (modified by thermal treatment)	$4\cdot5\pm0\cdot3$[c] (RT)	$-(2\cdot9\pm0\cdot3)$[c] (RT)	$-7\cdot5\pm1$[c] (RT)

[a] From Sawaoka and Kawaï (1967), [b] from Sawaoka *et al.* (1967), [c] from Kaminow and Jones (1960).

TABLE IX. Variation of exchange interaction with volume deduced from paramagnetic susceptibility and dilatation experiments, according to expression (32)

	$Fe_2O_3.NiO$	$Fe_2O_3.CoO$	$Fe_2O_3.FeO$	Fe_2O_3	$5Fe_2O_3.3Y_2O_3$
$-\gamma \times 10^4$	2·77[a]	1·17[a]	1·37[a]	1·27[a]	1[b]
$\alpha_T \times 10^6$	11[c]	12[c]	14[d]	12[d]	10[e]
$\dfrac{\partial \log n}{\partial \log V} = \dfrac{\gamma}{3\alpha_T}$	−8·4	−3·25	−3·26	−3·53	−3·36

[a] From Néel (1951), [b] from Pauthenet (1958), [c] from Weil (1950), [d] from Chenevard (1921), [e] from Geller and Gilleo (1957).

1. *Nickel-zinc Ferrite* ($Ni_{1-x}Zn_xFe_2O_4$)

The structure of the nickel ferrite is inverse. Each of the two sublattices A and B possess eight Fe^{3+} ions so that the spontaneous magnetization corresponds to the Ni^{2+} ions alone. The Zn^{2+} ions of substitution do not possess a magnetic moment; they have a preference for A sites, so that when the zinc concentration increases the magnetic moment increases and the Curie temperature decreases. The nickel ferrite $NiFe_2O_4$ has a Curie temperature of 889°K at atmospheric pressure (Foiles and Tomizuka, 1965), and it increases with pressure at the rate of $1{\cdot}16 \pm 0{\cdot}07$°K/kb (Table X). One observes in all the cases, whatever the concentration of nickel between 100 and 20%, an augmentation of the Curie temperature with pressure. The compressibility

TABLE X. Variation with pressure of the Curie temperature of various ferrites

	$NiFe_2O_4$	$Ni_{0\cdot8}Zn_{0\cdot2}Fe_2O_4$	$Ni_{0\cdot5}Zn_{0\cdot5}Fe_2O_4$
θ^{P}_{Fi} (°K)	889[a]	761[a]	550[a]
$\dfrac{\partial \theta_P}{\partial p}$ (°K/b)×10³	1·16±0·07[a]	0·73±0·06[a]	0·99±0·04[a]
	$Ni_{0\cdot3}Zn_{0\cdot7}Fe_2O_4$	$Ni_{0\cdot2}Zn_{0\cdot8}Fe_2O_4$	$Mn_{0\cdot5}Zn_{0\cdot5}Fe_2O_4$
θ^{P}_{Fi} (°K)	481[a] 318[c]	206[b]	363[d]
$\dfrac{\partial \theta^{P}_{Fi}}{\partial p}$ (°K/b)×10³	0·83±0·05[a] 0·87±0·06[c]	1·56±0·05[b]	0·9 ±0·04[d] 1·04±0·05[b]

[a] From Foiles and Tomizuka (1965), [b] from Kume *et al.* (1966), [c] from Werner (1959), [d] from Patrick (1954).

of $NiFe_2O_4$ at room temperature is about $0{\cdot}51 \times 10^{-3}$/kb (Waldron, 1955); using this value one obtains

$$\frac{\partial \log \theta^P_{Fi}}{\partial \log V} \simeq \frac{\partial \log |n_{AB}|}{\partial \log V} = -2{\cdot}4.$$

If one assumes the room temperature and Curie temperature compressibilities to be approximately the same, the corresponding value computed by means of expression (32) of the difference between the theoretical and experimental Curie constant is equal to $-8{\cdot}4$ (Table IX), a value much larger than that obtained directly.

2. *Manganese Zinc Ferrites*

The manganese and zinc ions occupy A sites; the non-magnetic Zn^{2+} ions substitute for magnetic Fe^{3+} ions which, at low concentration, contribute to an increase of the spontaneous magnetization and a diminution of the Curie temperature. For zinc concentrations greater than 42% the exchange interactions between A and B sublattice no longer have a large amplitude compared to the exchange interactions within the B sublattice, so that the spontaneous magnetization decreases at the same time as the Curie temperature (Guillaud and Greveaux, 1950).

The Curie temperature of the ferrite $Mn_{0{\cdot}5}Zn_{0{\cdot}5}Fe_2O_4$ increases linearly with pressure (Table X). This variation corresponds to the simultaneous modifications of the exchange interactions between sublattices A and B, and within sublattice B itself.

3. *Other Results*

The ferrite of iron (or magnetite) Fe_3O_4, or cobalt $CoFe_2O_4$ or nickel $NiFe_2O_4$, have experimental Curie constants greater than the theoretical Curie constants C'. This difference can be interpreted (eqn 32) as due to the thermal variation of the exchange interactions because of the thermal dilatation. The predominant exchange interactions are those which exist between the sublattices A and B. In all the cases (Table IX) n_{AB} diminishes when the temperature (and the volume) increases, in agreement with the positive sign of the magnetic volume anomaly for Fe_3O_4 (Chenevard, 1921), $CoFe_2O_4$, $NiFe_2O_4$ and $CuFe_2O_4$ (Weil, 1950).

In Table VIII are presented the principal results obtained from the studies of the effects of pressure on the magnetocrystalline anisotropy constants. The discussion of the experimental results is made difficult by the absence of specific experimental data on the pressure variation of the oxygen ion positions, of the spin-orbit coupling constant and of the radii of the electronic shells.

F. PEROVSKITE-TYPE COMPOUNDS

The perovskites have the chemical formula ABX_3, where A is a cation (or an atom) of large size (K, La, Sr or a rare earth element), B is a cation (or an atom) of smaller size (Cr, Mn, Fe, Co, Ga, Ti, V . . .) and X an anion (or atom) for example O or F. Although most of the perovskites may be ionic, some are metallic. The ideal perovskite structure is cubic; the B ions form a simple cubic sublattice with the anions at the edge centres of the cube and the A ions are at the body centre of the cube. The real structure is derived from the ideal structure by a small distortion.

The ionic perovskite $La_{0\cdot75}Sr_{0\cdot25}MnO_3$ has a Curie temperature of 353°K, which is elevated with pressure at the rate of $(0{\cdot}6 \pm 0{\cdot}04)$°K/kb (Patrick, 1954). The metallic perovskite $GaCMn_3$ (Howe and Myers, 1957) has a Curie temperature of 246°K, which increases at the rate of 1·3°K/kb. This compound also has a first-order transition at 150°K between ferromagnetic and antiferromagnetic states which decrease with pressure at the rate of $-2{\cdot}5$°K/kb (Bouchaud *et al.*, 1966). It is accompanied by a sudden increase in the volume of 1·5%. The variation with pressure of the magnetic transition temperature, as well as the volume discontinuity, has been explained by means of the Kittel (1960) theory of exchange inversions (Guillot and Pauthenet, 1966).

G. HEXAGONAL AND RHOMBOHEDRAL COMPOUNDS

A number of compounds can be obtained from the NiAs structure by an ordered distribution of vacancies. In the oxides α-Fe_2O_3 and Cr_2O_3, a third of the cations are absent, the vacancies being ordered in such a way that each Fe^{3+} or Cr^{3+} ion has only one neighbour along the trigonal c-axis (Fig. 10) which produces a slight displacement of the cations out of the (111) plane perpendicular to the c-axis.

1. α-Fe_2O_3

Hematite, α-Fe_2O_3, is antiferromagnetic with a Néel temperature of 950°K. The order of the spins along the c-axis is $+ + - - + + - -$, their direction below a transition temperature θ_M, called the Morin (1950) temperature, being along the c-axis and above θ_M in the plane perpendicular to the c-axis (Shull *et al.*, 1951). This transition is associated with the disappearance of the weak superposed ferromagnetism (Dzialoshinskii, 1958; Moriya, 1960) which exists between θ_M and θ_N. At atmospheric pressure the Morin temperature θ_M is about 250°K (Néel and Pauthenet, 1952). According to Kawai and

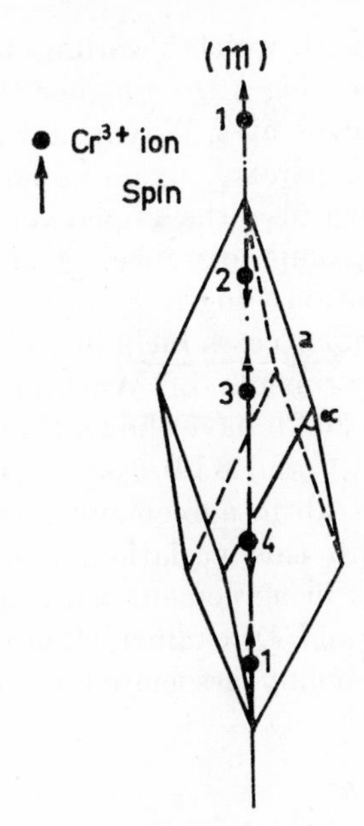

FIG. 10. Magnetic structure and axis of Cr_2O_3 (from Brockhouse, 1953).

Ono (1966), θ_M varies up to 2 kb at a rate of 10°K/kb. Neutron diffraction studies (Umebayashi *et al.*, 1966) up to 1·5 kb indicate a smaller rate of $3{\cdot}7 \pm 0{\cdot}2$°K/kb accompanied by a hysteresis in temperature of about 11°, independent of pressure. An identical value ($3{\cdot}6 \pm 0{\cdot}3$°K/kb) has been obtained by n.m.r. experiments at pressures up to 4·14 kb (Wayne and Anderson, 1967). These results are interpreted by the theory of Artman *et al.* (1965) according to which the Morin transition corresponds to a change of the sign of the total anisotropy energy K_T, the sum of an ionic anisotropy energy K_{SI} and a dipolar anisotropy energy K_{MD}. At absolute zero K_{MD} is negative and K_{SI} positive with an amplitude a little larger than K_{MD}. When the temperature increases, the amplitude of K_{SI} decreases more rapidly than that of K_{MD}. The Morin transition occurs when K_{SI} becomes equal in magnitude to K_{MD}. The variation of the Morin temperature with pressure is given by the relation

$$\frac{\partial \theta_M}{\partial p} = C\left(-\frac{\partial K_{SI}}{\partial p} + \frac{\partial K_{MD}}{\partial p}\right) \tag{43}$$

where C is a parameter independent of the pressure. The variation of θ_M, as a result of the modification of the dipolar anistropy energy

K_{DM}, is of the order 2°K/kb which according to Wayne and Anderson (1967) leads to a value of $\partial \log K_{SI}/\partial p$ higher than 10^{-3}/kb. Since the Fe^{3+} ions are in electronic S states, Wayne and Anderson attributed the effect observed to the " superexchange anisotropic energy " K_{SE} of the order of 10% of K_{SI} because the superexchange interactions vary rapidly with the interatomic distance. One must note, however, (Table VIII) that a variation $\partial \log K/\partial p$ of the order of 10^{-6} has been obtained in numerous similar cases including magnetic ions in S states.

The more recent experiments of Worlton *et al.* (1967) utilizing neutron diffraction up to 26 kb have shown that one must consider two distinct regions; between 0 and 6 kb θ_M varies linearly with pressure at the rate of $3{\cdot}8 \pm 0{\cdot}3$°K/kb in agreement with the preceding results, but between 6 and 26 kb this variation, again linear, is less rapid ($1{\cdot}0 \pm 0{\cdot}3$°K/kb). This is in agreement with Mössbauer effect studies under pressure of Lewis and Drickamer (1966) who have obtained at room temperature an anomaly associated with the Morin transition, between 30 and 60 kb.

2. Cr_2O_3

Cr_2O_3, like α-Fe_2O_3, is antiferromagnetic with a Néel temperature at about 307°K. The order of the spins along the c-axis is $+-+-$, with the spins pointing along the c-axis (Brockhouse, 1953) (Fig. 10). The linear magnetic dilatation anomaly is positive along the c-axis and negative along the a-axis (Greenwald, 1956) with the resultant magnetic volume anomaly being negative.

The pressure variation of the Néel temperature obtained under pressures up to 13·3 kb by neutron diffraction is, however, negative and equal to $-(1{\cdot}6 \pm 0{\cdot}3)$°K/kb (Worlton *et al.*, 1967).

3. $FeCl_2$ and $FeBr_2$

$FeCl_2$ and $FeBr_2$ have a hexagonal crystallographic structure of the $CdCl_2$ type. The magnetic structure is antiferromagnetic, having ferromagnetic planes and magnetic moments parallel to the c-axis with nearest neighbouring planes being coupled antiferromagnetically. At atmospheric pressure, $FeCl_2$ possesses a Néel temperature at 24°K (Starr *et al.*, 1940) and $FeBr_2$ at 11°K (Jacobs and Lawrence, 1965). A polymorphic transformation occurs for $FeCl_2$ at 4°K and 2 kb, probably between the forms $CdCl_2{}^{I}$ and $CdI_2{}^{II}$ (Fig. 11) (Narath and Schirber, 1966). $FeCl_2$ and $FeBr_2$ are both metamagnetic, having a threshold field of transition H_c proportional to the magnitude of the

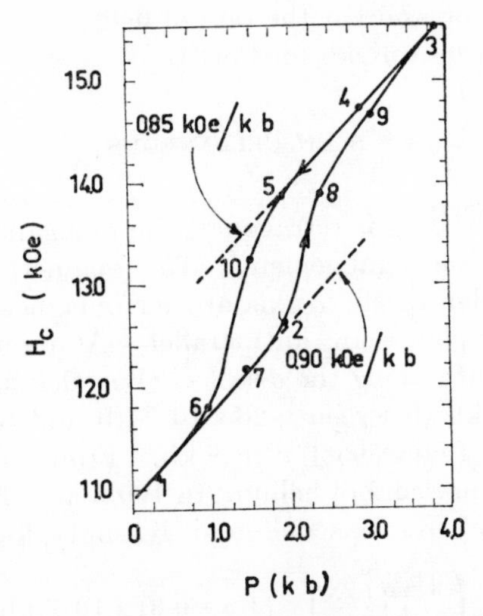

FIG. 11. Plot of metamagnetic critical field (H_c) for $FeCl_2$ as a function of hydrostatic pressure at 4·0°K. The experimental points are numbered in the sequence in which they were taken (from Narath and Schirber, 1966).

antiferromagnetic exchange interaction A_2 between ferromagnetic layers, so that one can write :

$$\frac{\partial \log H_c}{\partial \log V} = \frac{\partial \log |A_2|}{\partial \log V}. \tag{44}$$

At 4°K, the threshold field of $FeCl_2{}^{I}$ is 10,920 Oe at atmospheric pressure. It increases linearly with pressure as

$$(\partial \log H_c/\partial p)_{I} = (8{\cdot}7 \pm 0{\cdot}5)\ 10^{-2}/\text{kb}.$$

The threshold field of $FeCl_2{}^{II}$, extrapolated to atmospheric pressure is 12,240 Oe and one gets $(\partial \log H_c/\partial p)_{II} = (6{\cdot}9 \pm 0{\cdot}5) \times 10^{-2}/\text{kb}$. At atmospheric pressure, the threshold field of $FeBr_2$ is 23,800 Oe, and its pressure variation is $(\partial \log H_c/\partial p) = (6 \pm 0{\cdot}5) \times 10^{-2}/\text{kb}$. The three values obtained are close; in all cases $|A_2|$ increases with pressure, while at the same time the interatomic distance decreases.

4. *$MnCO_3$ and $CoCO_3$*

$MnCO_3$ and $CoCO_3$ have a rhombohedral structure with the magnetic moments of the manganese and cobalt ions ordered antiferromagnetically. There is a weak ferromagnetism superimposed on this antiferromagnetism. According to Astrov *et al.* (1961), who used the method of compression by freezing of Lazarev and Kan (1944), θ_N

increases with the pressure at the rate of $0{\cdot}42 \pm 0{\cdot}1$°K/kb in the case of $MnCO_3$, and $0{\cdot}10 \pm 0{\cdot}01$°K/kb for $CoCO_3$.

H. MISCELLANEOUS

1. *MnF_2*

The structure of MnF_2 is tetragonal; the manganese Mn^{2+} ions form an antiferromagnetic arrangement. The magnetic moments in the planes perpendicular to the c-axis are ferromagnetically aligned; the nearest neighbour planes are antiparallel. At room temperature the linear compressibility along the c-axis is (Benedek and Kushida, 1960) about $0{\cdot}31 \times 10^{-3}$/kb (between $0{\cdot}26 \times 10^{-3}$/kb and $0{\cdot}33 \times 10^{-3}$/kb) and along the a-axis $(0{\cdot}65 \pm 0{\cdot}03) \times 10^{-3}$/kb. From n.m.r. experiments under hydrostatic pressure of helium (to 100 b at 4°K; 700 b at 20·4°K and 1000 b at 35·7°K) Benedek and Kushida found the variation

$$\left(\frac{1}{\theta_N}\right)\left(\frac{\partial \theta_N}{\partial p}\right) = (4{\cdot}5 \pm 0{\cdot}3) \times 10^{-3}/\text{kb}$$

corresponding to

$$\frac{\partial \theta_N}{\partial p} = (0{\cdot}3 \pm 0{\cdot}2)\,°\text{K/kb}.$$

This value has been less accurately specified by Heller and Benedek (1962). The Néel temperature of MnF_2 which is $67{\cdot}336 \pm 0{\cdot}003$°K at atmospheric pressure increases at a rate of $0{\cdot}309 \pm 0{\cdot}003$°K/kb compared to the value 0·8°K/kb for polycrystalline MnF_2 obtained by Astrov *et al.* (1960) from magnetic susceptibility measurements under pressure produced by freezing water.

The magnetic anomaly of length (Gibbons, 1959) is negative along the c-axis, in agreement with the positive sign observed for the variation of the Néel temperature with pressure. It is positive, but small, along the a-axis. One can assume that the variations of the exchange interactions with the parameter c are dominant. Because of the difference between the linear compressibility along the c and a axes, the superexchange interaction angles may vary with pressure.

2. *$FeCl_2.2H_2O$ and $CoCl_2.2H_2O$*

The dihydrates $FeCl_2.2H_2O$ and $CoCl_2.2H_2O$ are isomorphic, the unit cell having monoclinic symmetry (Narath, 1964, 1965). They are antiferromagnetically ordered below 23°K ($FeCl_2.2H_2O$) and 17·2°K ($CoCl_2.2H_2O$). Like $FeCl_2$ and $FeBr_2$, these compounds are strongly anisotropic, and their magnetic configuration is made up of ferromagnetic chains coupled by two negative exchange interactions (A_1 and

A_2) to neighbouring chains. When a magnetic field is applied parallel to the direction of the spontaneous magnetization of the sublattices, the competition of the exchange interactions A_1 and A_2 result in two successive transitions between antiferro- and ferrimagnetic and ferri- and ferromagnetic configurations. The two threshold fields are given by the following relations:

$$\begin{aligned} H_{c,1} &= (g\mu_B)^{-1}(-A_1z_1 + 2A_2z_2) \\ H_{c,2} &= (g\mu_B)^{-1}(-A_1z_1 - A_2z_2). \end{aligned} \tag{45}$$

From their variations with pressure (Table XI) one can deduce (Narath and Schirber, 1966) the variations of A_1 and A_2. It should be noted that $|A_1|$ and $|A_2|$ increase with pressure when the volume of the sample decreases.

TABLE XI. Pressure effects on $FeCl_2.2H_2O$ and $CoCl_2.2H_2O$ (from Narath and Schirber, 1966)

	$FeCl_2.2H_2O$	$CoCl_2.2H_2O$
$H_{c,1(p=1)}$Oe	39200	31340
$(\partial \log H_{c,1}/\partial p) \times 10^5$/b	$1 \cdot 71 \pm 0 \cdot 08$	$1 \cdot 05 \pm 0 \cdot 05$
$H_{c,2(p=1)}$Oe	45600	44900
$(\partial \log H_{c,2}/\partial p) \times 10^5$/b	$2 \cdot 32 \pm 0 \cdot 11$	$2 \cdot 60 \pm 0 \cdot 18$
$-z_1A_1$ °K	24·8	17·4
$(\partial \log \lvert A_1 \rvert/\partial p) \times 10^5$/b	$2 \cdot 1 \pm 0 \cdot 1$	$2 \cdot 2 \pm 0 \cdot 1$
$-z_2A_2$ °K	1·2	1·9
$(\partial \log \lvert A_2 \rvert/\partial p) \times 10^5$/b	$5 \cdot 9 \pm 0 \cdot 1$	$6 \cdot 2 \pm 0 \cdot 6$

3. *MnTe*

In the NiAs structure, the anions form a hexagonal close-packed arrangement with cations occupying the octahedral voids. MnTe which has the NiAs structure is antiferromagnetic and the orientation of the magnetic moments is indicated in Fig. 12. The magnetic moment of the manganese ions is about $5\mu_\beta$, and the Néel temperature at atmospheric pressure is 310°K. The electrical conductivity of MnTe is that of a semi-conductor with hole conduction in the valence band. At room temperature, the scattering of these holes by the disordered magnetic moments of manganese is preponderant so that the value of the Néel temperature θ_N can be determined from the peak of the thermal variation of resistivity curve (Fig. 13). θ_N increases linearly with pressure at the rate of $(2 \cdot 0 \pm 0 \cdot 4)$°K/kb (Grazhdankina, 1957, 1965) in agreement with the more recent value (Ozawa *et al.*, 1966) 2·66°K/kb

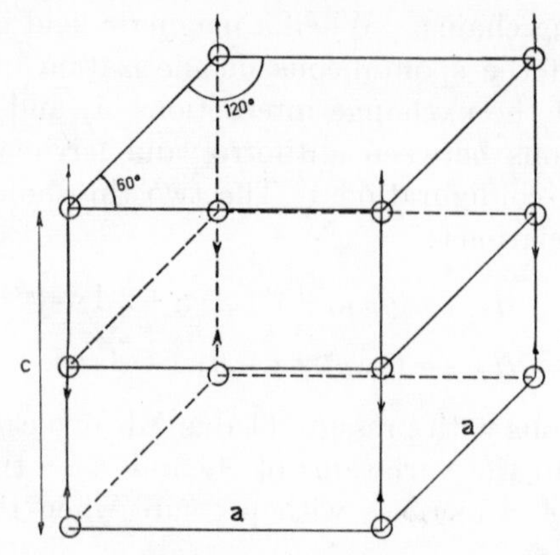

FIG. 12. Magnetic structure of the NiAs type compound MnTe.

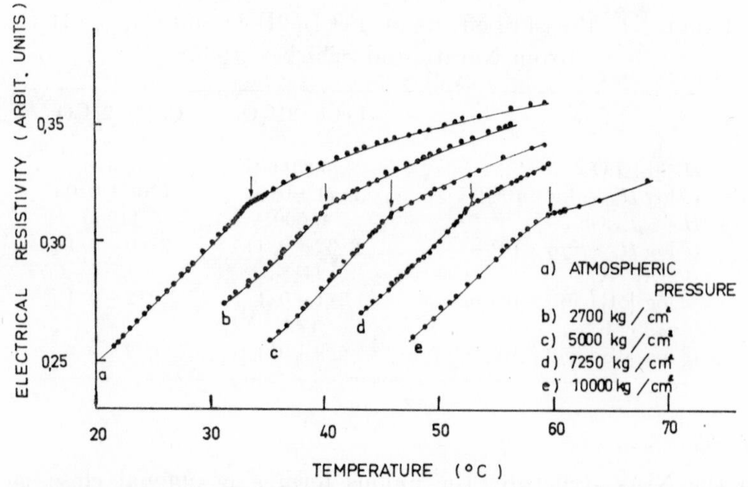

FIG. 13. The temperature dependence of the electrical resistivity of MnTe at atmospheric and high pressures (from Ozawa *et al.*, 1966).

up to 10 kb. The sign of the variation obtained is confirmed by the negative value of the volume anomaly at absolute zero (Greenwald, 1953) and by the positive value of the difference between the experimental and theoretical Curie constant (Uchida *et al.*, 1957).

V. METALS

A. INTRODUCTION

In the past, the understanding of magnetism in metals has been approached from two extreme points of view. Firstly, the electrons responsible for the magnetic behaviour are either in free or nearly-free electron energy bands (that is the itinerant model) or secondly, they are located in discrete atomic-like orbital energy states (that is localized model), with more or less direct interaction (Heisenberg type) between them. More recently, a theory of indirect exchange interaction (Rudermann and Kittel, 1954; Kasuya, 1956; Yosida, 1957) has been proposed in which the interaction between the atomic moments occurs indirectly *via* the conduction electrons. However, none of these approaches is sufficiently developed to give an adequate interpretation of the effects of hydrostatic pressures on the magnetic properties of metals. Instead, it is necessary to fall back on an application of thermodynamics to empirical relations from which various conclusions can be drawn about the variation of certain important quantities as a function of pressure, volume and/or interatomic distance. Or, alternatively, various types of interaction curves, using various quantities, have been constructed and used in attempts to correlate the results from various metals and alloys. The work in high pressures on magnetic metals and alloys has been motivated predominately by the desire to generate more information, from which, it is expected, some correlation of results will emerge.

It has not been possible to categorize the metals according to their crystallographic structures as in the case of the non-metals. Instead, it appeared more logical to classify them in accordance with their electronic structures, that is rare earth metals and first transition metals. These have been discussed in this order because greater emphasis has been placed on the rare earths than on the first transition elements and alloys during the period which this review covers. Similarly the intermetallic compounds and alloys do not lend themselves to a logical classification, and therefore they have neen lumped into Section V-D as individuals rather than groups.

B. RARE EARTH METALS AND ALLOYS

1. *Introduction*

The rare earth metals have been described by Néel (1938) and De Gennes (1958, 1962) as an assembly of ions, generally trivalent, with incomplete 4*f* shells, immersed in a " sea of conduction electrons ". These 4*f* electrons, which occupy an inner shell, are responsible for the highly localized magnetic moments of the atoms in the metals. Interactions between these atomic moments result in a large variety of ordered magnetic spin structures, that is collinear, helical, conical and fan structures, which in some cases can be altered by varying the temperature of the substances, by application of sufficiently high magnetic fields and/or by the application of hydrostatic pressure.

The arrangement of the spins in preferential planes or along preferential directions results from the balance of magnetocrystalline anisotropy and exchange forces (Elliott, 1965). In an effort to understand the dependence of these interactions on the interatomic spacing, experiments have been initiated which investigate the variation of the magnetic transition temperatures, the saturation magnetizations and magnetocrystalline anisotropy as a function of pressure and temperature. In some cases, it has been possible to compare the results for these quantities with those obtained by other types of measurements. These either help to refute or corroborate the high pressure results.

2. *Gadolinium*

The rare earth metal, gadolinium, possesses a hexagonal close-packed structure with atoms in electronic S states. It undergoes with decreasing temperature a paramagnetic–ferromagnetic transition at $\theta_c = 291{\cdot}8°K$.

Patrick (1954) was the first to measure the change in its Curie temperature under pressures up to 8 kb. He found that $\partial\theta_c/\partial p = -1{\cdot}2°K/kb$ (Table XII). This result is compared with what would be expected

TABLE XII. Effects of pressure on the Curie temperature and magnetization of gadolinium

Polycrystalline

θ_c (°K)	$\partial\theta_c/\partial p$ (°K/kb)	$(\sigma_s^{-1}\, \partial\sigma_s/\partial p) \times 10^3$/kb	References
289	$-1{\cdot}2$		Patrick (1954)
294·5	$-1{\cdot}53 \pm 0{\cdot}05$	$-1{\cdot}1 \pm 0{\cdot}3$	Bloch and Pauthenet (1962, 1965a)
290·8	$-1{\cdot}4$		Kouvel and Hartelius (1964a)
292·9	$-1{\cdot}6 \pm 0{\cdot}05$		Robinson *et al.* (1964)
292	$-1{\cdot}72 \pm 0{\cdot}07$		McWhan and Stevens (1965)
	$-1{\cdot}34 \pm 0{\cdot}06$		Livshitz and Genshaft (1965)
290·6	$-1{\cdot}48 \pm 0{\cdot}02$		Bartholin and Bloch (1967)
293	$-1{\cdot}8 \pm 0{\cdot}1$		Kawai *et al.* (1967a)

Monocrystalline

θ_c (°K)	$\partial\theta_c/\partial p$ (°K/kb)	$(\sigma_s^{-1}\, \partial\sigma_s/\partial p) \times 10^3$/kb	References
293	$-1{\cdot}56$		Austin and Mishra (1967)
293·05	$-1{\cdot}40 \pm 0{\cdot}02$		Bartholin and Bloch (1967)

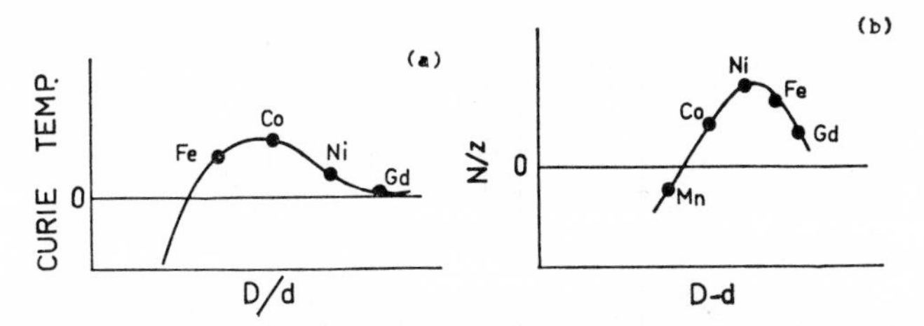

FIG. 14. (a) Bethe–Slater interaction curve, (b) Néel interaction curve.

from a Bethe–Slater interaction curve (Fig. 14a) and the Néel interaction (Fig. 14b). In both cases, the $\partial\theta_c/\partial p$ has the wrong sign for the interatomic distance, D, and atomic diameter, d, available.

Several years later, Bloch and Pauthenet (1962) repeated the measurements on polycrystalline gadolinium up to 6 kb and found that the change in the Curie temperature with pressure was linear with a slope of $-1{\cdot}53$°K/kb. In a subsequent paper (Bloch and Pauthenet, 1965a) these data were extended with measurements of the relative change of the spontaneous magnetization as a function of pressure at 77°K and 29 kOe, $(1/\sigma_s)(\partial\sigma_s/\partial p) = (-1{\cdot}1 \pm 0{\cdot}3) \times 10^{-3}$/kb. This value of the pressure dependence of the spontaneous magnetization is in conflict with the result of Kondorskii and Vinokurova (1965) who obtained $(1/\sigma_s)(\partial\sigma_s/\partial p) = (-1{\cdot}9 \pm 0{\cdot}4) \times 10^{-4}$/kb at 77°K. It is also in disagreement with the indirect result obtained from the magnetic field dependence of the volume magnetostriction (Coleman and Pavlovic, 1964) as deduced from eqn (8).

The rate of change with pressure of the Curie point can be calculated with eqn (10) using experimental $\partial\sigma_s/\partial T$ and $(1/\sigma_0)(\partial\sigma_0/\partial p)$ values. This becomes $(\partial\theta_c/\partial p) \simeq -1{\cdot}63$°K/kb in accordance with the direct result given above (Table XII). Employing the eqns (35, 36 and 37), derived from the theory of indirect exchange and inserting experimental results for K,

$$\left(\frac{1}{\theta_c}\frac{\partial\theta_c}{\partial p}\right)_{H,\,T} \quad \text{and} \quad \left(\frac{1}{\rho_m}\frac{\partial\rho_n}{\partial p}\right)_{H,\,T},$$

Bloch and Pauthenet (1964) computed the values for $(\partial \log \Gamma/\partial \log V)_{H,\,T}$ $(\partial \log m^*/\partial \log V)_{H,\,T}$ and $(1/\Delta M)(\partial\Delta M/\partial p)_{H,\,T}$ appearing in Table XIII. These results indicate that the coupling constant is a rapidly varying function of the volume. Also, if one considers the difference between the experimental saturation moment and the theoretical spin-only saturation moment to be due to the free electron moments, the magnetization of the conduction electrons changes with pressure

TABLE XIII. Pressure and volume variations of some physical quantities of Gd, Tb, Dy, Ho

Element	$\left(\frac{\partial \log \Gamma}{\partial \log V}\right)_{H,T}$	$\left(\frac{\partial \log m^*}{\partial \log V}\right)_{H,T}$	$\left(\frac{1}{\Delta M}\frac{\partial \Delta M}{\partial p}\right)_{H,T}$	$\left(\frac{1}{\rho_{mag}}\frac{\partial \rho_{mag}}{\partial p}\right)$ 10^{-3}/kb	Compressibility 10^{-3}/kb
Gd	2·6	−1·7	−4·1	−8·4[a] (310°K)[b]	−2·66[a]
	1·6[a]	0·2[a]			
Tb	1·2	0·3	−4·1		
Dy	1·9	−1·1	−3·9	−2·5[a, b]	−2·75
	2·0[a]	−1·4[a]			
Ho	2·3	−1·2	−4·6		

[a] From Austin and Mishra (1967), [b] *c*-axis values.

at the rate of $\partial \Delta M/\partial p = -2{\cdot}2 \times 10^{-3} \mu_\beta/\text{kb}$, compared to the experimental result of $(\partial \sigma_0/\partial p)_{77^\circ \text{K}} = (-4 \pm 2) \times 10^{-3} \mu_\beta/\text{kb}$. It appears that the pressure variation of magnetization for Gd can be attributed to the variation in the magnetization of the conduction electrons.

The previously cited investigations employed an a.c. initial permeability method to determine the Curie point at each pressure of measurement. This technique is better suited to determining the shift of the Curie point than for an absolute evaluation of the Curie point at any fixed pressure. Kouvel and Hartelius (1964a) have preferred to employ quasi-static magnetic measurements in magnetic fields up to about 12 kOe under pressures from 0 to about 5 kb. Under all conditions of field and temperature, the magnetization was found to vary linearly with pressure. Figures 15 and 16 illustrate some typical

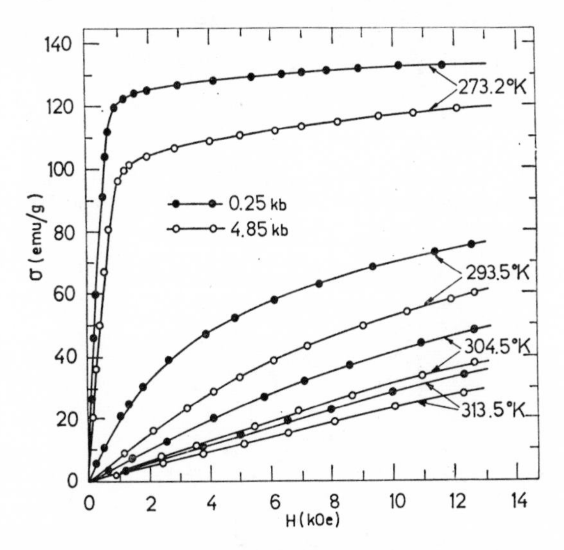

FIG. 15. The magnetization (σ) of gadolinium as a function of the applied field (H) at various temperatures and hydrostatic pressures (from Kouvel and Hartelius, 1964a).

curves of σ *versus* H and σ^2 *versus* H/σ. The advantage of the σ^2 *versus* H/σ curve arises from the simple behaviour predicted by the molecular field model. According to this model, the isotherms of σ^2 *versus* H/σ for a ferromagnetic material near its Curie point (at constant pressure) should form a succession of parallel lines that intersect the σ^2 or H/σ axis depending on whether the temperature is below or above the Curie point (θ_c). Moreover for $T > \theta_c$ the H/σ intercept corresponds to $1/\chi_0$, the reciprocal of the initial susceptibility. If

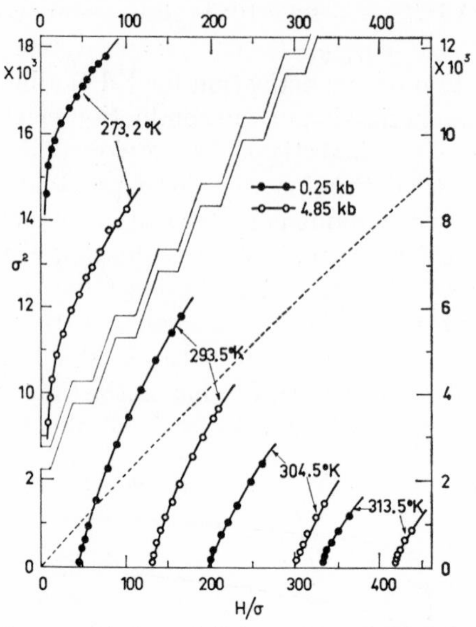

FIG. 16. Curves of σ^2 *versus* H/σ for gadolinium. Note shift of σ^2 scale for 273·2°K data. Dashed curve computed for $q=2$, $S=7/2$ (from Kouvel and Hartelius, 1964a).

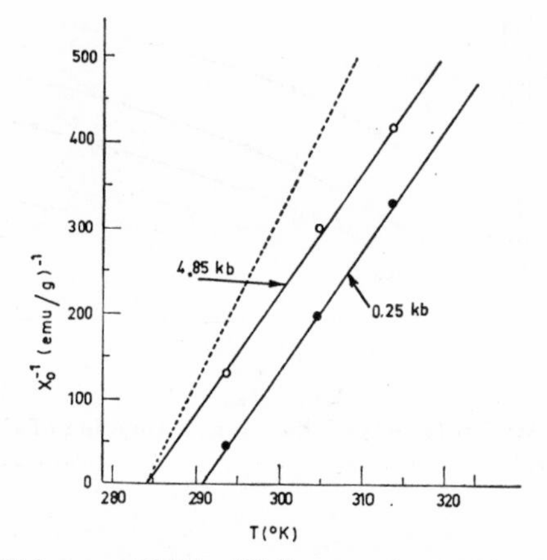

FIG. 17. Inverse initial susceptibility (X_0^{-1}) *versus* temperature for gadolinium under different pressures. Dotted curve computed for $g=2$, $S=7/2$ (from Kouvel and Hartelius, 1964a).

$1/\chi_0$ is plotted against T (Fig. 17), the intersection with the T axis gives θ_c which is true even if $1/\chi_0$ *versus* T is not a straight line as given by the molecular field theory. The Curie temperature for Gd was $\theta_c = 290{\cdot}8$°K and its variation with pressure $\partial\theta_c/\partial p = -1{\cdot}4$°K/kb. These are compared in Table XII with other results. Studies executed at higher pressures (up to 40 kb) than previous works were performed by Robinson *et al.* (1964). An a.c. initial permeability method was employed for determining magnetic transition temperatures with a solid AgCl pressure medium. Up to a pressure of 21·5 kb, their results, presented in Table XII, are in good agreement with previous results; beyond 21·5 kb, a transition to a non-ferromagnetic state was detected. Above the 21·5 kb transition, isobaric temperature runs show two new peaks in the initial permeability, but no explanation for these peaks was possible. This high pressure phase persisted in the samples after relaxation of the pressure. The initial ferromagnetic phase could be restored by a thermal anneal at 775°C for 17 min. From these results and their incompatibility with the Bethe–Slater interaction curve based on the first transition metals, Robinson and co-workers (1964) concluded that some other interaction curve was necessary. Their new interaction curve is discussed in part 11 of this section.

Although the magnetic structure of the high pressure phase was not investigated, it is speculated that this high pressure phase might have a helimagnetic structure (Jayaraman and Sherwood, 1964a).

A detailed investigation of the effect of pressure on the magnetic properties and crystal structure of four of the ferromagnetic rare earths was made at pressures much higher than the previous studies (up to 85 kb) by McWhan and Stevens (1965). Figure 18 shows the secondary voltage (from a transformer made of Gd) *versus* the temperature at a number of pressures. It should be noticed that as the pressure is increased a second broad hump appears. Between 20 and 25 kb three peaks grow in and the anomalies from the low pressure phase decrease

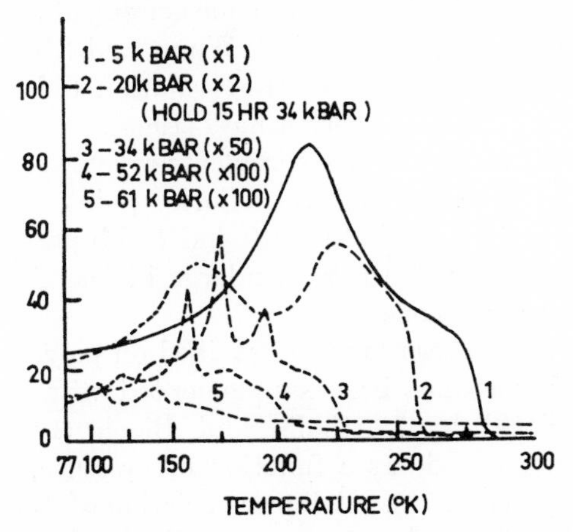

FIG. 18. Secondary voltage *versus* temperature for Gd at different pressures (from McWhan and Stevens, 1965).

in amplitude. The first of the three peaks is much smaller than the other two and disappears by 38 kb. Holding a Gd sample at 34 kb for 15 hr caused the peaks from the low pressure phase to disappear leaving only the two larger peaks. Around 50 kb, the first of these peaks decreases in amplitude and another grows in at a still lower temperature. Finally between 50 and 85 kb, isobaric temperature cycles show two peaks. As pointed out by the authors, it was difficult to know unambiguously which peaks belong to which phase and whether they represent equilibrium conditions. Nevertheless, ordering temperatures and the pressure variation in the ordering temperatures were assigned, and these are summarized in Table XIV, while the Gd^{I} result is compared in Table XII with the results of other investigations. On the assumption that the magnetic transitions are of second order, $\partial\theta/\partial p$ can be related by the thermodynamic equation (6) to the anomalies in the coefficient of thermal expansion ($\Delta\alpha$), compressibility (ΔK), and specific heat at constant pressure (ΔC_p). Values for $\Delta\alpha$ and ΔK have been calculated from the experimental values of ΔC_p and $\partial\theta/\partial p$ and are presented in Table XIV. The coefficient of thermal expansion data of Barson and co-workers (1957) for Gd, confirm the sign and magnitude of $\Delta\alpha$ at the ordering temperature. Using Bridgman's compressibility (K) for Gd, the logarithmic dependence of the ordering temperature with volume $\partial \log \theta/\partial \log V$ was calculated. This result appears in Table XIV and is compared with previous results.

Further high pressure investigations on Gd up to 35 kb by Livshitz and Genshaft (1965) revealed the $\partial\theta_c/\partial p$ value presented in Table XII. It is of the proper sign (negative) and of the right order of magnitude. However, the discontinuities in the measurements at about 21·5 kb which Robinson *et al.* (1964) and McWhan and Stevens (1965, 1967) found and ascribed to a phase transformation did not occur in the measurements of Livshitz and Genshaft (1965). Finally, Kawai *et al.* (1967a) employing resistance measurements reported for polycrystalline Gd a rate of variation of the Curie point with pressure of $\partial\theta_c/\partial p = -(1{\cdot}8 \pm 0{\cdot}1)$°K/kb between 0 and 18 kb pressure. This value although of the proper sign is rather large in magnitude compared to the previous determinations.

All of the preceding results were obtained on polycrystalline samples of Gd. These have since been supplemented with measurements on monocrystals of Gd by Bartholin and Bloch (1967, 1968). Their measurements were made as a function of pressure, temperature and crystal orientation (Fig. 19). A linear variation of θ_c with pressure was found to have the slope of $-(1{\cdot}48 \pm 0{\cdot}02)$°K/kb (Table XII) which is in excellent agreement with the results for polycrystalline samples.

TABLE XIV. The variation of ordering temperatures with pressure and second order transition anomalies of Gadolinium. The magnetic ordering temperatures θ_x are determined from specific heat measurements at usual pressure. (From McWhan and Stevens, 1965)

						Second order transition anomalies		
							Calculated	
	θ_x (°K)	$\theta_{p=1}$ (°K)	$\partial\theta/\partial p$ (°K/kb)	$\frac{\partial \ln \theta}{\partial \ln V}$	Pressure range (kb)	Exp. ΔC_p (cal/deg/ mole)	$\Delta\alpha \times 10^6$°K	$\Delta K \times 10^6$/kb
Gd (I)	291·8[a]	292 ± 2	−1·63 ± 0·07	2·2	5–52	6[a]	−72	124
		(255)	(−1·4)	2·0[g, h, i]	25–38			
				1·6				
(II)		247 ± 4	−1·46 ± 0·11		25–52			
(II) (III)		218 ± 2	−1·19 ± 0·04		25–84			
(III)		181 ± 5	−1·19 ± 0·08		34–84			
Tb (I)	227·7[b]	227 ± 1	−1·07 ± 0·03	1·8	4–71	11[b]	−110	118
(II)		196 ± 5	−0·83 ± 0·08		35·85			
		167 ± 4	−0·85 ± 0·06		35–85			
Dy (I)	174[c]	179 ± 2	−0·66 ± 0·04	1·4	5–77	85[c]	−67	44
(II)		166 ± 5	−0·67 ± 0·07	1·3[f,j]	49–85		−64[f]	40[f]
		145 ± 6	−0·74 ± 0·08	1·2	49–85			10[k]
Ho (I)	131·6[d]	133 ± 1	−0·48 ± 0·01	1·4	5–82	5·4[d]	−42	20
		(126 ± 14)	(−0·5 ± 0·2)		10–82			

[a] Griffel *et al.* (1954), [b] Jennings *et al.* (1957), [c] Griffel *et al.* (1956), [d] Gerstein *et al.* (1957), [f] Souers and Jura (1964), [g] Patrick (1954), [h] Bloch and Pauthenet (1962), [i] Robinson *et al.* (1964), [j] Landry and Stevenson (1963), [k] Tatsumoto *et al.* (1968) (experimental).

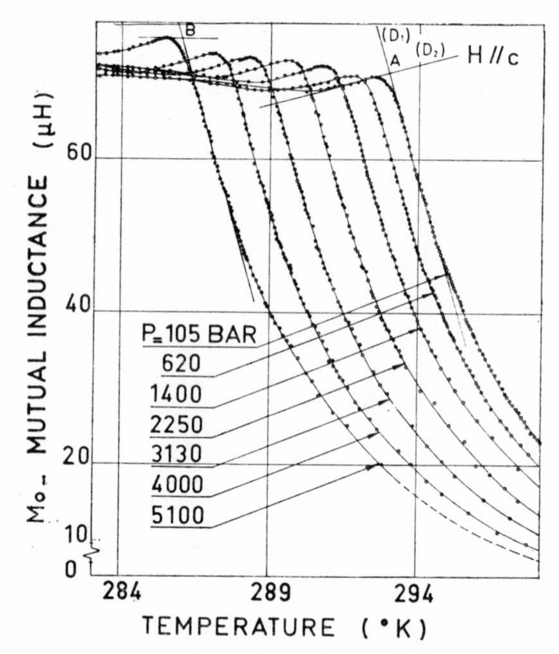

FIG. 19. Variation of M_0 with temperature for a single crystal of gadolinium at different pressures. The field H is applied along the c-axis (M_0 is proportional to the a c-susceptibility) (from Bartholin and Bloch, 1968).

The magnetic length anomaly along the c-axis at $\theta = 0°K$ was computed to be $(e_{cc})_0 = 1{\cdot}2 \times 10^{-2}$ for Gd in disagreement with $0{\cdot}3 \times 10^{-2}$ from X-ray diffraction and dilatation measurements (Darnell, 1963a; 1963b; Darnell and Moore, 1963). Electrical resistance investigations on single crystals of Gd by Austin and Mishra (1967) up to 8 kb and from 273 to 313°K indicated a shift in the Curie point having the rate of approximately $\partial\theta_c/\partial p = -1{\cdot}56°K/kb$ (Table XII).

Volume magnetostriction studies were made on single crystals of Gd by Coleman and Pavlovic (1964). Using eqn (8), the pressure coefficient of the magnetization was calculated at 77°K.

$$\left(\frac{1}{\sigma}\frac{\partial\sigma}{\partial p}\right)_{77°K} = -1{\cdot}1 \quad \times 10^{-4}/\text{kb}$$

which is an order of magnitude smaller than the direct result by Bloch and Pauthenet (1965b) of $(-1{\cdot}1 \pm 0{\cdot}3) \times 10^{-3}/\text{kb}$. Similarly from eqn (10), the rate of variation of the Curie temperature with pressure $\partial\theta/\partial p = -1{\cdot}24 \pm 0{\cdot}1°K/kb$ was calculated (Coleman and Pavlovic, 1964). Although this quantity is low, it is satisfactory considering the many assumptions made and the many possibilities for errors to enter into the calculation.

As high pressure techniques have improved and become more commonplace, a greater variety of physical quantities have been investigated. One such quantity receiving more attention recently is that of magnetocrystalline anisotropy. Birss and Hegarty (1967) have developed a torque magnetometer for the determination of the magnetocrystalline anisotropy constants under high pressures. Gd was investigated at pressures up to 2·7 kb and in magnetic fields of 12 kOe at 283°K. Figure 20 shows a plot of " A_2 " (in Birss *et al.* notation), the

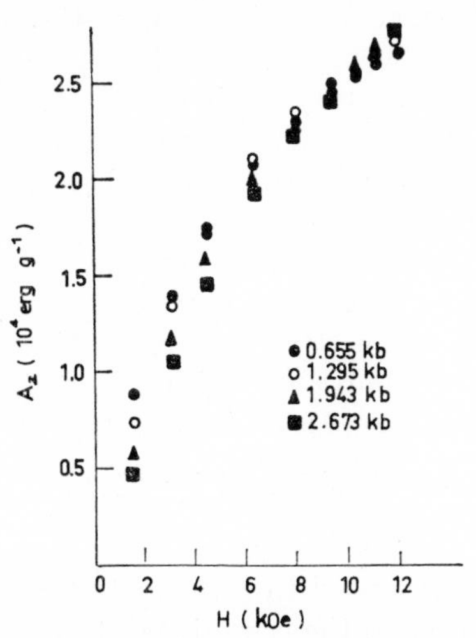

Fig. 20. Variation of A_2, the component of torque with two-fold periodicity (referred to unit mass of the (1120) gadolinium disk), with applied field H at 283·2°K and various pressures (from Birss and Hegarty, 1967).

coefficient of the Fourier-analysed second harmonic (2θ) component of torque (referred to unit mass) *versus* magnetic field H at various values of pressure. The general form of the field dependence and magnitude of the results agrees fairly well with that found by Graham (1965). The magnitude of the relative change with pressure at this temperature in the first anisotropy constant is very much larger than for the transition metals reported by Veerman and Rathenau (1965). Furthermore, it can be realized from the figure that A_2 increases with increasing pressure at high fields.

3. *Terbium*

Terbium, a rare earth metal, presents a more complicated magnetic behaviour than Gd. It has a paramagnetic–antiferromagnetic transition at a Néel temperature of $\theta_N = 227{\cdot}7\,^\circ$K and an antiferromagnetic–ferromagnetic transition with a temperature of $\theta_F^{AF} = 221{\cdot}0\,^\circ$K. The crystallographic structure of Tb is hexagonal close packed. In the antiferromagnetic domain, the structure of the magnetic moments is helimagnetic with all spins in planes perpendicular to the c-axis ferromagnetically aligned in the plane. In each successive plane along the c-axis the spin direction is rotated through an angle ω with respect to the adjacent planes.

Bloch and Pauthenet (1965a) investigated the pressure dependence of the Néel temperature as well as the pressure dependence of the spontaneous magnetization at 77°K in a magnetic field of 29 kOe. The Néel temperature was $\theta_N = 237\,^\circ$K and the slope of its linear pressure dependence was $-(0{\cdot}82 \pm 0{\cdot}1)\,^\circ$K/kb (Table XV), while the relative variation of the spontaneous magnetization with pressure was $(1/\sigma_0)\,(\partial\sigma_s/\partial p) = (-3{\cdot}8 \pm 0{\cdot}4) \times 10^{-3}$/kb (Table XV). A similar study of the magnetization by Vinokurova and Kondorskii (1965) indicated that $(1/\sigma_s)\,(\partial\sigma_s/\partial p) = -7{\cdot}3 \times 10^{-3}$/kb, considerably larger than that of Bloch and Pauthenet. Using the preceding results in eqn (10), the pressure dependence of the saturated magnetization at absolute zero $(1/\sigma_0)\,(\partial\sigma_0/\partial p) = (-3{\cdot}6 \pm 0{\cdot}5) \times 10^{-3}$/kb was found. In Table XIII are presented the values for $(\partial \log \Gamma/\partial \log V)_{H,\,T}$, $(\partial \log m^*/\partial \log V)_{H,\,T}$ and $\left(\dfrac{1}{\Delta M}\dfrac{\partial \Delta M}{\partial p}\right)_{H,\,T}$ of Tb obtained by Bloch and Pauthenet (1965a) from the indirect exchange theory.

The range of pressures used to investigate the shift of the magnetic transition temperature and the crystal structure of Tb was extended to 90 kb by McWhan and Stevens (1965). They report a phase transition at 35 kb from Tb^{I} to Tb^{II}. In this structural transition Tb goes from a h.c.p. to a Sm-type structure. In both regions it was not possible to distinguish the θ_F^{AF} point from the Néel point because of the small difference between the two ordering temperatures. Their results for the ordering temperatures are presented in Tables XIV and XV for both phases. The data is shown on Fig. 21 for Tb as well as the other rare earths. As in the case of Gd, the transition to the high pressure phase is accompanied by a decrease in the ordering temperature. The pressure at which this transition occurs in Tb is greater than the pressure at which it occurs for Gd. However, the decrease in the ordering temperature for Tb is less than for Gd. The anomalies in the coefficient of thermal expansion ($\Delta\alpha$) and compressibility ΔK have been computed as discussed above on the assumption that the transition is of second order (Table XIV).

Within the pressure range 0–25 kb Robinson *et al.* (1966) found

TABLE XV. Pressure dependence of the ordering temperatures of terbium

Polycrystals

Tb	θ_N (°K)	$\partial\theta_N/\partial p$ (°K/kb)	$\frac{1}{\sigma_s}\cdot\frac{\partial\sigma_s}{\partial p}$ ($\times 10^3$/kb)	θ_c (°K)	$\partial\theta_c/\partial p$ (°K/kb)	References
	231	$-0\cdot82\pm0\cdot1$	$-(3\cdot8\pm0\cdot4)$ (77°K)			Bloch and Pauthenet (1965a)
(I)	227 ± 1	$-1\cdot07\pm0\cdot03$				McWhan and Stevens (1965)
(II)	196 ± 5	$-0\cdot83\pm0\cdot08$				
(III)	167 ± 4	$-0\cdot85\pm0\cdot06$				
			$-7\cdot3$ (77°K)			Vinokurova and Kondorskii (1965)
	$\sim226\cdot4$	$-1\cdot0$				Robinson *et al.* (1966)
A	231	$-1\cdot05\pm0\cdot1$		223	$-1\cdot24\pm0\cdot10$	Wazzan *et al.* (1967)
B	231	$-1\cdot14\pm0\cdot1$		222	$-1\cdot08\pm0\cdot10$	
	$227\cdot1$	$-0\cdot86$	$\frac{1}{n_{eff}}\frac{\partial n_{eff}}{\partial p}$[b] $=-0\cdot96$	$219\cdot3$	$-1\cdot10$	Tatsumoto *et al.* (1968)
				$\theta_p^a=230$°K	$\frac{\partial\theta_p^a}{\partial p}=-0\cdot76$	
	229	$-1\cdot3\pm0\cdot2$				Kawai *et al.* (1967a)

Monocrystals

Tb	θ_N (°K)	$\partial\theta_N/\partial p$ (°K/kb)	$\frac{1}{\sigma_s}\cdot\frac{\partial\sigma_s}{\partial p}$ ($\times 10^3$/kb)	θ_c (°K)	$\partial\theta_c/\partial p$ (°K/kb)	References
	$229\cdot3_8$ (T ↑) $229\cdot2_0$ (T ↓)	$-0\cdot84\pm0\cdot02$		$220\cdot7_6$ (T ↑) $219\cdot3_2$ (T ↓)	$-1\cdot24\pm0\cdot02$	Bartholin and Bloch (1968)
		$-0\cdot76\pm0\cdot05$			$-1\cdot06\pm0\cdot1$	Umebayashi *et al.* (1968)

[a] θ_p is the paramagnetic Curie temperature, [b] n_{eff} is the effective number of Bohr magnetons.

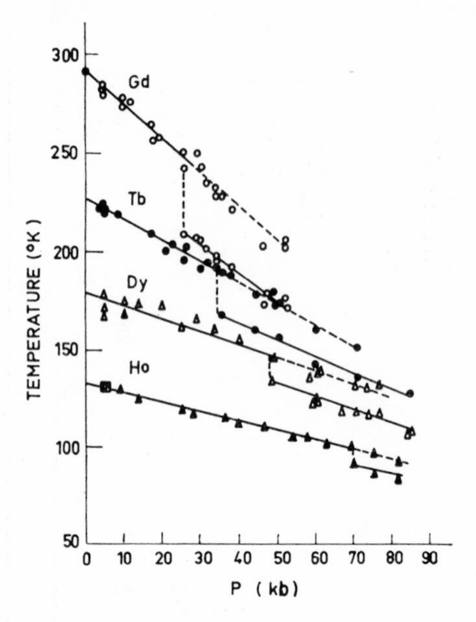

Fig. 21. Change in ordering temperature with pressure of Gd, Tb, Dy and Ho (from McWhan and Stevens, 1965).

$\theta_N = 226{\cdot}4°K$ and $\partial\theta_N/\partial p = -1{\cdot}0°K/kb$. At pressures close to atmospheric pressures, they were able to resolve θ_F^{AF} and θ_N but found that the two were indistinguishable at higher pressures ($>0{\cdot}2$ kb). Wazzan *et al.* (1967) investigated the pressure dependence (up to 14·2 kb) of the magnetic transition temperatures in terbium determined from electrical resistance measurements. Figure 22(a) shows a resistance *versus* temperature curve for Tb at 7·1 kb. The resistance varies linearly at both ends of the temperature range; the lower temperature at which the resistance begins to deviate from the linear behaviour is taken to be θ_F^{AF} whereas the upper temperature is taken to be θ_N. These results give a linear variation of θ_F^{AF} and θ_N *versus* pressure. The slopes of these curves and their values at $p = 1$ b are labelled as " *A* " in Table XV. A second method of determining θ_F^{AF} and θ_N from the data was used. The slope of the resistance *versus* temperature curve was obtained and plotted against temperature as illustrated in Fig. 22(b) and θ_F^{AF} and θ_N determined as shown in the figure. The results " *B* " in Table XV were determined in this way. Good agreement exists between " *A* " and " *B* " and with other results. This clearly indicates that the magnetic technique of Robinson *et al.* (1966) and McWhan and Stevens (1965) was not sufficiently accurate to differentiate between θ_F^{AF} and

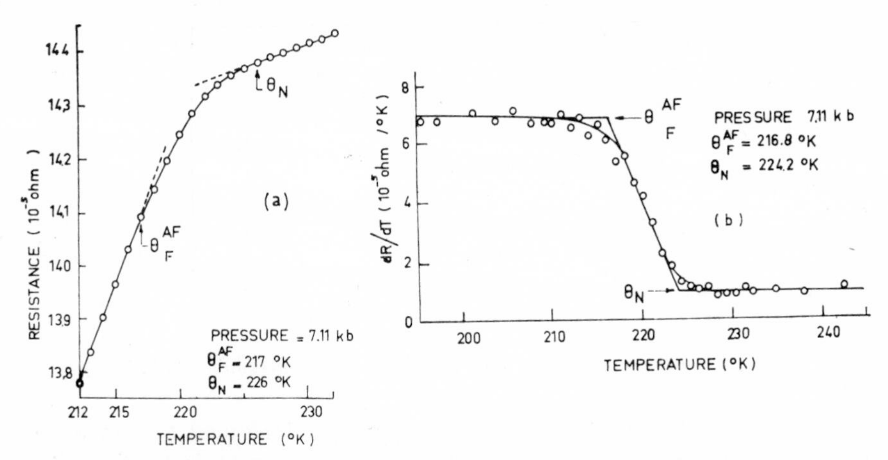

FIG. 22. (a) Electrical resistance of a polycrystalline terbium sample as a function of temperature at a pressure of 7·1 kb. (b) dR/dT *versus* temperature at a pressure of 7·1 kb (from Wazzan *et al.*, 1967).

θ_N. These studies led them to conclude (1) that the function $\theta_N/[(g-1)^2 J(J+1)]$ is a function of the hexagonal c/a ratio and (2) that the effect of pressure upon the Rudermann–Kittel interaction or upon $\theta_N/[(g-1)^2 J(J+1)]$ appears to be largely due to the effect of pressure upon the number of states and density of states per energy level in the Brillouin zone which is directly related to the effect of pressure upon the c/a ratio (these points are discussed more fully later in part 11 of this section).

Measurements of weak field a.c. susceptibilities have been made on Tb under hydrostatic pressures up to about 6 kb by Tatsumoto *et al.* (1968). From these data the pressure effects on the magnetic transition temperatures have been determined (Table XV). In addition, it was possible for them to obtain the paramagnetic Curie temperature $\theta_p = 230°K$, its pressure variation $\partial\theta_p/\partial p = -0{\cdot}76°K/kb$ and the relative variation with pressure of the effective number of Bohr magnetons n_{eff} in the paramagnetic state, $\dfrac{1}{n_{eff}}\dfrac{\partial n_{eff}}{\partial p} = -9{\cdot}6\times10^{-4}/kb$. The linear compressibility was measured as a function of temperature (Fig. 23). The anomaly observed at the transition temperature θ_N was $\Delta K_1 = 0{\cdot}4\times10^{-4}/kb$ which agrees well with the thermodynamic Ehrenfest relation for second-order transitions (eqn 6). The anomaly at θ_F^{AF} appears to indicate a first-order transition which is not inconsistent with the reported lattice distortion and the thermal hysteresis observed.

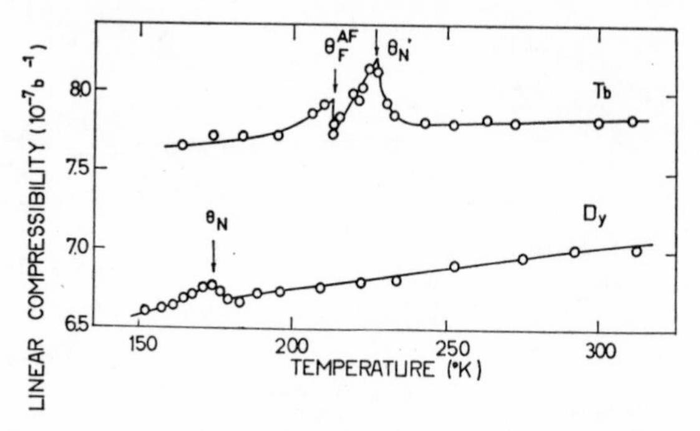

FIG. 23. Linear compressibility for Tb and Dy (from Tatsumoto *et al.*, 1968).

More recently, Bartholin and Bloch (1968) determined the pressure coefficient of the transition temperatures of monocrystals of Tb up to 6 kb. Again, as for Gd, data was taken (Fig. 24) which clearly presents the paramagnetic–antiferromagnetic and antiferromagnetic–ferromagnetic transitions and their shifts with pressure. The variation of the Curie and Néel temperatures was linear with pressure, having a negative

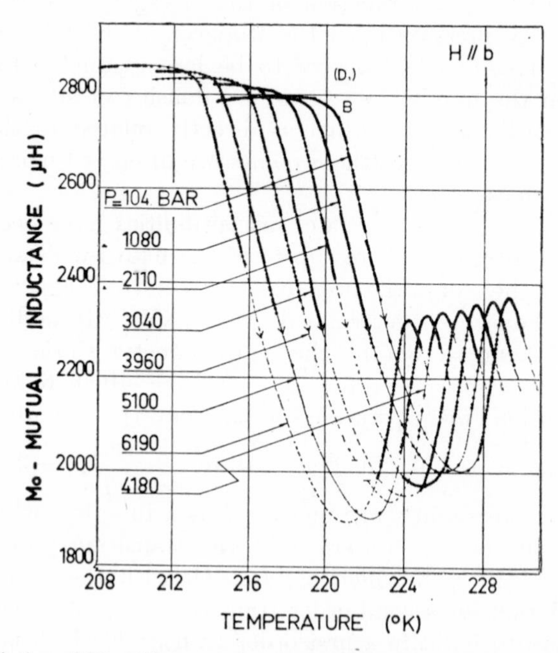

FIG. 24. Variation of M_0 with increasing temperature for a single crystal of terbium at different pressures. The field H is applied along the b-axis (from Bartholin and Bloch, 1968).

slope (Table XV). A slight difference of 0·15°K was found between the Néel temperature for increasing and decreasing temperature. A similar but larger temperature hysteresis was observed for the θ_F^{AF} temperature. It was constant and independent of pressure and amounted to 1·44°K. These results were used to evaluate the variation of the nearest neighbour and next-nearest neighbour exchange interactions as a function of the c-axis parameter; $\partial(A_1/k)/\partial \log c = 75$°K (nearest neighbour) and $\partial(A_2/k)/\partial \log c = 0$ (next-nearest neighbour) (eqn (29) in Section IIIC). By applying these values of $\partial(A_i/k)/\partial \log c$, a calculation of the magnetic length anomaly, e_{cc} at 0°K was made; obtaining for terbium $e_{cc} = 0{\cdot}90 \times 10^{-2}$ which is in poor agreement with the value of $0{\cdot}2$–$0{\cdot}3 \times 10^{-3}$ from X-ray diffraction and dilatometric measurements (Darnell, 1963a, b; Darnell and Moore, 1963).

Quite contrary to previous results which hinted that the two transition temperatures approached each other with pressure, Bartholin and Bloch (1968) found that they diverge at the rate of $(0{\cdot}40 \pm 0{\cdot}04)$°K/kb.

The pressure dependence of the helical turn angle, ω, of Tb single crystals was measured by neutron diffraction up to 6 kb at temperatures above 80°K (Umebayashi *et al.*, 1968). At constant T–θ_N, the relative changes of ω with pressure $1/\omega\,(\partial\omega/\partial p)$ was slightly temperature dependent, but a representative average value was 20×10^{-3}/kb. This experimental value compares well with values obtained from calculations based on the theory of Miwa (1965). A more extensive comparison is given in Table XVI. The values for the pressure coefficient of the Néel temperature, as well as of the antiferro–ferromagnetic transition temperature, of Tb were found to be in reasonable agreement with other reported values (Table XV). The temperature range of the antiferromagnetic phase expanded with pressure at the rate of $(0{\cdot}3 \pm 0{\cdot}12)$°K/kb in excellent agreement with that of Bartholin and Bloch (1968).

TABLE XVI. Change of interlayer turn angle with pressure (From Umebayashi *et al.*, 1968)

	Terbium		Holmium	
$T - \theta_N$	$-4{\cdot}0$°K	$-9{\cdot}0$°K	$7{\cdot}0$°K	$36{\cdot}5$°K
ω (at 1 atm)	20·0°	17·7°	50°	45°
$\frac{1}{\omega}\frac{\partial\omega}{\partial p} \times 10^3$/kb				
Observed	20 ± 2	23 ± 2	$1 \pm 0{\cdot}2$	$2{\cdot}3 \pm 0{\cdot}4$
Calculated	19	25	0·6	0·9

4. *Dysprosium*

Dysprosium (Dy) like Tb has two transition temperatures. One, a Néel temperature at about 178·5°K, represents the temperature of the paramagnetic–antiferromagnetic transition (θ_N). The other, at about 85°K, is the temperature of the antiferro–ferromagnetic transition (θ_F^{AF}). The antiferromagnetic phase has a helicoidal spin structure of the type described for Tb. At the Néel temperature, the angle, ω, between the spin directions in two adjacent planes is 43·2° and decreases with temperature to 26·5° at the antiferro–ferromagnetic transition. The antiferromagnetic phase can be made to transform to a collinear ferromagnetic spin structure by the application of a large magnetic field, the magnitude of which depends on the temperature.

Swenson *et al.* (1960) were the first to make a determination of the pressure variation of the antiferro–ferromagnetic transition temperature of Dy; $\partial\theta_F^{AF}/\partial p = -1$°K/kb. This work was followed by that of Landry and Stevenson (1963) who obtained $\partial\theta_N/\partial p = -0{\cdot}56 \pm 0{\cdot}01$°K/kb (Table XVII) for the pressure variation of the Néel point. Measurements of the antiferro–ferromagnetic transition were not possible because of a large magnetic hysteresis; also these investigators pointed out that there is a broadening of the maximum with the application of pressure. It is concluded however that friction and pressure inhomogeneities in these experiments was small and thus this broadening must be attributed to the physical behaviour of the substance. The preceding result was in agreement with $\partial\theta_N/\partial p$ calculated using the Ehrenfest relation (eqn 6).

A very cursory experiment was performed by Belov *et al.* (1964) in order to determine the magnetization at atmospheric pressure and 1·9 kb in a magnetic field of 3·10 kOe. They found the antiferro–ferromagnetic transition temperature shifted at a rate of $-3{\cdot}6$°K/kb (Table XVII) which, compared with the other results given in Table XVII, is rather large although of the right sign.

The shift in the Néel temperature of polycrystalline Dy was again determined by Bloch and Pauthenet (1965a). Their result of $\partial\theta_N/\partial p = -0{\cdot}6 \pm 0{\cdot}1$°K/kb shown in Table XVII is in agreement with the results of Landry and Stevenson. In addition the magnetization at 77°K was measured in a magnetic field of 29 kOe, and the relative change in magnetization with pressure was $(-0{\cdot}15 \pm 0{\cdot}2) \times 10^{-3}$/kb. Employing the same thermodynamic relation, eqn (10), as in the cases of Gd and Tb, the relative dependence of the saturated magnetization at absolute zero was calculated $(1/\sigma_0)(\partial\sigma_0/\partial p)_{77°K} = (0{\cdot}1 \pm 0{\cdot}25) \times 10^{-3}$/kb. On the basis of the theory of indirect exchange the derivatives of Table XVII were computed. It is found that $\partial(\Delta M)/\partial p$ is of the same

TABLE XVII. Pressure data on dysprosium

Polycrystalline material

Dy	θ_N (°K)	$\partial\theta_N/\partial p$ (°K/kb)	θ_c (°K)	$\partial\theta_c/\partial p$ (°K/kb)	$\frac{1}{\sigma_s}\left(\frac{\partial\sigma_s}{\partial p}\right)$ (10^{-3}/kb)	References
		$-1\cdot0$				Swenson *et al.* (1960)
	~177	$-0\cdot56\pm0\cdot01$				Landry and Stevenson (1963)
				$-3\cdot6$		Belov *et al.* (1964)
	172	$-0\cdot6\ \pm0\cdot1$			$-0\cdot15\pm0\cdot2$ (77°K)	Bloch and Pauthenet (1965a)
		$-0\cdot62\pm0\cdot04$				Souers and Jura (1964)
I	179 ± 2	$-0\cdot62\pm0\cdot04$ ($P=5$–77 kb)				McWhan and Stevens (1965)
II	166 ± 5	$-0\cdot67\pm0\cdot07$ ($P=49$–85 kb)				
III	145 ± 6	$-0\cdot74\pm0\cdot08$ ($P=49$–85 kb)				
	$\sim177\cdot0$	$-0\cdot4$		$-0\cdot8$		Robinson *et al.* (1966)
	$175\cdot3$	$-0\cdot32$	$84\cdot7$	$-1\cdot16$		Okamoto *et al.* (1966)
		$-0\cdot44\pm0\cdot02$		$-1\cdot24\pm0\cdot10$		Milton and Scott (1967)
	172	$-0\cdot40\pm0\cdot05$				Kawai *et al.* (1967a)
	$178\cdot3$	$-0\cdot44$			$\frac{1}{n_{\mathrm{eff}}}\frac{\partial n_{\mathrm{eff}}}{\partial p}=-0\cdot67$ $n_{\mathrm{eff}}=10\cdot7\ \mu_B$	Tatsumoto *et al.* (1968)
Monocrystals						
	$180\cdot1_7$ (T ↑) $179\cdot9$ (T ↓)	$-0\cdot47\pm0\cdot01$	$89\cdot8_8$ (T ↑) $86\cdot9_0$ (T ↓)	$-1\cdot27\pm0\cdot02$		Bartholin and Bloch (1968)

order of magnitude as $(\partial\sigma_0/\partial p)_{77^\circ\text{K}}$. Their conclusion again is that the pressure variation of the magnetization of Dy seems to be due only to the variation in the magnetization of the conduction electrons.

Souers and Jura (1964) confirmed the preceding results by obtaining $\partial\theta_N/\partial p = -0{\cdot}62 \pm 0{\cdot}04^\circ$K/kb from electrical resistance data as a function of temperature and pressure. An application of the Ehrenfest equation (6) yielded $\Delta\alpha = -64 \times 10^{-6}/^\circ$C and $\Delta K = 40 \times 10^{-6}$/kb for the linear expansion and compressibility anomalies when the above values for $\partial\theta_N/\partial p$ and $\Delta C_p = 8{\cdot}5$ cal/(deg mole) were used. These are compared in Table XIV with others.

Studies by McWhan and Stevens (1965) of the crystal structure of Dy under pressure, suggest by analogy to Gd that it has a first-order transition from a hexagonal close-packed to a samarium-type structure at high pressures. The magnetic properties under pressure were also measured and yielded a variation of the Néel point $\theta_N = (179 \pm 2)^\circ$K with pressure of $\partial\theta_N/\partial p = (-0{\cdot}62 \pm 0{\cdot}04)^\circ$K/kb in good agreement with previous results (Table XVII). Apparent transitions in the high pressure phase were also measured (Tables XIV and XVII). The temperature *versus* pressure curve for Dy is presented in Fig. 21 where the discontinuity due to the transition to the high pressure phase is clearly shown. As in the case of Gd and Tb, McWhan and Stevens have considered the magnetic transition of Dy to be of second order and calculated the anomalies of $\Delta\alpha$ and ΔK (Table XV). It is reported that good correlation is obtained for $\Delta\alpha$ with experimental results. Finally the logarithmic dependence of the Néel point on the volume was calculated (Table XIV).

Qualitative magnetic measurements were executed by Robinson *et al.* (1966) on Dy under pressures ranging from 1 b–25 kb. Prominent cusps at the Néel temperature allowed an accurate determination of the shift in the Néel point $\partial\theta_N/\partial p = -0{\cdot}4^\circ$K/kb (Table XVII). However the antiferro–ferromagnetic transition peak broadened under pressure making the determination of the transition temperature $\theta_{\text{F}}^{\text{AF}}$ difficult. As a result, the plot of $\theta_{\text{F}}^{\text{AF}}$ initially increased (at an undetermined rate) then, above 5–7 kb, $\theta_{\text{F}}^{\text{AF}}$ decreased linearly with the slope $\partial\theta_{\text{F}}^{\text{AF}}/\partial p = -0{\cdot}8^\circ$K/kb. This result is small in comparison with the previous ones presented in Table XVII.

Unlike previous results which gave the relative changes in terms of the magnetization or electrical resistance of Dy, Okamoto *et al.* (1966) measured the pressure dependence of the transverse magneto-resistance in a magnetic field of 6 kOe under pressure up to 6 kb. The advantage obtained is that both the $\theta_{\text{F}}^{\text{AF}}$ and θ_N transitions appear as cusps in the data curves shown in Fig. 25. The data yielded a Néel temperature,

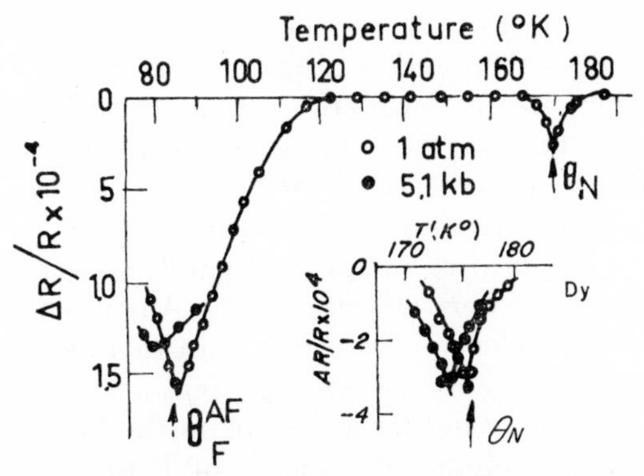

FIG. 25. Temperature dependence of the magnetoresistance of Dy under atmospheric pressure and 5·1 kb (from Okamoto *et al.*, 1966).

$\theta_N = 175{\cdot}3$°K, with $\partial\theta_N/\partial p = -0{\cdot}32$°K/kb and a $\theta_F^{AF} = 84{\cdot}7$°K with a $\partial\theta_F^{AF}/\partial p \simeq -1{\cdot}16$°K/kb (Table XVII). The data for θ_F^{AF} appeared to be linear at high pressures but between 1 b and 1·5 kb some curvature appeared, similar to that reported by Robinson *et al.* (1966). An estimate of $\partial\theta_N/\partial p$ can be made from eqn (6) and, again, it is in agreement with the previous results. Similar experiments by Kawai *et al.* (1967a) corroborate those given above.

The pressure dependence of the Néel temperature and antiferro–ferromagnetic transition were re-examined by Milton and Scott (1967), measuring the self-inductance of a coil which enclosed the sample. The Néel temperature peaks were sharp and reversible with temperature. However, it was reported that the Néel temperature *versus* pressure curve exhibited a slight curvature; as a result an initial pressure coefficient $(\partial\theta_N/\partial p) = -0{\cdot}44 \pm 0{\cdot}02$°K/kb, and an overall average initial pressure coefficient $(\partial\theta_N/\partial p)_{\text{av.}} = 0{\cdot}5$°K/kb, are reported. As usual, considerable difficulty was experienced in extracting the pressure coefficient of the antiferro–ferromagnetic transition temperature. These temperature peaks were broad and accompanied by considerable thermal hysteresis. To counteract the thermal hysteresis effect, it was necessary to cool the sample below 85°K and hold it at that temperature for several hours. Data was recorded on the warm-up cycle. Presumably the data obtained in this way were reproducible. The inflection point in the inductance *versus* temperature curve was employed to obtain the transition temperature. The pressure coefficient of the antiferro–ferromagnetic transition was $\partial\theta_F^{AF}/\partial p = (-1{\cdot}24 \pm 0{\cdot}10)$°K/kb (Table XVII).

Measurements of weak field a.c. susceptibilities have been made on Dy under hydrostatic pressures up to about 6 kb by Tatsumoto *et al.*

(1968). From these measurements, they find that $\partial\theta_N/\partial p = -0{\cdot}44\,°\mathrm{K/kb}$ in good agreement with previous results (Table XVII). Utilizing the pressure variation of the magnetic susceptibility $(1/\chi)(\partial\chi/\partial p)$, the relative change of the effective number of Bohr magnetons n_{eff} with pressure, $(1/n_{\mathrm{eff}})(\partial n_{\mathrm{eff}}/\partial p) = -6{\cdot}7 \times 10^{-4}/\mathrm{kb}$, was obtained. The linear compressibility K_1 was measured (using strain gauges) as a function of pressure (Fig. 23). The discontinuity of K_1 at θ_N is $\Delta K_1 = 0{\cdot}1 \times 10^{-4}/\mathrm{kb}$.

Single crystals of Dy were investigated by Austin and Mishra (1967) for the pressure dependence of resistivity up to 12 kb. The pressure coefficient of resistivity at 310°K was $(1/\rho_{\mathrm{mag}})(\partial\rho_{\mathrm{mag}}/\partial p) = -2{\cdot}5 \times 10^{-3}/\mathrm{kb}$ (Table XIII). By analysing these results on the basis of the indirect exchange model, the volume dependence of Γ the coupling constant and m^* the effective mass of the conduction electrons were computed. These results, shown in Table XIII, agree with those of Bloch and Pauthenet (1965a).

These results of Austin and Mishra were followed up by a careful investigation of the variation of the transition temperatures of Dy up to 6 kb by Bartholin and Bloch (1968). The Néel temperature decreased at the rate of $\partial\theta_N/\partial p = (-0{\cdot}41 \pm 0{\cdot}01)\,°\mathrm{K/kb}$ (Table XVII). At the antiferro–ferromagnetic transition the curves of $M_0(T)$ (mutual inductance) are characterized by a sudden discontinuity (Fig. 26) characteristic of a first-order transition. However, a thermal hysteresis was found for this discontinuity; the rate of change of the discontinuity with pressure, $\partial\theta_{\mathrm{F}}^{\mathrm{AF}}/\partial p = -(1{\cdot}27 \pm 0{\cdot}02)\,°\mathrm{K/kb}$, was independent of the way in which the transition temperature was reached. The magnetic c-axis length anomaly at $T = 0\,°\mathrm{K}$ was computed to be $(e_{cc})_0 = 0{\cdot}60 \times 10^{-2}$ for Dy in excellent agreement with the values $0{\cdot}6 \times 10^{-2}$–$0{\cdot}7 \times 10^{-2}$ estimated from X-ray diffraction and dilatometric measurements (Darnell, 1963a; Darnell and Moore, 1963). It should be recalled that the agreement for Gd and Tb was poor.

5. *Holmium*

As in the case of the preceding rare earth metals (with more than half filled $4f$ electron shells and a h.c.p. structure) holmium (Ho) undergoes two kinds of magnetic transitions. When its temperature is decreased, it passes from a paramagnetic phase at about 130°K to an antiferromagnetic phase and subsequently, at about 20°K, to a ferromagnetic phase. The antiferromagnetic phase possesses a helicoidal spin structure along the c-axis. The interlayer turn angle decreases from 50° at 130°K to 30° at 20°K. At 20°K, an increase in the 4th and 6th order axial anisotropy energy terms produces a transition to a warped conical ferromagnetic of mean cone angle 80° from the c-axis. The

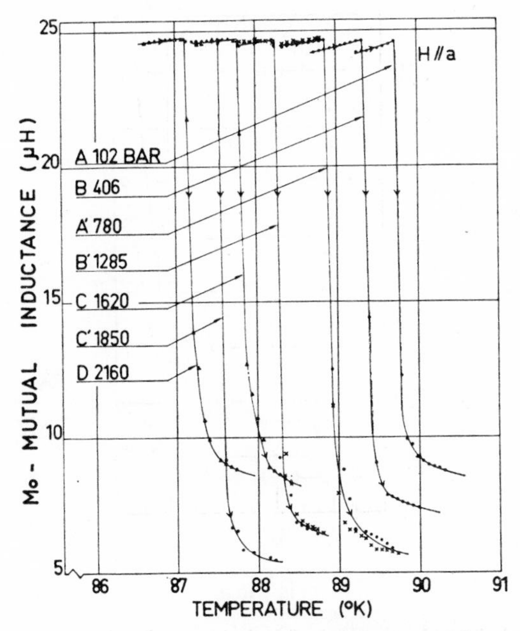

FIG. 26. Variation of M_0 with increasing temperature for a single crystal of dysprosium at different pressures. The field is applied along the a-axis (from Bartholin and Bloch, 1968).

spin structures are influenced by high magnetic fields and high pressures.

Bloch and Pauthenet (1965a) determined the shift in the Néel point of Ho; $\partial\theta_N/\partial p = -0{\cdot}45 \pm 0{\cdot}15$°K/kb at $\theta_N = 118$°K. Because of the metamagnetic transition in Ho, it was not possible to ascertain the pressure variation of the magnetization. However, they observed that at 77°K, the threshold field for Ho increased with pressure at the rate of (370 ± 50) Oe/kb. Use of the theory of indirect exchange by conduction electrons allowed them to obtain the pressure and volume dependence of the factors important in that theory; these appear in Table XIII.

Vinokurova and Kondorski (1964) found the dependence of $(1/\sigma)(\partial\sigma/\partial p)$ on the magnetic field at 77°K and 111°K shown on Fig. 27. At 77°K, the pressure coefficient is independent of the field up to 7 kOe, its value being $(1/\sigma)(\partial\sigma/\partial p) = -(84{\cdot}4 \pm 5{\cdot}6) \times 10^{-4}$/kb at $p = 2{\cdot}6$ kb. The sharp enhancement of the effect in strong field (Fig. 27) must be due to the partial transformation of the helicoidal spin structure to the ferromagnetic structure. At 111°K, the effect is smaller and independent of the magnetic field with $(1/\sigma)(\partial\sigma/\partial p) = (-44{\cdot}3 \pm 2{\cdot}9)$

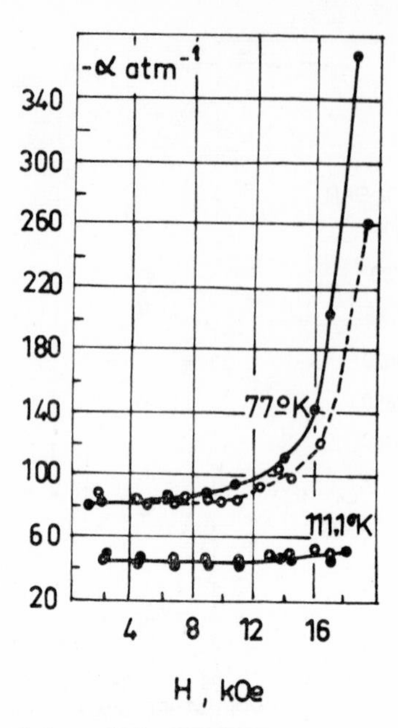

FIG. 27. The field dependence α (H) of Holmium at 77°K: ○ at p=2600 atm, ● at p=1800 atm; at 111·1°K: ○ at p=3700 atm, ● at p=1880 atm $(\alpha = (1/\sigma)(\partial\sigma/\partial p))$ (from Vinokurova and Kondorskii, 1964).

$\times 10^{-4}$/kb. Thus the magnetization of Ho decreases with pressure and is independent of field in the antiferromagnetic region.

Holmium was investigated by McWhan and Stevens (1965) in the same manner as were the previous three rare earths and they obtained a $(\partial\theta_N/\partial p) = -0{\cdot}48 \pm 0{\cdot}01$°K/kb (Table XIV) in good agreement with Bloch and Pauthenet. The curve of temperature *versus* pressure in Fig. 21 illustrates their data. The discontinuity appearing in the curve at about 70 kb marks the pressure at which Ho undergoes a transition to a high pressure phase of the Sm-type. Table XIV contains other information obtained by the use of eqn (6).

Investigations of the electrical resistance of Ho as a function of pressure were made by Kawai *et al.* (1967a). The rate of change of θ_N with pressure was $\partial\theta_N/\partial p = -0{\cdot}5 \pm 0{\cdot}1$°K/kb at $\theta_N = 117$°K.

Finally, a very detailed neutron diffraction study of single crystals of Ho was made by Umebayashi *et al.* (1968) up to pressures of 6 kb and at temperatures above 80°K. The variation of θ_N with pressure was found to be $\partial\theta_N/\partial p = -0{\cdot}33 \pm 0{\cdot}05$°K/kb. The pressure dependence of the helical turn angle, ω, was determined; at constant $(T - \theta_N)$, $(1/\omega)(\partial\omega/\partial p)$ was slightly temperature dependent, however, an average value was $1{\cdot}2\times10^{-3}$/kb. Calculation of $(1/\omega)(\partial\omega/\partial p)$ based on the theory of Miwa resulted in values, given in Table XVI, which are in good agreement with the other experimental results.

6. *Erbium*

Erbium (Er) is the last of the heavy rare earths to exhibit several well-defined magnetically ordered phases in a h.c.p. crystal structure. In zero applied field, Er exhibits three distinct magnetically-ordered phases (Cable *et al.*, 1961). As shown in Fig. 28, the *c*-axis component

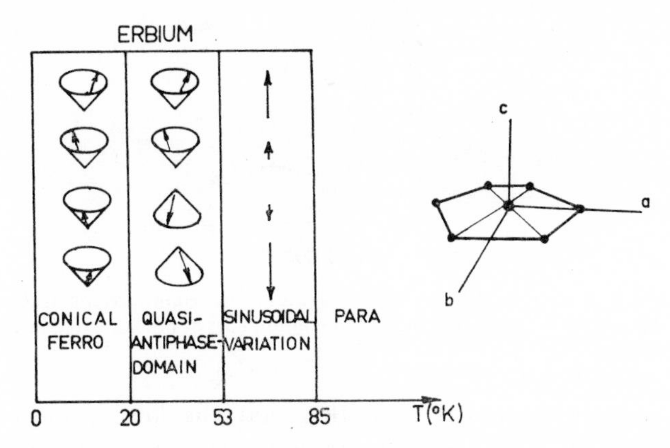

FIG. 28. Magnetic structure of Er in zero applied field as observed by neutron diffraction (from Rhyne and Legvold, 1965).

of the magnetization is observed to be sinusoidally modulated from 85 to 53°K, no ordering being found in the basal plane. Below 53°K, the sinusoidal variation changes to a square wave modulation and helical ordering is observed in the basal plane (a quasi–antiphase–domain structure). At about 20°K, the anisotropy energy spontaneously effects a transition to conical ferromagnetism in which the easy magnetic directions are generators of a cone. At 4·2°K, the *c*-axis component of the moment is 7·2 μ_β and the basal plane component is 4·7 μ_β which suggests that the moments lie on a cone of half angle of about 30°.

In spite of the interesting magnetic properties of Er only two studies of the effects of pressure on the magnetic properties have been reported. Milton and Scott (1967) established the Néel temperature at $\theta_N=85$°K with a pressure variation of $\partial\theta_N/\partial p=-0{\cdot}26\pm0{\cdot}01$°K/kb. Considerable difficulties were encountered in the determination of the lower temperature transitions because of thermal hysteresis effects illustrated in Fig. 29. Two peaks were observed on cooling and three on warming, and

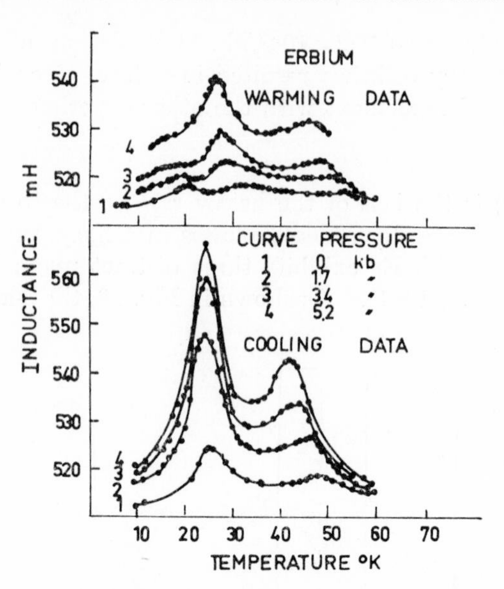

Fig. 29. Inductance readings *versus* temperature for an erbium sample at four different values of pressure. Both warming and cooling curves are shown (from Milton and Scott, 1967).

the investigators concluded from this that the low temperature and high temperature peaks on the warming curves correspond to the ferromagnetic point at 20°K and the spin transition at 53°K, respectively. The middle peak occurring at about 30°K on warming and the huge peak at 25°K on cooling are not understood and do not seem to correlate with other work. The temperatures of all the peaks are pressure dependent, except the cooling peak at 25°K which does not seem to depend on pressure up to 6 kb. Trial runs without a sample indicated that all the peaks are characteristic of Er. The peak at 20°K which is associated with the ferromagnetic transition is weak and vanishes with pressure above 3 kb. At lower pressures, it was possible to deduce a pressure shift of $\partial\theta_F^{AF}/\partial p = -0{\cdot}8 \pm 0{\cdot}2$°K/kb. The pressure shift of the upper peak is linear from the warming data but a slight curvature exists for the cooling data. The slope of the warming curve is $\partial\theta_F^{AF}/\partial p = -1{\cdot}3 \pm 0{\cdot}2$°K/kb.

The other work on Er by Vinokurova and Kondorskii (1964) involved the measurement of the relative change of the specific magnetization at 77°K and magnetic fields up to 17 kOe. The pressure coefficient was independent of the magnetic field and constant with pressure within experimental error: $(1/\sigma)(\partial\sigma/\partial p) = (-11{\cdot}63 \pm 0{\cdot}77) \times 10^{-3}$/kb ($p = 2{\cdot}8$ kb) and $(1/\sigma)(\partial\sigma/\partial p) = (-11{\cdot}73 \pm 0{\cdot}78) \times 10^{-3}$/kb ($p = 1{\cdot}8$ kb).

7. *Europium*

Europium (Eu) precedes the rare earth elements discussed above and, as such, is considered as one of the light rare earths. Unlike the heavy rare earths (Gd, Tb, . . .) which have hexagonal close-packed structures, Eu has a body-centred cubic (b.c.c.) structure. At 91°K it undergoes an antiferromagnetic transition; the antiferromagnetic spin structure is helicoidal with the magnetic moments lying in a cubic face with the rotation axis perpendicular to the moments. The magnetic properties of Eu do not fit a simple divalent model of seven localized $4f$ electrons or a trivalent model of six localized $4f$ electrons. The magnetic moment as determined from neutron diffraction studies is $5{\cdot}9\ \mu_\beta$ whereas a divalent model would give a value of $7\ \mu_\beta$. It also possesses a large atomic volume (29 cm^3/mole) and compressibility ($8{\cdot}5 \times 10^{-3}$/kb).

The situation for Eu is confused. On the one hand, McWhan *et al.* (1966a) find from resistance measurements that the Néel temperature is initially independent of pressure up to about 40 kb after which it takes a negative slope. On the other hand, Grazhdankina (1967) finds, from similar measurements, a positive pressure dependence of θ_N up to 12 kb. McWhan and co-workers find a qualitative justification for the form of their results by employing the theory of indirect interaction via the conduction electrons.

8. *Samarium*

Samarium (Sm) is one of the light rare earths having a structure consisting of a close-packed stacking arrangement of atomic layers in the sequence ABABCBCACABAB. It has a Néel temperature of about 14·8°K. Bridgman was the first to measure the resistivity and volume of Sm as a function of pressure; he found no unusual features. Recently, Stager and Drickamer (1964) have observed some minor deviations from normal behaviour in the resistivity of Sm in the region 50–500 kb.

The interest in Sm, here, is not that it was investigated under pressure but that under a pressure of about 40 kb and at temperatures above 300°C it transforms to another stable phase. This new phase was found to have a Néel temperature $\theta_N = 27$°K (Jayaraman and Sherwood, 1964b), and this pressure-induced phase, which has the double h.c.p. structure, is also close packed but has a different stacking sequence–ABACABAC.

9. *Gadolinium–Rare Earth Alloys*

Milstein and Robinson (1967a) chose the Gd–Dy binary alloy system for a study of the pressure effects on the magnetic properties. Alloys containing from 100 %–28 % Gd were prepared. The magnetic transitions were investigated for all alloys at atmospheric pressure and the temperature–composition diagram of Fig. 30 summarizes their findings. From this group of alloys three, *viz* 58·6 %, 50·8 % and 42·0 % Gd were selected for pressure studies. The transition temperatures *versus* pressure curves are given in Fig. 31 while the pertinent data from these

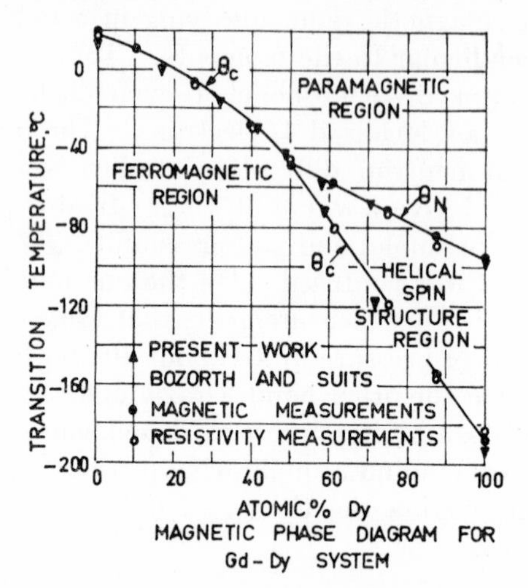

Fig. 30. Magnetic phase diagram for Gd–Dy system (from Milstein and Robinson, 1967a).

studies is compiled in Table XVIII. Noteworthy in these characteristics is the fact that a non-linearity exists in the ferromagnetic transition temperature and that this departure appears to increase as the concentration of Dy increases until in the 42 % Gd–58 % Dy sample a maximum appears. This peculiar behaviour is given a plausible explanation in part 11 of this section.

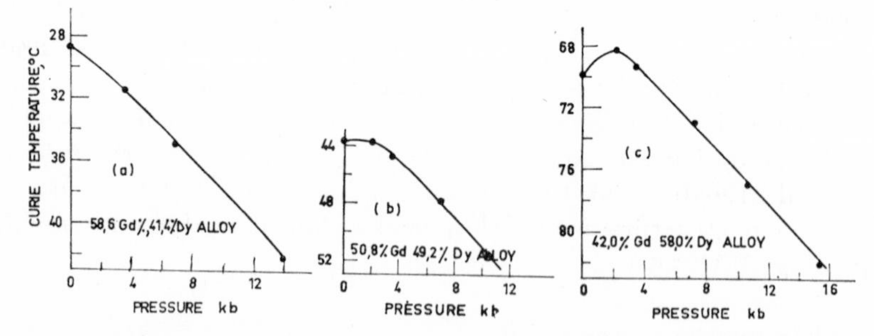

Fig. 31. Variation of the Curie temperature with pressure for alloys of (a) 58·6% Gd—41·4% Dy, (b) 50·8% Gd—49·2% Dy and (c) 42·0% Gd—58·0% Dy (from Milstein and Robinson, 1967a).

TABLE XVIII. Influence of pressure on the transition temperatures of Gd–Dy alloys. (From Milstein and Robinson, 1967a)

Composition of Gd–Dy alloy (% Gd)	Apparent Curie temp. θ_c (°K) in present study	Effect of pressure upon θ_c	Apparent Néel temp. θ_N (°K) in present study	Effect of pressure upon θ_N	Source of data
100	289	Decreases linearly at rate $-1 \cdot 60 \pm 0 \cdot 05$°K/kb			Reference[a]
58·6	244·2	Decreases at rate $-1 \cdot 04 \pm 0 \cdot 10$°K/kb			Present work
50·8	229·3	Appears not to change in 0–2 kb range, and then decreases at rate $-0 \cdot 98 \pm 0 \cdot 10$°K/kb			Present work
42·0	203	Appears to increase by about 2°K in range 0–2 kb and then decreases at rate $-1 \cdot 01 \pm 0 \cdot 10$°K/kb	216·6	Decreases linearly at rate $-0 \cdot 85 \pm 0 \cdot 08$°K/kb	Present work
0	80	Increases by about 6°K in range 0–5 kb and then decreases at rate $-0 \cdot 8$°K/kb. Results reported as tentative	~275·5	$-0 \cdot 4$°K/kb	Reference[b]

[a] Robinson *et al.* (1964), [b] Robinson *et al.* (1966).

Austin and Mishra (1967) prepared five alloys of Gd with lutecium (Lu) and yttrium (Y) and determined the shift in the Curie temperature with pressure. The results of the experiments are summarized in Table XIX. These two alloy systems were chosen because Lu has a full and Y has an empty $4f$ shell and alloying these elements with Gd is not likely to alter the electron concentration significantly considering

TABLE XIX. The Curie temperature and pressure variation of the Curie temperature of Gd–Lu and Gd–Y alloys. (From Austin and Mishra, 1967)

Alloy	θ_c (°K)	$\partial\theta_c/\partial p$ (°K/kb)	$\partial \log \theta_c/\partial p$ (10^3/kb)
Gd	293	−1·56	−5·3
$Gd_{0\cdot97}Lu_{0\cdot09}$	267	−1·46	−5·4
$Gd_{0\cdot80}Lu_{0\cdot20}$	259	−1·30	−5·0
$Gd_{0\cdot69}Lu_{0\cdot37}$	221	−1·25	−5·6
$Gd_{0\cdot85}Y_{0\cdot15}$	268	−1·30	−4·9
$Gd_{0\cdot70}Y_{0\cdot38}$	220	−1·23	−5·6

the stability of their configuration. Also, since Gd, Lu and Y have similar atomic volumes the unit cell volumes should be approximately the same; the difference between pure Gd and $Gd_{0\cdot57}Lu_{0\cdot43}$ is about 4·5%. Smidt and Daane (1963) have measured the resistivity of Gd–Lu and Er alloys as a function of temperature at atmospheric pressure, and these results support the hypothesis that a small decrease in the Fermi energy on alloying occurs, amounting to a 7% decrease for 30% Lu. Austin and Mishra have computed the volume dependence of θ_c in pure Gd using the data of Smidt and Daane (1963), the pressure data and Bridgman's compressibilities. The curves are presented in Fig. 32. It is evident that the variation in θ_c produced by alloying cannot be explained simply in terms of a change in the lattice parameter, which leads to the conclusion that it appears to be more appropriate to view the alloying as a process which dilutes the magnetic ions. The expression for the Curie temperature of a ferromagnetic metal with an indirect exchange mechanism is a function of $n(E_f)$, the density of states, k_f, the wave vector at the Fermi level, Γ, the average matrix element for the exchange interaction and R, the position vector of a $4f$ ion. It is concluded that R will change significantly on alloying but k_f and $n(E_f)$ hardly at all. Although a large change in θ_c will occur, it is expected that $\partial \log \theta_c/\partial p$ will be approximately the same for all alloys. Examination of Table XIX shows that the measured variation in $\partial\theta_c/\partial p$ is about 25% for Gd–Lu and Gd–Y alloys but $\partial \log \theta_c/\partial p$ is constant to ± 5%, in agreement with the conclusion.

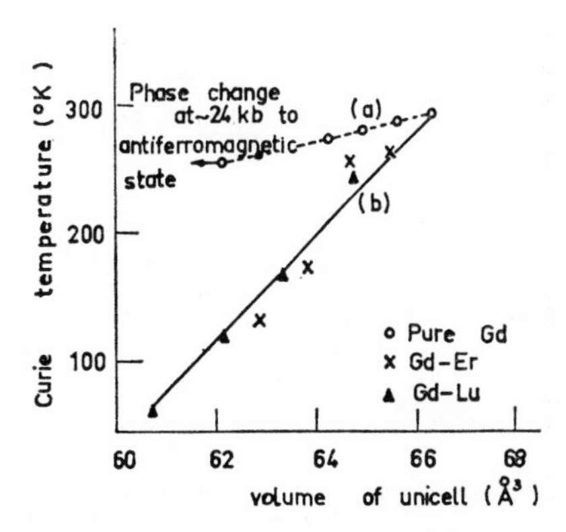

FIG. 32. Volume dependence of the Curie temperature in Gd, (a) from pressure data, (b) from alloy data (from Austin and Mishra, 1967).

McWhan and Stevens (1967) studied the pressure variation of the transition temperature of $Gd_{0\cdot45}$–$Y_{0\cdot55}$, $Gd_{0\cdot45}$–$Lu_{0\cdot55}$ and $Gd_{0\cdot343}$–$Er_{0\cdot657}$; the results appear at the bottom of Table XX. These alloys are part of a series having the same average De Gennes functions

$$\sum_i c_i(g_i-1)^2 J_i(J_i+1) = 7\cdot08,$$

which is the value for pure Dy; i designates a particular alloy constituent and c_i its concentration. The alloy $Gd_{0\cdot343}$–$Er_{0\cdot657}$ was particularly interesting because it was suspected that it might have three ordering temperatures similar to Er. Magnetization measurements of $Gd_{0\cdot343}$–$Er_{0\cdot657}$ *versus* temperature made at 1 atm confirm the existence of three transitions in this alloy. The susceptibility above 180°K follows a Curie–Weiss law with an effective moment $p = 8\cdot87\ \mu_\beta$ and $\theta_p = 150$°K. Assuming that the three transitions observed are similar to the three in Er, the pressure dependence of the paramagnetic to sinusoidal transition is $-0\cdot44$°K/kb and of the sinusoidal to antiphase domain transition is $-0\cdot36$°K/kb. These are comparable to the results for Er.

10. *Terbium—Rare Earth Alloys*

The alloy system Tb–Y has been thoroughly investigated by McWhan *et al.* (1966b) and McWhan and Stevens (1967). In general, all the transition points exhibited a similar behaviour as a function of pressure and temperature. The Néel temperature possesses little or no temperature hysteresis in cooling and warming cycles. However, the ferromagnetic transition temperature has a large hysteresis ($\simeq 9$°K), and the curves of initial susceptibility obtained on cooling and warming

TABLE XX. Pressure data for some rare earth elements and alloys. (From McWhan and Stevens, 1967)

Sample	Néel temperature θ_N (°K) (1 b)		$d\theta_N/dp$ (°K/kb) ±10%	p range (kb) 5 to	Phase transition		High pressure
	literature[a]	extrap.			$\theta_{N1}-\theta_{N2a}$ (°K)	at p (kb)	$\theta_{N2a}-\theta_{N2b}$
Tb	228	227	−1·08	34	18	35	29
95% Tb–5% Y	—	219	−0·80	30	17	30	28
90% Tb–10% Y	211	213	−0·77	30	17	40	26
85% Tb–15% Y	—	207	−0·70	47	17	52	26
80% Tb–20% Y	196	201	−0·63	45	15	60	24
70% Tb–30% Y	183	186	−0·52	45	15	87	21
60% Tb–40% Y	169	168	−0·41	55	—	—	—
50% Tb–50% Y	149	149	−0·32	53	—	—	—
40% Tb–60% Y	129	136	−0·32	70	—	—	—
30% Tb–70% Y	111	111	−0·28	85	—	—	—
Dy	179	182	−0·62	40	13	49	25
67·5% Tb–32·5% Y	—	182	−0·50	45	13	60	21
67·5% Tb–32·5% Y	—	175	−0·51	75	—	—	—
57% Tb–43% Er	—	181	−0·60	75	15	85	21
45% Gd–55% Y	—	161	−0·59	75	—	~85	—
45% Gd–55% Lu	—	147	−0·50	45	—	—	—
34·3% Gd–65·7% Er	—	144	−0·44	75	—	~85	—
		98	−0·36	60	—	—	—
Gd	291·8[b]	291	−1·63	25	38	25	20
Ho	133	133	−0·48	65	9	76	(15–20)

[a] Child *et al.* (1965), [b] Griffel *et al.* (1954).

exhibit large differences in magnitude and shape below the ferromagnetic transition temperature. The cooling curves at different pressures below 30 kb are similar in shape but the warming curves exhibit two broad maxima. This effect was also observed for Tb and Gd. At pressures above 30 kb, the initial susceptibility drops by a factor of ten, and the sharp rise indicative of the antiferro–ferromagnetic transition disappears. At lower temperatures, two new peaks form as in the rare earth elements Gd, Tb, Dy and Ho. These ordering temperatures occur in the high pressure Sm-type phase. Strangely, the pressure at which the peaks appear and their amplitudes depend strongly on the time the sample is held at a particular pressure. The results of these studies are presented in Table XX. The data from the high pressure phase cusps θ_{N2a} and θ_{N2b} are shown in Fig. 33. The lines in the

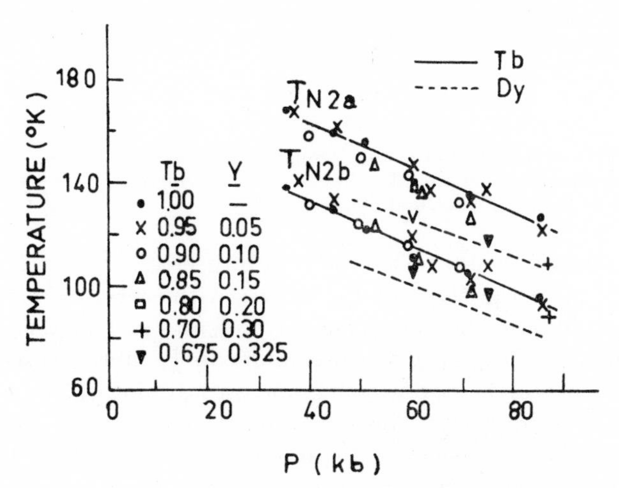

FIG. 33. Effect of pressure on the two peaks of the high pressure phases in the Tb–Y system. The lines are calculated from the data for pure Tb and Dy (from McWhan and Stevens, 1967).

figure are calculated for Tb and Dy. The results for the Tb–Y alloys fall fairly well between the calculated curves for the elements Tb and Dy in agreement with their average De Gennes functions which lie between those of Tb and Dy.

The effect of pressure on the ferromagnetic transition temperatures of the Tb–Y alloys determined from the warm-up curves is presented in Fig. 34. However, McWhan and Stevens warn that in view of the problems of hysteresis and broad peaks, any interpretation of the θ_c curves should be made with caution. The hysteresis observed in the inflection points increases with increasing Y concentration with $\Delta\theta_C = 9, 11$ and 15°K for the alloys with 5, 10 and 15% Y, respectively. The initial slopes and intercepts calculated from the smooth curves are $-0{\cdot}9$°K/kb (204°K), $-0{\cdot}8$(180°K), $-1{\cdot}6$ (160°K) and $-0{\cdot}2$ (128°K)

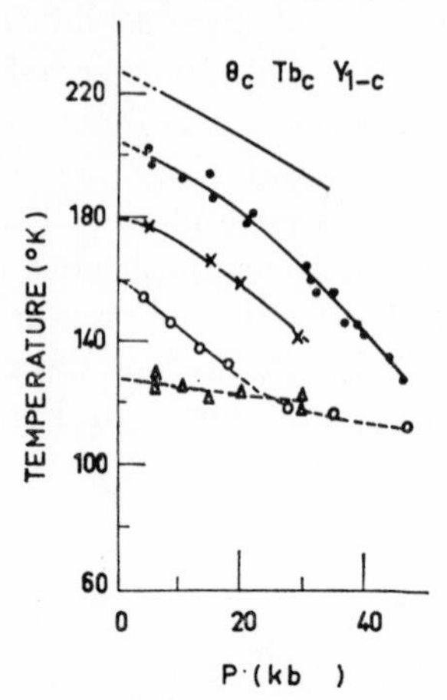

Fig. 34. Effect of pressure on the Curie temperature of the Tb–Y alloys. From top to bottom the curves are for alloys with Tb concentration of 1·00, 0·95, 0·90, 0·85, and 0·80 (from McWhan and Stevens, 1967).

for the alloys with 5, 10, 15 and 20% Y, respectively. There is a qualitative change in the slope of the curves with concentration which is complicated by the fact that at higher pressures in $Tb_{0{\cdot}85}\,Y_{0{\cdot}15}$, and at all pressures in $Tb_{0{\cdot}8}\,Y_{0{\cdot}2}$, the maximum in χ_i (initial susceptibility) was not reached at 77°K on cooling and on warming χ_i rose to a maximum and then dropped. This condition was simulated in $Tb_{0{\cdot}95}\,Y_{0{\cdot}05}$ and found to have no effect on the observed θ_c which suggests that the trend shown in Fig. 34 may be real.

A group of alloys were prepared by McWhan and Stevens (1967) having the average De Gennes function equal to that of Dy (7·08). These alloys are listed at the bottom of Table XX. All measurements on each sample were carried out in the same pressure sequence. The results for three alloys and Dy are plotted in Fig. 35. Qualitatively, all four materials show the same general features; namely, similar slopes and the appearance of two new peaks at high pressures. The pressure at which the polymorphic transition in Dy has been shown to occur at is 45 kb. The transformation in $Tb_{0{\cdot}675}\,Y_{0{\cdot}325}$ occurs between 45 and 60 kb, but in $Tb_{0{\cdot}675}\,Y_{0{\cdot}325}$ and $Tb_{0{\cdot}57}\,Er_{0{\cdot}43}$ it does not occur below 75 kb. As reported previously for the rare earths the high-pressure phase transitions are sluggish, and this is so too for the alloys.

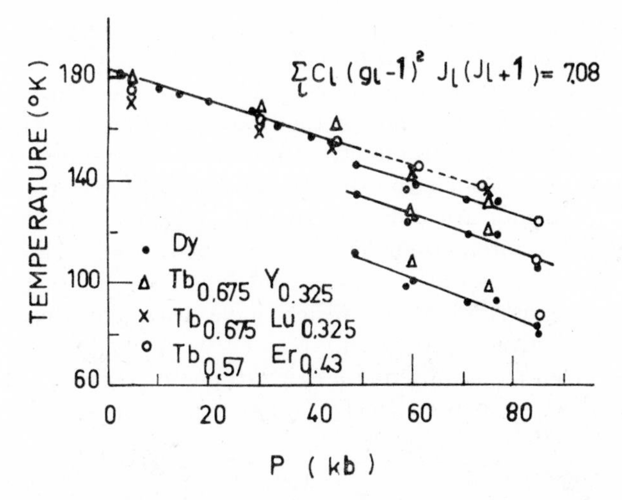

FIG. 35. Comparison of the effect of pressure on the ordering temperatures of different alloys with an average De Gennes function equal to that of Dy (from McWhan and Stevens, 1967).

Employing the molecular field approximation with a first-order correction for the anisotropy energy, McWhan and Stevens were able to calculate the change in the exchange interaction with the lattice parameter. Their result was $(k^{-1})\, \mathrm{d}J(Q)/\mathrm{d}a = (42 \pm 6)$°K/A, and this result and the aggregate of results presented above have been considered in some theoretical discussions concerning the mechanisms which might occur when pressure is applied. This is more extensively discussed in the following sub-section.

11. *Theoretical Speculations*

The work described above leading to an exposé of the effects of pressure on the magnetic properties of pure metallic elements and alloys has been motivated by the desire to reveal the dependence of the interactions in these materials on the interatomic distances. These results are usually correlated in terms of an interaction (exchange) curve which is the plot of some quantity proportional to the interaction against some quantity, usually a function of the interatomic distance, which is considered to be an independent variable. Most often, this is the paramagnetic Curie temperature or magnetic ordering temperature *versus* the ratio of the interatomic distance (D) to the unfilled electron shell diameter (d). Such a curve constructed for the first transition elements is shown in Fig. 14(a). Early measurements of the shift of the Curie point of Gd with pressure give a $\partial\theta_c/\partial p$ which is negative. Thus Gd should appear on the curve of Fig. 14(a) to the left of Fe, but the large D/d ratio for Gd would place it to the right of Ni. This suggests that there is an incompatibility associated with placing Gd on this curve constructed for the first transition elements. Other differences (such as electronic structure, crystal structure, and so on) serve to lead to the conclusion that the first transition elements and the rare earth elements are different species, indeed, would lead one to suppose that the interactions have a different nature. Thus, it is well to review briefly some of the ideas concerning the interactions in the rare earth metals and alloys which have been contributed and/or supported by high pressure work.

The first attempt to construct a new interaction curve was by Robinson *et al.* (1964) who used Néel's (1938) expression for the exchange energy in the rare earths

$$J_e = \frac{3k\theta J}{2ZS^2(J+1)}$$

where k is Boltzman's constant, θ is the magnetic ordering temperature, J is the total angular momentum quantum number, S is the total spin quantum number, and Z is the coordination number. Figure 34 shows a plot of J_e/k *versus* the interatomic distance D divided by the diameter $2R$ of the $4f$ electron shell (as calculated by the Slater orbital method) for the ferromagnetic rare earths using both ferromagnetic and paramagnetic Curie temperatures, θ_c and θ_p, respectively. A point for Gd under 19·6 kb pressure is included. The effect of pressure on the Curie temperature of Gd is found to be consistent with this interaction curve.

Subsequently, Bloch and Pauthenet (1965a) measured the shift of θ_c for Gd and of θ_N for Tb, Dy and Ho as well as their relative magnetizations as a function of pressure. They analysed the data on the basis of an indirect exchange mechanism with the conduction electrons as intermediaries. As inferred from the quantities $\partial(\log \Gamma)/\partial(\log V)$ in Table XIII, Γ, the coupling constant, is a fairly strong function of the volume or interatomic distances.

A short time later, McWhan and Stevens (1965) reported results on the same group of elements in agreement with Bloch and Pauthenet. They pointed out that the Robinson *et al.* (1964) interaction curve (Fig. 36) predicts a slope $\partial\,\theta_{\mathrm{F}}^{\mathrm{AF}}/\partial p$ for Dy which is not in agreement with their result. Consequently, they constructed a new interaction curve in which was plotted the observed ordering temperatures divided by the De Gennes function $(g-1)^2 J(J+1)$ *versus* the ratio of the interatomic distance, R, to the diameter of the unfilled $4f$ shell r shown in Fig. 37. The data falls fairly well into two curves representing the

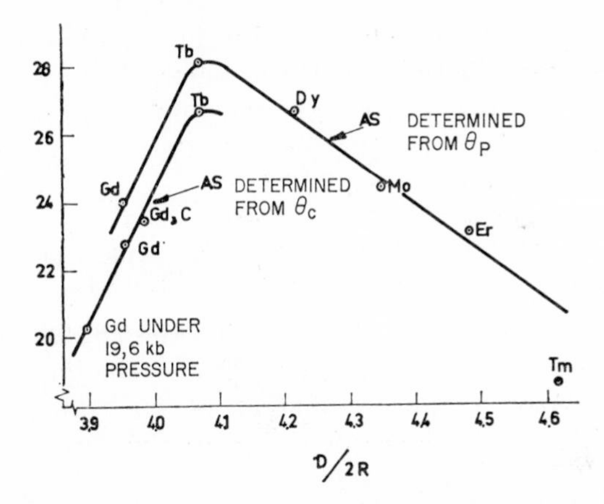

FIG. 36. Exchange interaction curve for the ferromagnetic rare earth metals. Two curves are shown, one based on θ_p, the paramagnetic Curie temperature, and the other on θ_c, the ferromagnetic Curie temperature (from Robinson *et al.*, 1964).

low and high pressure phases. In view of the many assumptions and approximations involved, the fit to a smooth curve is reasonable. Also, these curves show, at least qualitatively, that the exchange interaction increases smoothly on going from Gd to Ho, that is with increasing R/r. In this same report McWhan and Stevens indicated that their high pressure X-ray diffraction results reveal that the c/a ratio in the h.c.p. phase increases with pressure towards the ideal value of 1·63.

As a continuation of this work McWhan and Stevens (1966, 1967) considered the magnetic properties of a series of rare earth alloys at

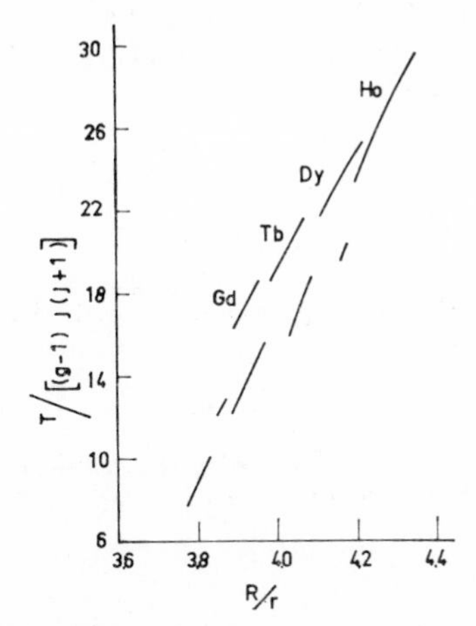

FIG. 37. Experimental exchange interaction curve for rare earth metal Gd through Ho (from McWhan and Stevens, 1965).

high pressure. The results were considered on the basis of the indirect exchange mechanism *via* the conduction electrons where, in addition to considering the exchange energy, they also introduced the anisotropy energy. They obtained the following expression for the exchange interaction

$$J(Q) = \tfrac{3}{2}\chi^{-1}\left[\theta_N(V) + \tfrac{1}{3}(\theta_{p\parallel} - \theta_{p\perp})\left(\frac{V_0}{V}\right)\right] \tag{46}$$

where χ is the De Gennes function. Using the $\partial\theta_N/\partial p$ and the anisotropy of the paramagnetic Curie temperatures, the quantity $J(Q)$ was evaluated and plotted as a function of the dimensionless ratio $(r_{4f}^2)^{1/2}/V^{1/3}$. These curves are presented in Fig. 38 and, surprisingly,

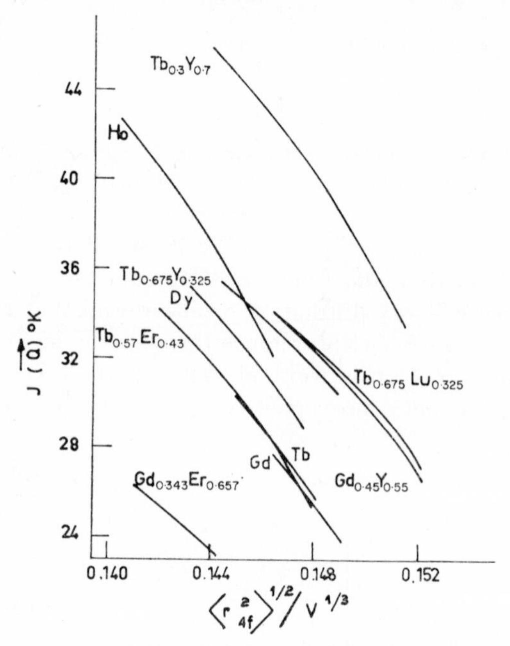

Fig. 38. The variation of the exchange interaction with $\langle r_{4f}^2\rangle^{1/2}/V^{1/3}$ for several rare earth elements and alloys. The average slope for all the alloys is $k^{-1}\,\mathrm{d}J(Q)/\mathrm{d}(\mathrm{K}r_{4f}^2) = -1300 \pm 200°\mathrm{K}$ (from McWhan and Stevens, 1967).

the slopes are quite similar for a wide variety of materials. The average slope is $k^{-1}\,\mathrm{d}J(Q)/\mathrm{d}a = (42 \pm 6)°\mathrm{K}/\text{Å}$ (a being the lattice parameter) which is approximately the same for all the materials. Hence, the observed variation in $\partial\theta_N/\partial p$ results mainly from the variation in X. As $\partial\theta_N/\partial p$ scales with X and not $X^{2/3}$ (Koehler, 1965), it would appear that the 2/3 law is not a universal one.

In the same paper McWhan and Stevens reported that a re-examination of their older X-ray diffraction data on the rare earths suggests that the previous report is erroneous and that the change in c/a with pressure is in fact very small. Milstein and Robinson (1967a) caught between the two papers of McWhan and Stevens proceeded to construct yet another interaction curve from new data for a series of Gd–Dy alloys on the premise that the c/a ratio varies and controls the interaction. Figure 39(a) shows the behaviour of the transition temperatures divided by

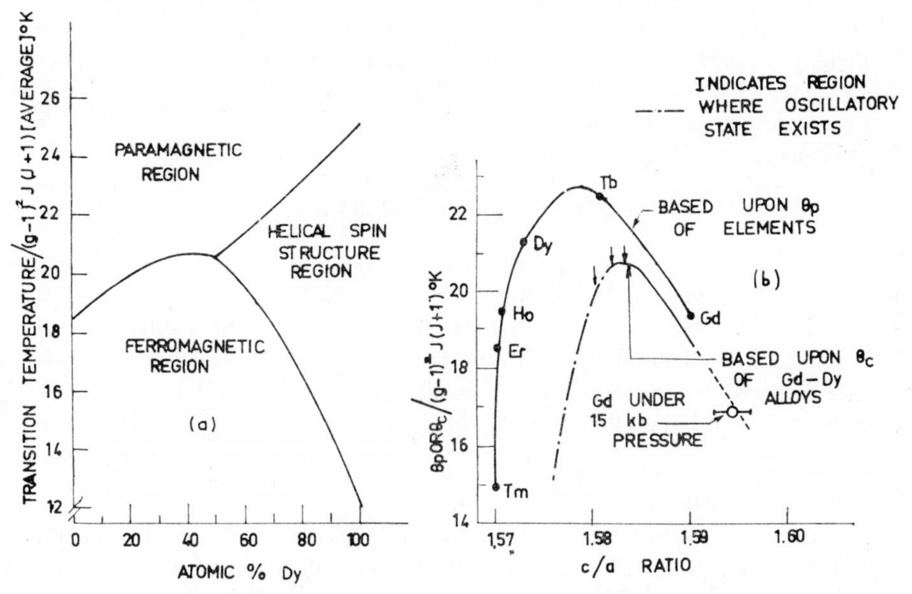

FIG. 39. (a) Magnetic transition temperature divided by average value of De Gennes factor *versus* composition for Gd–Dy system. (b) Upper curve shows paramagnetic Curie temperatures divided by De Gennes factors for heavy rare earth elements. Lower curve is ferromagnetic Curie temperatures divided by average values of De Gennes factors for Gd–Dy alloy system. Arrows indicate position of alloys subjected to pressure in present work, with the Gd content increasing in passing from left to right (from Milstein and Robinson, 1967a).

the De Gennes function *versus* composition of the Gd–Dy alloys. It should be noticed the $\theta_c/[(g-1)^2 J\,(J-1)]$ passes through a maximum at about 40% Dy, which is similar to the phenomenon observed for $\theta_p/[(g-1)^2\,J\,(J+1)]$ for the elements Gd through Tm. It has been observed that for Dy-rich alloys the θ_c apparently increased and then decreased with pressure (Fig. 31). From this it is concluded that the pertinent effect that pressure has upon the crystal structure is similar to the effect upon the crystal structure observed in passing through the series of elements in the order Tm to Gd. The reasons cited for the importance of the parameter c/a ratio are (1) the c/a ratio increases from Tm to Gd and (2) the c/a ratio of the rare earths increases with pressure as reported by McWhan and Stevens. In Fig. 39(b) are plotted (1) $\theta_p/[(g-1)^2 J(J+1)]$ *versus* the room temperature c/a values for the rare earths from Gd through Tm, (2) $\theta_c/[(g-1)^2 J(J+1)]$ (average) *versus* c/a (assuming c/a varies linearly with composition) for Gd–Dy alloys at atmospheric pressure, and (3) a data point for Gd metal under 15 kb pressure. The Gd data point should be compared with the extrapolation of the empirical results, and there is good agreement. In addition, with this curve, it is possible to give a rather straightforward explanation for the non-linear behaviour of the transition temperature *versus* pressure curves.

Finally, each factor in the exchange interaction relation

$$J(R) = \frac{3ZV^2\pi}{2E_F} \sum_R \Phi(2k_F R)$$

was examined individually in order to ascertain which might contribute a behaviour such as that shown in Fig. 39(b). The exchange integral V was eliminated since a calculation of V was made (Milstein and Robinson, 1967b) and it was found that V would be relatively insensitive to changes of the interatomic distance. E_F should be relatively constant for all the elements. $\sum_R \Phi(2k_F R)$ could be responsible for the behaviour above but no explicit connection between the variation of $\sum \Phi(2k_F R)$ with c/a could be found.

Arguing in a similar manner to that of Milstein and Robinson, Wazzan and co-workers (1967) conclude that it is reasonable that the variation of $\theta_N/[(g-1)^2 J(J+1)]$ with pressure must be largely due to the variation of the c/a ratio with pressure. Their interaction curve is shown in Fig. 40. The introduction into this plot of the values of $\theta_N/[(g-1)^2 J(J+1)]$ for Tb at various pressures, that is f(c/a), shows good agreement with the extrapolated curve.

Obviously, more work is required to ascertain definitely the effects of pressure on the c/a ratio as well as other important properties. Even so, perhaps too much is being expected of such a naive picture as an interaction curve.

C. FIRST TRANSITION SERIES METALS AND ALLOYS

1. *Introduction*

An examination of the following pages, which contain a review of the pressure studies of the magnetic properties of the first transition metals and alloys, indicates that there has not been as great an amount of

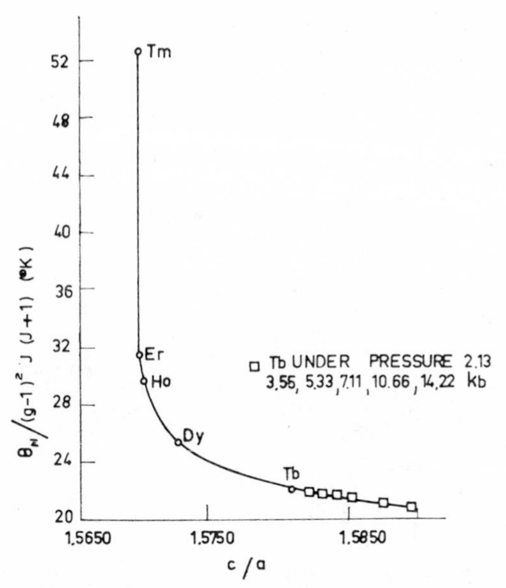

FIG. 40. Plot of $\theta_N/[(g-1)^2\, J(J+1)]$ as a function of the c/a ratio for the elements Tb–Tm (from Wazzan *et al.*, 1967).

work on these materials as on the rare earth metals and alloys. Undoubtedly, the lack of interest has been due to the fact that these metals and alloys have small pressure effects and that these effects are not so easily interpretable by means of existing theoretical models. Nevertheless, the papers reviewed below reveal a greater variety of studies and indicate some progress in experimental techniques and theoretical speculations.

2. *Iron*

In 1962, Tatsumoto *et al.* (1962a, b) measured the pressure coefficient of the saturation magnetization of iron. These results were made at room temperature under hydrostatic pressures up to 11 kb. It was found that the pressure coefficient $(1/\sigma_s)(\partial\sigma_s/\partial p) = -3{\cdot}10\times10^{-4}$/kb is in good agreement with the previous result of Kouvel and Wilson (1961), $-2{\cdot}74\times10^{-4}$/kb. It is in agreement also with the pressure coefficient determined indirectly from recent volume magnetostriction measurements: $-2{\cdot}85\times10^{-4}$/kb (Hasuo, 1964), $-2{\cdot}67\times10^{-4}$ (Stoelinga *et al.*, 1965), $-2{\cdot}49\times10^{-4}$ (Fawcett and White, 1967) and $-2{\cdot}50\times10^{-4}$ (Williams and Pavlovic, 1968). Bloch and Pauthenet (1965b) have shown that the pressure coefficient of saturation magnetization at absolute zero can be calculated by utilizing eqn (8), if one knows $(1/\sigma_s)(\partial\sigma_s/\partial p)$, $(1/\sigma_s)(\partial\sigma_s/\partial T)$ and $(1/\theta_f)(\partial\theta_f/\partial p)$. For iron it is found that the second term on the right-hand side of eqn (8) is negligible so that $(1/\sigma_0)(\partial\sigma_0/\partial p) \simeq (1/\sigma_s)(\partial\sigma_s/\partial p)$.

Patrick (1954) had measured the shift in the Curie point of iron with pressures up to 9 kb and found $\partial\theta_c/\partial p = 0 \pm 0{\cdot}1$°K/kb, that is no shift. More recently, this experiment was repeated by Léger *et al.* (1966b) who found $\partial\theta_c/\partial p = 0 \pm 0{\cdot}03$°K/kb, confirming Patrick's result. This result persisted up to about 17 kb, where the α Fe–γ Fe phase transformation occurs. This α Fe–γ Fe transformation was thoroughly investigated up to 65 kb (Fig. 41).

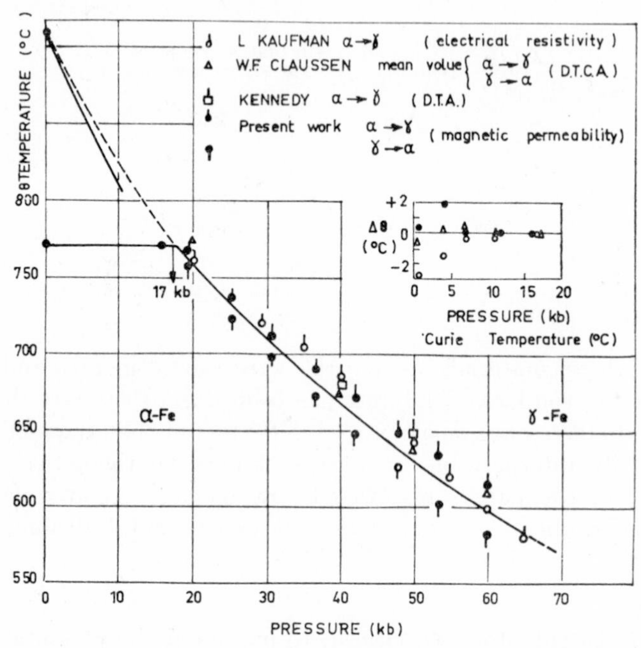

Fig. 41. Variation of the Curie temperature and the (α→γ) transition temperature in iron as a function of pressure (from Leger *et al.*, 1966b).

An important contribution to studies at high pressures were the measurements of Veerman and Rathenau (1965) of the pressure dependence of the first magnetocrystalline anisotropy constants of the transition elements at room temperature up to about 3 kb. Their results indicate for iron $(1/K_1)(\partial K_1/\partial p) = -4{\cdot}0 \times 10^{-3}$/kb, where K_1 is the constant per unit mass. Compare this with $-5{\cdot}6 \times 10^{-3}$/kb reported by Kouvel (1963). Veerman and Rathenau employed the relation

$$\frac{1}{K_1}\frac{\partial K_1}{\partial p} = m_p \frac{1}{\sigma}\frac{\partial \sigma}{\partial p} \tag{47}$$

to calculate the pressure coefficient of magnetization. Assuming $m_p = 8$, $(1/\sigma)(\partial\sigma/\partial p) = -5 \times 10^{-4}$/kb was obtained which is two or three times higher than the previously determined values. It should be noted that the value of $m_p = 8$ is not firm, although using even an $m_p = 10$ does not improve the situation much. Further measurements of $(1/K_1)(\partial K_1/\partial p)$ have been presented by Kawai and Sawaoka (1968) at room temperature and 77°K,

$$\frac{1}{K_1}\frac{\partial K_1}{\partial p} = -4{\cdot}6 \times 10^{-3}/\text{kb } (RT) \qquad \text{and} \qquad -8{\cdot}0 \times 10^{-3}/\text{kb } (77°\text{K}).$$

The room temperature result agrees in magnitude and sign with the previous results.

The pressure dependence of still another physical property has been determined by Franse *et al.* (1967). This is the pressure dependence of the anisotropic magnetostriction constants h_1 and h_2 at room temperature under pressures up to 3 kb. The pressure coefficient of the magnetostriction constants for iron were reported as

$$\frac{1}{h_1}\frac{\partial h_1}{\partial p} = 2{\cdot}2 \times 10^{-6}/\text{kb}$$

and

$$\frac{1}{h_2}\frac{\partial h_2}{\partial p} = 0{\cdot}3 \times 10^{-6}/\text{kb}.$$

3. *Nickel*

As part of their programme, Tatsumoto *et al.* (1962a, b) also determined the pressure coefficient of the saturation magnetization of nickel at room temperature; $(1/\sigma_s)(\partial\sigma_s/\partial p) = -2{\cdot}43 \times 10^{-4}$/kb. The result is in good agreement with the $-2{\cdot}85 \times 10^{-4}$/kb of Ebert and Kussman (1937) but not at all in agreement with the result of Kouvel and Wilson (1961), $(+1{\cdot}3 \pm 0{\cdot}7) \times 10^{-4}$/kb. However, Kouvel and Wilson point out that a later measurement on a single crystal of nickel was very close to the value of Ebert and Kussman. It should be recalled that over the past fifteen years there has existed a controversy concerning the sign of the pressure coefficient of nickel. All volume magnetostriction measurements up to 1954 when converted to $(1/\sigma)(\partial\sigma/\partial p)$ by means of eqn (6) give values of about -2×10^{-4}/kb, except for the result of Azumi and Goldman (1954) which converts to $+1{\cdot}15 \times 10^{-4}$/kb. Since that time numerous volume magnetostriction measurements have been made which give:

Laurens and Alberts (1964 $(1/\sigma_s)(\partial\sigma_s/\partial p) = -1{\cdot}6 \times 10^{-4}$/kb
Stoelinga *et al.* (1965) $= -1{\cdot}23 \times 10^{-4}$/kb
Fawcett and White (1967) $(T = 4{\cdot}2°\text{K}) = +1{\cdot}44 \times 10^{-4}$/kb.

The first two have a negative sign while the last is positive. The Fawcett and White result was obtained at $T = 4{\cdot}2°$ and there is some question as to the validity of comparing it with room temperature results, especially since it is known that Ni is not a well-behaved metal at low temperatures.

The shift in the Curie temperature of nickel with pressure has been found by Bloch and Pauthenet (1965b) to be $\partial\theta_c/\partial p = 0{\cdot}32 \pm 0{\cdot}02$°K/kb at $\theta_c = 633{\cdot}1$°K which was confirmed by Léger *et al.* (1966a). Employing this result, they show that $(1/\sigma_0)(\partial\sigma_0/\partial p) = -2{\cdot}8 \times 10^{-4}$/kb and is constant between 4·2° and 2·93°K. Further corroboration of this result has come from Okamoto *et al.* (1967a) who obtained $\partial\theta_c/\partial p = 0{\cdot}37$°K/kb.

Wisniewski (1962) has measured the pressure dependence of the saturation magnetostriction of polycrystalline nickel up to 6 kb, and he reports that it increases monotonously with pressure at the rate of about $0{\cdot}4 \times 10^{-6}$/kb. However, Franse *et al.* (1967) have determined the variation with pressure of the first two magnetostriction constants h_1 and h_2 and they find that $\partial h_1/\partial p = -0{\cdot}2 \times 10^{-6}$/kb and $\partial h_2/\partial p = 0$.

Veerman and Rathenau (1964) report the pressure coefficient of K_1 at room temperature for nickel $(1/K_1)(\partial K_1/\partial p) = (-7{\cdot}0 \pm 1) \times 10^{-3}$/kb. Again, utilizing the eqn (47) and assuming $m_p = 25$, they obtain $(1/\sigma_s)(\partial\sigma_s/\partial p) = -2{\cdot}8 \times 10^{-4}$/kb. Similar measurements have been made by Kawai and Sawaoka (1967b; 1968) at room temperature and 77°K, $(1/K_1)(\partial K_1/\partial p) = -7{\cdot}5 \times 10^{-3}$/kb ($RT$) and $-4{\cdot}8 \times 10^{-3}$ (77°K).

4. *Cobalt*

The early measurements of the pressure dependence of the magnetization of cobalt were uncertain because of the difficulties of obtaining magnetic saturation. Kouvel and Hartelius (1964b) have circumvented the difficulties by using a single crystal magnetized along the easy direction in a closed magnetic circuit so as to eliminate the demagnetization effects. Figure 42(a) illustrates one of their magnetization curves up to about 14 kOe, while Fig. 42(b) shows the value of $(1/\sigma)(\partial\sigma/\partial p)$ *versus* H. Above 3 kOe saturation is attained and $(1/\sigma_s)(\partial\sigma_s/\partial p) = -2{\cdot}18 \times 10^{-4}$/kb; compare this result with $-3{\cdot}5$, $-4{\cdot}2$ and $-3{\cdot}4 \times 10^{-4}$/kb obtained by Kornetzki (1934), Bozorth (1954) and Fawcett and White (1967), respectively. It was noted that since these measurements were made at room temperature, the $(1/\sigma_s)(\partial\sigma_s/\partial p)$ value could be attributed to a pressure dependence of σ_0 and/or of θ_c. Additional arguments lead to the conclusion that essentially the whole of $(1/\sigma_s)(\partial\sigma_s/\partial p)$ can be identified with $(1/\sigma_0)(\partial\sigma_0/\partial p)$ and therefore with the relative change of the atomic moment at 0°K with pressure.

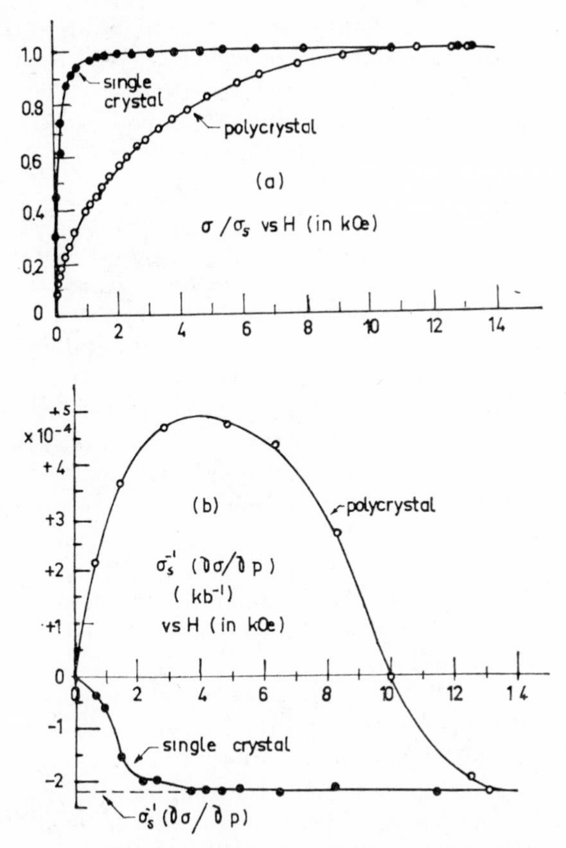

FIG. 42. (a) σ/σ_s and (b) $\sigma^{s}1\ (\partial\sigma/\partial p)$ as functions of applied field for single crystal and polycrystal of cobalt (from Kouvel and Hartelius, 1964b).

Additional measurements were made on a polycrystalline sample of cobalt and the results are illustrated in Fig. 42(a) and (b). Employing a graphical method (Kouvel, 1963) a measure of the pressure coefficient of the magnetocrystalline anisotropy constant $(1/K)(\partial K/\partial p) = -23{\cdot}2 \times 10^{-4}$/kb was obtained where K is an unknown linear combination of K_1 and K_2 since both constants are large in cobalt. In a subsequent paper Kouvel and Hartelius (1964c) were able to obtain $(1/K_1)(\partial K_1/\partial p) = -18 \times 10^{-4}$/kb and $(1/K_2)(\partial K_2/\partial p) = -85 \times 10^{-4}$/kb from a single crystal of cobalt magnetized at an angle of 72° with the hexagonal c-axis.

Experiments performed by Léger *et al.* (1967a) to measure the shift in the Curie temperature of cobalt up to 60 kb pressure yield $\partial\theta_c/\partial p = 0 \pm 0{\cdot}05$°K/kb, that is independent of pressure.

A tentative value of the pressure coefficient of the first magnetocrystalline anisotropy constant K_1 of cobalt is given by Veerman and Rathenau (1964); $(1/K_1)(\partial K_1/\partial p) \simeq -3{\cdot}5 \times 10^{-4}$/kb. This result is in better agreement with the polycrystalline sample value of Kouvel and Hartelius than with the single crystal result.

5. *Chromium*

Chromium is a metal which has an antiferromagnetic transition at about 310°K. Its magnetic structure (Corliss *et al.*, 1959; Bacon, 1961; Shirane and Takei, 1962) is one in which the magnetic moments $\boldsymbol{\mu}$ of the nearest neighbour atoms are oriented antiparallel in the [100] directions and whose magnitudes vary sinusoidally with a long period incommensurate with the periodicity of the crystal lattice. Thus $\boldsymbol{\mu}$ may be described as $\boldsymbol{\mu} = \boldsymbol{\mu}_0 \cos \mathbf{Q}.\mathbf{r}$, where $\boldsymbol{\mu}_0$ is the maximum value of $\boldsymbol{\mu}$ and $\mathbf{Q}$ is the wave vector. The direction of $\mathbf{Q}$ is along one of the cube axes and perpendicular to $\boldsymbol{\mu}$ above a spin flip temperature θ_{SF} of 122°K (transverse wave) but parallel below θ_{SF} (longitudinal wave). These particular transitions have been found to be greatly influenced by the additions of small amounts of alloy materials as well as by mechanical strain. It follows that they should also be influenced by the application of hydrostatic pressure.

The first neutron diffraction investigation of the effects of pressure on the antiferromagnetic transition was performed by Litvin and Ponyatovsky (1964). They found that the Néel temperature is lowered with pressure (up to 3·2 kb) at a mean rate of $-5{\cdot}9$°K/kb.

Electrical resistance measurements of Cr were made under pressure up to 82 kb and from 4·2° or 77° to 298°K by McWhan and Rice (1967). The volume or pressure dependence of the Néel temperature (Fig. 43) has a strong curvature. This curve is described by the exponential equation

$$\ln\left(\frac{\theta_N(V)}{\theta_N(V_0)}\right) = \frac{B_0}{\theta_{N_0}}\left(\frac{\partial \theta_N}{\partial p}\right)_{p\to 0}\left(\frac{V_0 - V}{V_0}\right) = -26{\cdot}5\left(\frac{V_0 - V}{V_0}\right)$$

where $\theta_{N0} = 312$°K and $(\partial\theta_N/\partial p)_{p\to 0} = -5{\cdot}1$°K/kb. This type of dependence seems to differ from that found in the one-band model but is compatible with the two-band model of antiferromagnetism. It is demonstrated that the temperature dependence of the magnetic contribution to the electrical resistivity can be explained by taking into account only the variation in the number of effective carriers from the introduction of a band gap (2Δ) due to the magnetic ordering.

More recently, Umebayashi *et al.* (1968) have made a thorough neutron diffraction study of the spin density waves in a monocrystal of

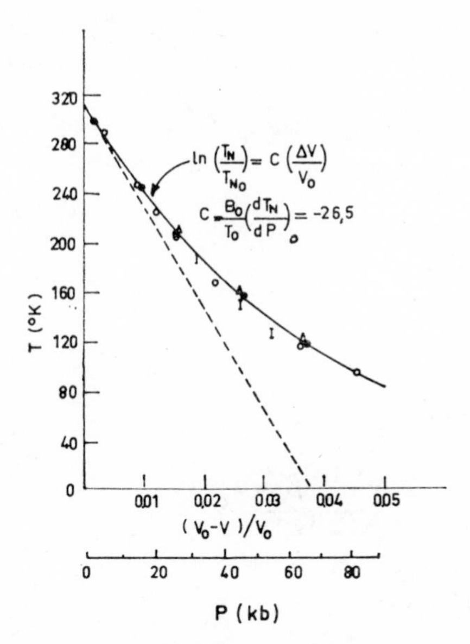

FIG. 43. Volume dependence of the Néel temperature of chromium. Vertical bars, sample 1; crosses, sample 2; and solid circles, sample 3 (from McWhan and Rice, 1967).

Cr at pressures up to 6 kb. The Néel temperature at a pressure of 0·05 kb was established at $\theta_N = 313{\cdot}5 \pm 1{\cdot}5$°K and diminished with pressure at a rate of $(-5{\cdot}3 \pm 0{\cdot}3)$°K/kb, while the spin flip temperature θ_{SF} decreased at a rate of $(-5{\cdot}8 \pm 0{\cdot}2)$°K/kb. The decrease of the Néel point is in agreement with the value of McWhan and Rice above, as well as with the other results of $(-5{\cdot}1 \pm 0{\cdot}2)$°K/kb (Mitsui and Tomizuka, 1965) and $-5{\cdot}4$°K/kb (Werner *et al.*, 1967). However, the spin flip variation with pressure is low compared to the more recent result, $\partial\theta_{SF}/\partial p \simeq 10$°K/kb determined by Werner *et al.* (1968).

6. *Iron Alloys*

Of the iron alloys which have been investigated, the system receiving the greatest amount of attention is the iron–nickel system. Kouvel (1963) has adequately discussed the work prior to 1962. Since then, the pressure dependence of the Curie temperatures of alloys ranging in composition from 30 % nickel in iron to 100 % nickel have been determined by Léger *et al.* (1967a) up to 60 kb. The Curie temperature *versus* pressure curves were not linear, although it is not stated whether this is a true curvature or an error fluctuation. However, in a previous study, on a 36 % Ni–63 % Fe alloy, Léger *et al.* (1966b) reported a definite curvature to the pressure curve. Nevertheless, the average slopes become smaller with nickel content, change sign at about 70% Ni and increase to the positive value of pure Ni. In Fig. 44 are plotted the results of Patrick (1954), Kaneko (1960) and the Léger *et al.* results using the average slopes of the first four alloys of low nickel content together with the Curie temperature curve as a function of composition. These are in fair agreement with previous results.

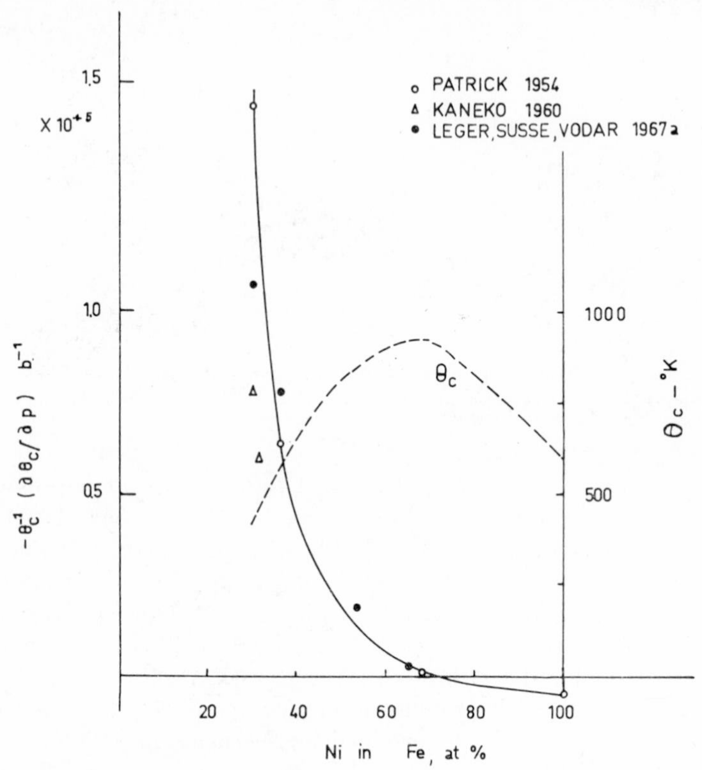

FIG. 44. The Curie temperatures and the relative Curie temperature variation with pressure of Fe–Ni alloys.

Two other iron base alloys reported to have a behaviour like the invar alloys—52% Fe–36% Ni–12% Cr (alloy No. 1) and 37·8% Fe–53·5% Co–8·7 Cr (alloy No. 2) were investigated up to 25 kb by Livshitz and Genshaft (1964). They measured the shift of the Curie temperature with pressure and found a non-linear relationship. The data was fitted to the parabolic expression,

$$\Delta\theta_c = -ap - bp^2$$

where the constants a and b have the values given in Table XXI.

TABLE XXI. Constants a and b in the relation
$\Delta\theta_c = -ap - bp^2$

Alloy	a (°K/kb)	b (°K/kb)
No. 1	$2\cdot6 \pm 0\cdot3$	$0\cdot05 \pm 0\cdot01$
No. 2	$6\cdot7 \pm 0\cdot6$	$0\cdot27 \pm 0\cdot01$

Another invar-like behaviour has been observed by Kussman and Jessen (1962) in γ iron–palladium alloys containing about 30 atm % Pd. In order to make a comparison with the Fe–Ni invar alloys, Wayne (1967) measured the pressure dependence of the Curie temperature of γ–Fe_{1-x} Pd_x alloys for $0{\cdot}29 \leqslant x \leqslant 0{\cdot}35$ up to pressures of 20 kb. The values of $\partial\theta_c/\partial p$ vary continuously with increasing Pd content from $-(2{\cdot}75 \pm 0{\cdot}05)$°K/kb to $-(1{\cdot}63 \pm 0{\cdot}05)$°K/kb. Such a variation is similar to that of the γ Fe–Ni alloys.

Fujimori (1966) reports that an antiferromagnetic alloy of iron with 30 atm % Mn has a pressure dependence of its Néel point which amounts to $\partial\theta_N/\partial p = -2{\cdot}5$°K/kb. Veerman and Rathenau (1964) have reported the results of experiments which lead to the pressure dependence of the magnetocrystalline anisotropy for some alloys of iron with Be, Si and Al together with earlier values arrived at by other workers in an indirect manner.

The iron–rhodium alloys in the vicinity of the composition $Fe_{0\cdot5}$–$Rh_{0\cdot5}$ exhibit an interesting magnetic behaviour (Fallot and Hocart, 1939). The first alloy to be studied under pressure (Zakharov *et al.*, 1964) is antiferromagnetic at room temperature and atmospheric pressure; as its temperature increases, it undergoes an antiferromagnetic–ferromagnetic transition at about $\theta_{F}^{AF} = 350$°K and, finally, at $\theta_c = 660$°K it has a ferromagnetic–paramagnetic transition. Its CsCl type BCC structure (α) remains intact throughout, only the magnetic structure changes accompanied by uniform unit cell dilatation. The $(\alpha_{AF} - \alpha_F)$ transition is a first-order transformation ($\Delta V \neq 0$) while the second transformation $(\alpha_F - \alpha_p)$ is second order ($\Delta V = 0$, $\Delta S = 0$) and corresponds to a Curie point.

This was investigated for the pressure dependence of the $(\alpha_{AF} - \alpha_F)$ transition up to about 25 kb. A linear increase in θ_{AF}^{F} with pressure was observed having the slope $\partial\theta_{AF}^{F}/\partial p = 4{\cdot}3$°K/kb, with a 10° to 15° thermal hysteresis. In applying the Kittel theory of exchange inversion, Zakharov and co-workers were able to calculate the compressibility of Fe – Rh; $K = 1{\cdot}1 \times 10^{-4}$/kb. This result appears to be low for this material. Later, in a very beautiful set of experiments, Ponyatovskii *et al.* (1967) combined measurements of both the $(\alpha_{AF}-\alpha_F)$ transition and the $(\alpha_F-\alpha_p)$ transition temperature with the thermal expansion at the $(\alpha_{AF}-\alpha_F)$ transition with pressure to construct the $T-p$ phase diagram shown in Fig. 45. For the $(\alpha_{AF}-\alpha_F)$ transition the slope of the characteristic is $\partial\theta^F_{AF}/\partial p = 4{\cdot}7\,°K/kb$ while for the $(\alpha_F-\alpha_p)$ transition it is $\partial\theta_c/\partial p = -0{\cdot}6\,°K/kb$. Since the lines in the plot are linear, they were extrapolated to higher pressures until they intersected. This point of intersection at about 50 kb would be a magnetic triple point where it would be possible for one space structure to exist simultaneously with three different magnetic structures (antiferromagnetic, ferromagnetic and paramagnetic). It was also possible to extract from the data at room temperature the compressibility $K = 4{\cdot}8\times10^{-4}/kb$ and to show that at high temperatures the compressibility of the ferromagnetic phase is slightly higher than that of the antiferromagnetic phase. In conclusion, Ponyatovskii and co-workers state it appears

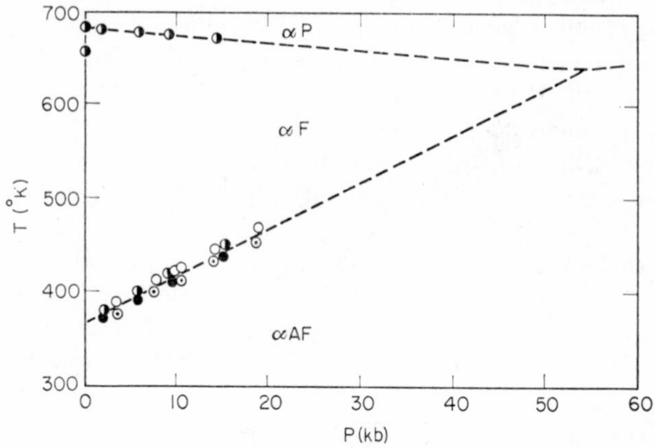

FIG. 45. Phase diagram of the Fe–Rh alloy. Magnetometric measurements: ◑, ●; tensometric measurements: ⊙, ○ (from Ponyatovskii *et al.*, 1967).

desirable to study Fe–Rh alloys which have a narrower temperature range of stability of the θ_F phase and possibly triple points at pressures below 30 kb. It was suggested that this might be achieved by reducing the iron content in the binary alloy or else by introducing a third element.

In 1966 Bloch (1966a) reported results on a $Fe_{0{\cdot}47}Rh_{0{\cdot}53}$ alloy. The shift in the $(\alpha_{AF}-\alpha_F)$ transition temperature *versus* pressure was linear and the slope $\partial\theta^F_{AF}/\partial p = 5{\cdot}1\,°K/kb$ which confirms the previous results. By invoking the ideas of the Kittel theory of exchange inversion he derives a theoretical expression for $\partial\theta^F_{AF}/\partial p$ and obtains the result $\partial\theta^F_{AF}/\partial p = 4{\cdot}75\,°K/kb$ in excellent agreement with the experimental result. Although this confirmation would strongly recommend the validity of this theory, Kouvel (1966) disagrees and argues in favour of a theory in which the moments of the Rh atoms play the dominant role.

Léger *et al.* (1967b) have studied the alloy $Fe_{0\cdot47}$–$Rh_{0\cdot53}$ up to 60 kb. The variation of the $(\alpha_{AF}-\alpha_F)$ and $(\alpha_F-\alpha_p)$ transition temperatures as a function of pressure are shown on Fig. 46. At low temperatures, the

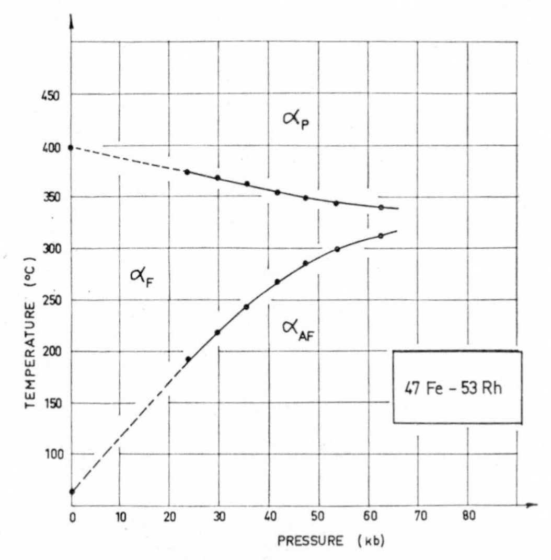

FIG. 46. Magnetic phase diagram of a 47% Fe–53% Rh alloy (from Léger *et al.*, 1967b).

investigators find that $\partial\theta^{F}_{AF}/\partial p = (5\cdot34 \pm 0\cdot2)$°K/kb and $\partial\theta_c/\partial p = -(1\cdot04 \pm 0\cdot05)$°K/kb; both results being in good agreement with previous results. Also it is observed that the extrapolations of Ponyatovskii *et al.* (1967) are incorrect because the $T-P$ characteristics have a curvature above 30 kb so that even at 60 kb the curves have not yet intersected. The magnetic triple point predicted to be at 50 kb actually occurs above 80 kb, if at all. Concomitantly, Léger *et al.* (1967b) prepared the alloy $Fe_{0\cdot50}$ $Rh_{0\cdot46}$ $Ir_{0\cdot04}$, expecting to reduce the stability range of the ferromagnetic phase and lower the pressure at which the magnetic triple point might occur. The results of this experiment are shown in Fig. 47. The triple point has been attained and occurs at a pressure of $(31\cdot3 \pm 1\cdot5)$ kb and a temperature of $(326\cdot7 \pm 3)$°C. The lack of discontinuities in the curves of $(\alpha_{AF}-\alpha_F)$ and $\alpha_{AF}-\alpha_p$ indicates that the thermodynamic relations $\sum\Delta V = 0$ and $\sum\Delta S = 0$ hold at the triple point. In conclusion it is inferred from the

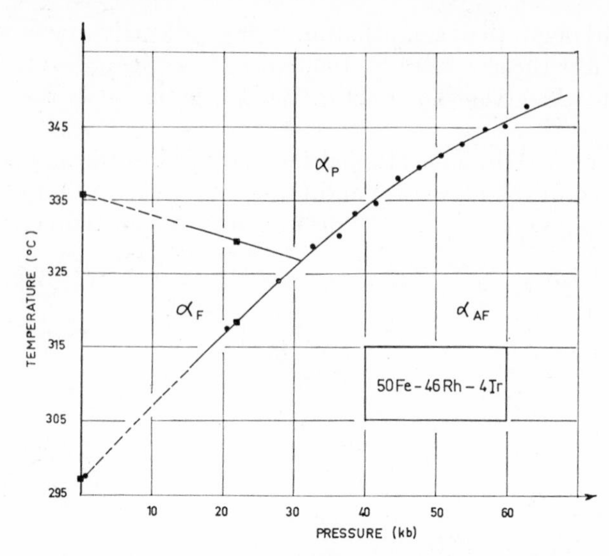

FIG. 47. Magnetic phase diagram of a 50% Fe–46% Rh–4% Ir alloy (from Léger *et al.*, 1967b).

similarity between the curves of the $Fe_{0\cdot50}Rh_{0\cdot46}Ir_{0\cdot04}$ alloy and the $Fe_{0\cdot47}Rh_{0\cdot53}$ alloy that a triple point does exist for $Fe_{0\cdot47}Rh_{0\cdot53}$ at higher pressures.

7. *Nickel Alloys*

The Ni–Cu alloy system has been the most extensively studied nickel system. It is a face centred cubic solid solution across the entire phase diagram and represents the addition of a non-magnetic atom, Cu, to a ferromagnetic metal, Ni. The first experiments to determine the effects of pressure on these alloys had been performed by Ebert and Kussman (1937), Kondorskii and Sedov (1960) and Patrick (1954). Since that time a few other investigations have been made.

The pressure dependence of the Curie temperature of Ni, $Ni_{0\cdot80}$–$Cu_{0\cdot20}$ and $Ni_{0\cdot70}$–$Cu_{0\cdot30}$ was investigated by Bloch and Pauthenet (1965b). They reported that $\partial\theta_c/\partial p$ decreases with copper content in a monotonic fashion. Using the molecular field approximation eqn (22) for the volume dependence of the molecular field constant, n, the $\partial \log n/\partial \log V$ are calculated for the materials given above, and it varies from $-1\cdot95$ to $-2\cdot15$, that is the molecular field constant varies with the interatomic distance to a power of about -6. To supplement these results, Fujiwara *et al.* (1966) have determined the pressure coefficient of the magnetization and Okamoto *et al.* (1967a) the pressure coefficient of the Curie temperature of a series of Ni–Cu alloys up to 50 % copper. On the whole, they find that $(1/\sigma_s)(\partial\sigma_s/\partial p)$ is negative at low temperatures and increases with temperatures until in the vicinity of θ_c it changes from negative to positive. From these results, the pressure coefficient of magnetization at 0°K is obtained and it is found to be in good agreement with the earlier results of Kondorskii and Sedov (1960). The $\partial\theta_c/\partial p$ for the Ni–Cu alloys decreases with copper content from 0·31°K/kb for pure Ni to zero at about 34 atm % Cu and reverses sign for higher copper contents.

Three alloys of Ni–Si (containing 7, 8 and 9 % silicon) were studied by Bloch and Pauthenet (1965b) to determine the pressure dependence of their Curie temperatures. Again, as in the Ni–Cu alloys, $\partial\theta_c/\partial p$ decreased with silicon content but at a rate three to four times faster.

Several Pd–Ni alloys were selected by Fujiwara *et al.* (1967) for the investigation of the variation of the magnetic moment with pressure. These were chosen because a comparison of their properties with those of the Ni–Cu alloys would be of interest since Pd is a non-ferromagnetic metal with holes in its *d*-band while Cu does not have holes in its *d*-band. Qualitatively, the behaviour of the Ni–Pd alloys is much like that of the Ni–Cu alloys, that is at low temperatures $(1/\sigma_s)(\partial\sigma_s/\partial p)$ is negative and increases with temperature becoming positive close to the Curie temperature. It is suggested that for the Ni–Pd alloys the sign of $(1/\sigma_0)(\partial\sigma_0/\partial p)$ is negative, but that $\partial\theta_c/\partial p$ is positive for the Ni-rich alloys and negative for the Pd-rich alloys.

8. *Chromium Alloys*

Neutron diffraction studies on dilute chromium alloys have revealed that spin density waves (SDW) in chromium are sensitively modified by the presence of transition metal impurities. This modification is manifest by the occurrence of a stable anti-ferromagnetic structure (AF). In alloys containing about 1 atm % Mn or 2 atm % Fe, however, a first-order transition between SDW and AF states has been observed. It has been shown that the low temperature phase of Cr–(Mn) is a SDW state whereas that of Cr–(Fe) is an AF state. Syono and Ishikawa (1967) have investigated the pressure effect on these first-order transitions of both alloys in order to clarify the mechanism of the transition. In the case of the Cr (Fe) alloy, the transition was detected using electrical resistivity measurements, while for the Cr (Mn) alloy the anomaly in the thermal expansion was used because for this alloy, no anomaly in the resistivity exists. A linear relation between the applied pressure and the transition temperature holds up to 2·4 kb for a Cr–(2·3 atm % Fe) alloy. The slope $\partial\theta_{\mathrm{SDW}}^{\mathrm{AF}}/\partial p$ was determined as

$-(21{\cdot}1 \pm 0{\cdot}2)$°K/kb with no thermal hysteresis. In contrast with the transition in Cr (Fe), a large thermal hysteresis was observed for a Cr–(0·43 atm % Mn) alloy. Assuming that the centre of the hysteresis loop is the mean transition temperature, the pressure dependence of the transition temperature was estimated to be $\partial\theta_{AF}^{SDW}/\partial p = +(22 \pm 1)$°K/kb. A discussion is put forth to explain these results on the bases of Fermi surface arguments.

D. INTERMETALLIC COMPOUNDS AND ALLOYS

1. *Introduction*

In this section on intermetallic compounds and alloys, the materials are not grouped in any order based on a predominant or related property. Most of these are individuals with no particular connection or correlation with any other. Because manganese is a major constituent of a large number of them, these have been presented first in an alphabetical order according to the next most abundant element. Although many of these materials exhibit interesting properties the pressure investigations of them have been sketchy and only a few have received considerable attention.

2. *Manganese–Arsenide*

The metallic compound MnAs possesses, at low temperature, the classic NiAs structure ($B8_1$). The magnetic moments are parallel to each other and perpendicular to the *c*-axis of the crystal. It has a first-order transition temperature θ_c between the ferro- and paramagnetic phases. At atmospheric pressure, the value of θ_c is about 317°K for increasing temperatures and 307°K for decreasing temperatures, and the transition is accompanied by a latent heat of 1·79 cal/g. The magnetic volume anomaly at θ_c is positive and amounts to about 1·86 % (de Blois and Rodbell, 1963; Goodenough and Kafalas, 1967). Utilizing the relation (5) one expects a variation of the magnetic transition temperature with pressure of -12°K/kb. The paramagnetic phase which exists immediately above θ_c has an orthorhombic structure of MnP (B31) type. At 400°K (θ_t) occurs a new transition of second order between an MnP phase and a NiAs phase.

The experiments of Rodbell (1961) to 1·5 kb have given the value $(\partial\theta_c/\partial p)\,(T\uparrow) = -12$°K/kb; those of Kamigaichi *et al.* (1965) to 2·5 kb, $-17{\cdot}5$°K/kb, and of Samara and Giardini (1965) also to 2·5 kb, $-12{\cdot}3$°K/kb. The more recent experiments of Grazhdankina and Bersenev (1966) give $-(16{\cdot}2 \pm 0{\cdot}3)$°K/kb up to 2·5 kb, followed by an abrupt decrease attributed to a phase change. The phase thus obtained would be antiferromagnetic with a Néel temperature which rises with pressure at the rate of $(2{\cdot}26 \pm 0{\cdot}07)$°K/kb. The stable domain of the NiAs phase was more precisely specified by Goodenough and Kafalas (1967). The expression $p = -5{\cdot}6230 + 0{\cdot}1072\ \theta_c(T\uparrow) - 0{\cdot}000283\ \theta_c^2(T\uparrow)$ describes the results obtained where $\theta_c\ (T\uparrow)$ is the value measured with increasing temperature in Kelvin degrees and the pressure p is expressed in kb. The derivative of this relation gives the initial value $(\partial\theta_c/\partial p)\ (T\uparrow) = -13{\cdot}8$°K/kb. The stable domain of the NiAs phase is between 0 and 4·6 kb.

The transition temperature $\theta_c\ (T\downarrow)$ obtained with decreasing temperature varies with pressure at the rate of -24°K/kb up to 1 kb (de Blois and Rodbell, 1963) and $-(21 \pm 0{\cdot}5)$°K/kb up to 2 kb (Kamigaichi *et al.*, 1965), in qualitative agreement with the results obtained by Goodenough and Kafalas which indicate that a pressure of 2 kb is sufficient to stabilize the B31 phase.

Recent experiments (Menyuk *et al.*, to be published), using a vibrating coil magnetometer allows a better description of the magnetic phase diagram of MnAs (Fig. 48). It also provides detailed information on

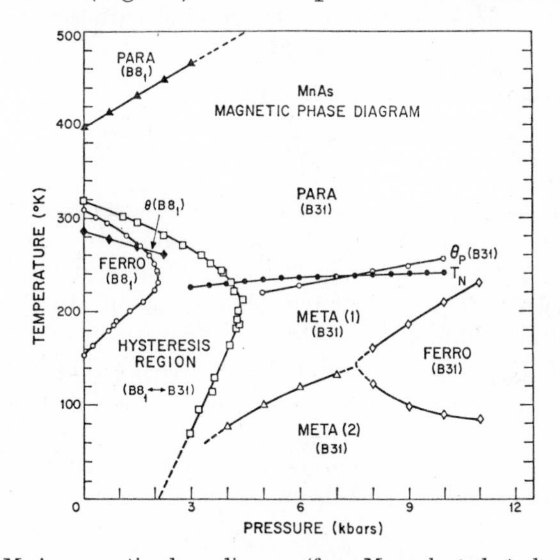

FIG. 48. MnAs magnetic phase diagram (from Menyuk *et al.*, to be published).

the first order crystallographic transition. The magnetic behaviour of orthorombic (B31) MnAs immediately below T_N (θ_N) is similar to that observed in dysprosium or holmium. However, the analogy is limited since increasing pressure decreases θ_N in the rare earth metals which is opposite to the effect observed in MnAs. In region meta (1)

$\partial H_c/\partial T > 0$, while in region meta (2) $\partial H_c/\partial T < 0$, where H_c is defined as the critical applied field below which the material appears to exhibit normal antiferromagnetic behaviour. The boundary of the ferromagnetic (B31) region is defined as the one at which $H_c \to 0$. The values of θ_p (B31) and θ (B81) are the paramagnetic Curie points.

The existence and the nature of the first-order transition between the ferro- and paramagnetic states has been given numerous interpretations. As given by the theory of Kittel (1960), the transition would be due to the variation with distance (because of thermal dilatation) of the exchange interactions which are positive below the critical parameter a_c and negative above a_c. This model does not give an interpretation, however, of the properties of MnAs which are paramagnetic and not antiferromagnetic above θ_c. The model of Bean and Rodbell (1962) allows a better description to be given of the experimental situation. The variation of the exchange interactions with temperature (due to the thermal expansion) can give an interpretation of the ferro-paramagnetic transition of first order since these exchange interactions vary rapidly with the volume of the sample. This transition temperature is a function of the volume variation given by the relation

$$\theta_c = \theta_{c,0}[1 + \beta(V - V_0)/V_0] \tag{48}$$

where $\theta_{c,0}$ and V_0 would be the transition temperature and volume in the absence of magnetoelastic interactions.

There exists for a given compressibility and coefficient of dilatation, a critical value of β above which the transition is of first order. The investigation of the compounds $MnAs_{1-x}P_x$ and $MnAs_{1-y}Sb_y$ have, however, shown that one can have, in the case of MnAs, a variation with the volume (due to the thermal dilatation) of the magnetic moment itself (Goodenough and Kafalas, 1967). The NiAs phase is characterized by the manganese ions with larger spins than in the MnP phase. The magnetic moment increases with the volume. The variation of θ_c results then not only from the modification of the exchange interactions but also from the modification of the magnetic moment itself. Band theory gives an interpretation of the increase of the magnetic moment with volume.

3. *Mn–Au Intermetallic Compounds*

The magnetic properties of the intermetallic compound $MnAu_2$ were revealed by Meyer and Taglang (1956). It was shown that $MnAu_2$ is an antiferromagnetic material with weak negative exchange interactions within the same lattice. A magnetic field of about 10 kOe is sufficient to decouple the antiferromagnetic interactions and to orient the magnetic moments along the field direction. Thus, in intense fields, $MnAu_2$ behaves like a ferromagnet. Later, Herpin *et al.* (1958, 1959) and Herpin (1961, 1962) found that $MnAu_2$ had a body-centred tetragonal cell of Mn atoms. These Mn atoms have a helical arrangement of magnetic spins aligned ferromagnetically in planes perpendicular to the c-axis and with a rotation of 51° between successive planes. The spin angle ω between planes decreases with temperature. The Néel temperature is $\theta_N = 364\,°K$. They confirmed the fact that a static magnetic field perpendicular to the c-axis beyond a threshold of 9·6 kOe caused a transition to a ferromagnetic state.

The electrical resistivity and the magnetization of $MnAu_2$ as a function of pressure have been measured by Grazhdankina and Rodionov (1963). It is reported that at room temperature $(1/R)(\partial R/\partial p) = -7{\cdot}75 \times 10^{-3}/\text{kb}$. The Néel temperature occurs at $\theta_N = 364{\cdot}6\,°K$ while its shift with pressure is $\partial\theta_N/\partial p = (0{\cdot}69 \pm 0{\cdot}05)\,°K/\text{kb}$. On the other hand, the investigators give the threshold magnetic field as $H_{\text{th}} = 8$ kOe at room temperature. With the application of hydrostatic pressure this threshold field is displaced at the rate of $\partial H_{\text{th}}/\partial p = -(0{\cdot}68 \pm 0{\cdot}07)$ kOe/kb. These results confirm the previous ones of Klitzing and Gielessen (1958). Extrapolation of the H_{th} *versus* pressure characteristic indicated that at a pressure of about 12 kb H_{th} went to zero. Some years later Bloch (1966a) performed similar experiments in which he found $\theta_N = 367°$, $\partial\theta_N/\partial p = 0{\cdot}47\,°K/\text{kb}$, $\partial H_{\text{th}}/\partial p = -0{\cdot}82$ kOe/kb and $(1/\chi_{\text{AF}})(\partial\chi_{\text{AF}}/\partial p) = 74 \times 10^{-3}/\text{kb}$. Using the compressibility at room temperature $K = 0{\cdot}6 \times 10^{-3}/\text{kb}$, he found from the experimental results that $\partial \log H_{\text{th}}/\partial \log V = 118$ and $\partial \log \chi_{\text{AF}}/\partial \log V = -123$.

At the same time as the above work, Smith *et al.* (1966) undertook a neutron diffraction investigation of $MnAu_2$ under pressures ranging from 1–10 kb, at temperatures from 77–295 °K. Their results indicate that the helicoidal ferromagnetic transition must occur very abruptly. The spin angle ω was found to decrease slowly from 51° at $p = 1$ b to 42° at $p = 9$ kb, beyond which it drops to zero. Furthermore a comparison of the pressure data and the thermal data at atmospheric pressure shows that a pressure induced volume change is much more effective in reducing the spin angle than a temperature-induced contraction of the same magnitude. It is concluded that if it can be assumed that the interplane interaction is the most important in stabilizing this layer structure, then small changes of the interatomic distance within the plane should have little effect on the magnetic structure and the results would indicate that interactions which are functions only of inter-plane separation are inadequate to describe the variation of magnetic structure with pressure and temperature.

Another intermetallic compound of manganese and gold, $MnAu_4$, was studied by Hirone *et al.* (1963) at pressures up to 3 kb and temperatures between room temperature and 100°C The Curie point was $(331 \cdot 8 \pm 0 \cdot 9)$°K and its pressure shift $\partial\theta_c/\partial p = 2 \cdot 7$°K/kb. From magnetization data, the relative pressure change of the absolute saturation magnetization at $T = 0$°K was found to be $(1/\sigma_0)(\partial\sigma_0/\partial p) = -7 \cdot 3 \times 10^{-3}$/kb, by assuming $\alpha = 1 \cdot 4 \times 10^{-5}$, $K = 0 \cdot 7 \times 10^{-3}$/kb and eqn (8). To confirm this result, the forced magnetostriction was measured at room temperature; $\partial\omega/\partial H = -3 \cdot 27 \times 10^{-6}$/kOe to be compared to $\partial\omega/\partial H = -3 \cdot 4 \times 10^{-6}$/kOe obtained from eqn (8) and the direct magnetization measurements.

4. *MnBi*

At low temperatures, MnBi, normally, has the NiAs structure. It undergoes a first-order transformation at 633°K accompanied by an abrupt change in the crystallographic parameters (Willis and Rooksby, 1954; Roberts, 1956). The high temperature structure is of the $NiIn_2$ type; the low temperature phase is ferromagnetic, the magnetic moment direction being that of the *c*-axis above 84°K and perpendicular to the *c*-axis below 84°K. The high temperature phase can be obtained by quenching from above 650°K. It is ferrimagnetic with a Curie point of 470°K. Pressure stabilizes the high temperature phase which has a pressure variation of its Curie temperature of $-0 \cdot 7$°K/kb (Fig. 49).

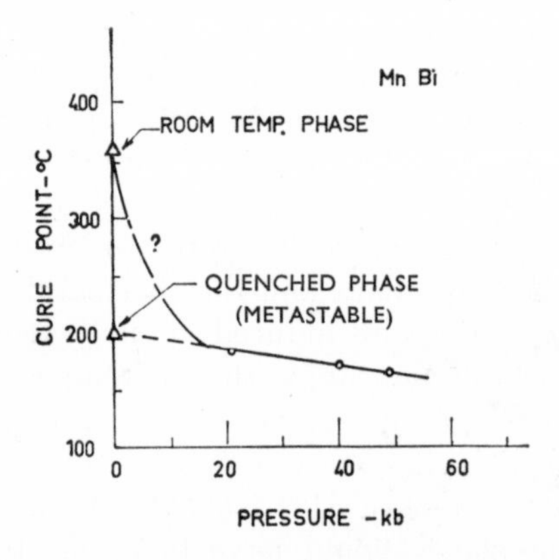

FIG. 49. Shift of the Curie point of MnBi with pressure (from Samara and Giardini, 1965).

5. *Mn_5Ge_3*

The compound Mn_5Ge_3 has a hexagonal B81 type crystallographic structure. It is ferromagnetic at low temperature with a Curie point at 293 °K (Fontaine and Pauthenet, 1962). The magnetic moments are directed along the *c*-axis, and according to their crystallographic positions two sites are to be distinguished in which the manganese atoms have the moments $(3 \pm 0{\cdot}1)\ \mu_\beta$ (4 sites–4 *d*) and $(2 \pm 0{\cdot}1)\ \mu_\beta$ (6 sites–6 *g*) (Ciszewski, 1963). The Curie temperature of Mn_5Ge_3 is elevated linearly with pressure up to 6 kb at the rate of 0·42°K/kb (Bloch and Pauthenet, 1962).

6. *MnP*

It has been reported by Huber and Ridgley (1964) that MnP is metamagnetic at temperatures below 50°K, while above 50°K it is ferromagnetic with a Curie point of 291°K. The magnetic anisotropy energy and the Curie temperature of MnP were measured as a function of pressure by Kawai *et al.* (1967b). Using a torque magnetometer they found the relative magnetic anisotropy energy variation with pressure to be $(1/E_a)\,(\partial E_a/\partial p) \simeq -17 \times 10^{-3}$/kb at 77°K and the Curie temperature shift to be $(\partial \theta_c/\partial p) = -1{\cdot}3$°K/kb.

7. *MnSb*

MnSb has a NiAs type structure and is ferromagnetic with the magnetic moments of the manganese ions directed along the crystallographic *c*-axis. The Curie temperature is (587 ± 1)°K (Guillaud, 1949). The effect of pressures ranging to 4 kb has been experimentally determined by Ido *et al.* (1967). The Curie temperature is lowered at the rate of $-(3{\cdot}2 \pm 0{\cdot}5)$°K/kb. Experiments to 50 kb (Samara and Giardini, 1965) have likewise shown (Fig. 50) an initial

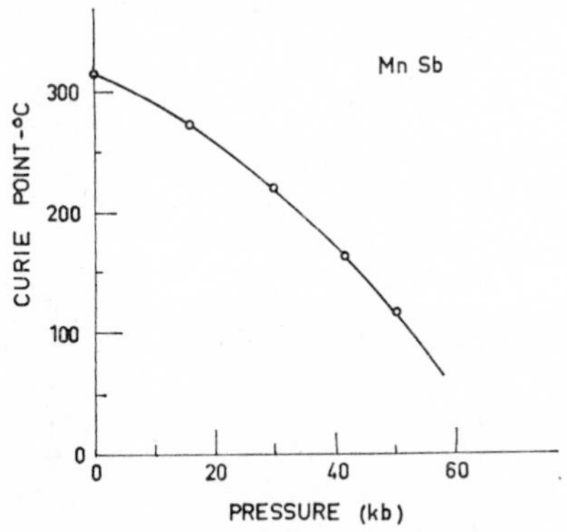

FIG. 50. Shift of the Curie point of MnSb with pressure (from Samara and Giardini, 1965).

variation of the order of $-3°K/kb$. One should notice however that the amplitude of this variation increases rapidly with pressure. No anomaly characterizes the thermal variation of parameter c at the ordering temperature (Willis and Rooksby, 1954). On the other hand, the parameter a undergoes a positive magnetic anomaly, in agreement with the negative sign of the Curie temperature variation with pressure. The pressure modifies simultaneously the exchange interactions and the absolute saturation magnetic moments. The forced magnetostriction experiments (Hirone *et al.*, 1965) give a relative variation of $(1/\sigma_0)(\partial\sigma_0/\partial p) = -5{\cdot}9 \times 10^{-3}/\text{kb}$, a value particularly important.

8. $Mn_{2-x}Cr_xSb$

Chromium-modified Mn_2Sb exhibits a transition with decreasing temperature from a ferrimagnetic (Fi) to an antiferromagnetic (AF) state. The substitution of chromium atoms for Mn atoms of the compositions $Mn_{2-x}Cr_xSb$ $(0{\cdot}02 < x < 0{\cdot}16)$ causes the lattice to contract and thus influence the transition temperature. Kamigaichi *et al.* (1965) have determined the pressure dependence of the (Fi–AF) transition temperature θ_{AF}^{Fi} for $Mn_{1\cdot 9}Cr_{0\cdot 10}Sb$. A large pressure hysteresis was observed for the transition temperature amounting to $\Delta\theta_{AF}^{Fi} \simeq 9° \pm 2°C$; however, the pressure dependence of θ_{AF}^{Fi} for increasing and decreasing pressure was the same, about $5{\cdot}0 \pm 0{\cdot}4°K/kb$. Doerner and Flippen (1965) performed the same experiments but do not report a pressure hysteresis. However, the pressure derivatives of the transition temperature of several compositions was 2·17, 2·45 and 3·4°K/kb for compositions having $x = 0{\cdot}05$; 0·10 and 0·16, respectively.

9. *Heusler alloys*

The original Heusler alloys Cu_2MnAl exhibited ferromagnetism even though none of the constituent elements were ferromagnetic. It was later found that Mn is the essential ingredient and that the Cu and Al could be replaced by other similar elements. Table XXII presents the

TABLE XXII

Heusler alloy	θ_c (°K)	$\partial\theta_c/\partial p$ (°K/kb)	$\frac{1}{\sigma_0}\frac{\partial\sigma_0}{\partial p}$ (kb)$^{-1}$	
Cu_2MnIn	449	1·50	$-2{\cdot}1 \times 10^{-3}$	Hirone *et al.* (1963)
Ni_2MnIn	228	0·56		Austin and Mishra (1967)
Ni_2MnSn	200	0·62		Austin and Mishra (1967)
Pd_2MnSn	185	0·75		Austin and Mishra (1967)
Pd_2MnSb	251	0·12		Austin and Mishra (1967)
Au_2MnAl	252	−1·10		Austin and Mishra (1967)

pressure dependence of the Curie temperature and absolute saturation magnetization for a number of these alloys.

10. *CrTe*

The structure of CrTe is of the NiAs type. However stoechiometric CrTe cannot be prepared and does not exist in equilibrium at room temperature (Chevreton *et al.*, 1963). One obtains, at the time of preparation, a phase with a deficiency of chromium, $Cr_{1-x}Te$. $Cr_{1-x}Te$ is ferromagnetic with a Curie temperature at 338°K and a metallic resistivity at room temperature equal to about $5 \times 10^{-4}\ \Omega-\text{cm}$. The easy axis of magnetization is the *c*-axis. From measurements of the galvano-magnetic effect under pressure, Grazhdankina *et al.* (1961) have found a moment at absolute saturation independent of pressure. The pressure variation of the Curie temperature will result, then, essentially from the modification of the exchange interaction. Between 1 and 4·6 kb the Curie temperature decreases linearly with pressure at a rate of $-(6 \pm 0{\cdot}3)$°K/kb (Grazhdankina *et al.*, 1961) in good agreement with the more recent value (Ido *et al.*, 1967) $-5{\cdot}6$°K/kb. Since the compressibility of CrTe is about $(2{\cdot}2 \pm 0{\cdot}3) \times 10^{-3}$/kb (Grazhdankina *et al.*, 1961) one obtains $\partial \log \theta_f / \partial \log V = 8 \pm 1{\cdot}5$.

11. *Cr_3Te_4*

The compound Cr_3Te_4 has a structure which is derived from that of NiAs. In order to employ an analogy it is convenient to write the formula $Cr_3 \square Te_4$ where $\square$ designates a vacancy. Because the neighbourhood of the anions are no longer symmetric, a monoclinic deformation takes place (Bertaut *et al.*, 1966). The magnetization is in the plan of the base, probably along the [100] and equivalent directions. The Curie temperature is 329°K, and it diminishes linearly with pressure between 1 b and 6 kb, at the rate $-7{\cdot}35$°K/kb (Bloch, unpublished).

12. *$CrS_{1{\cdot}17}$*

$CrS_{1{\cdot}17}$ has a trigonal structure intermediate between that of NiAs and of Cd $(OH)_2$ (Jellinek, 1957). Its behaviour is ferrimagnetic (Yuzuri *et al.*, 1957) with two types of ions, Cr^{2+} and Cr^{3+}, coexisting in the lattice. The Curie temperature is close to 307°K. At 153°K, a transition takes place between the ferri- and antiferromagnetic states. The effect of pressure up to 6 kb has been studied by Kamigaichi *et al.* (1966); the Curie temperature lowers with pressure by $-2{\cdot}6$°K/kb, whereas the ferri–antiferromagnetic transition decreases at the rate of $-3{\cdot}5$°K/kb.

13. *α-FeS*

Haraldsen (1941) has shown that FeS experiences two magnetic transitions. The one (α) at about 413°K is characterized by an abrupt variation of the volume and of the susceptibility. The other (β) is at a much higher temperature. X-ray diffraction investigations (Hägg and Sucksdorff, 1933) have shown that θ_α separates two domains, one stable above θ_α, which has the NiAs type structure, the other stable below θ_α, with a superstructure characterized by a displacement of the iron and sulphur atoms with respect to the regular NiAs structure (Bertaut, 1954). The transition at θ_α is of the first order between two states of different antiferromagnetic order. Below θ_α, the magnetic moments are directed along the c-axis of the crystal, and above θ_α they are perpendicular to this axis. The temperature θ_β is the Néel temperature associated with the second antiferromagnetic phase. Its value is near 593°K. The experiments to 11 kb (Ozawa and Anzai, 1966) show that θ_α decreases with pressure at the rate of $-(0{\cdot}8 \pm 0{\cdot}1)$°K/kb.

The volume anomaly ΔV at the transition is $-(0{\cdot}15 \pm 0{\cdot}07)$ cm^3/mole and the enthalpy variation ΔH is 550 cal/mole. From the eqn (5) one gets (Ozawa and Anzai, 1966) $\partial\theta_\alpha/\partial p = -(2 \pm 1)$°K/kb, in qualitative agreement with the result obtained directly.

14. *Sc_3In*

The intermetallic compound Sc_3In was found by Matthias *et al.* (1961) to be ferromagnetic below $\theta_c = 6{\cdot}1$°K. Gardner *et al.* (1968) investigated the pressure variation of θ_c and found it linear with $\partial\theta_c/\partial p = 0{\cdot}195$°K/kb and $(\partial \log \theta_c/\partial \log V) = -13{\cdot}9$ if $K = 23 \times 10^{-4}$/kb is assumed to be the linear compressibility and equal to that for scandium (Montfort and Swenson, 1965).

VI. Conclusion

The investigations of the magnetic properties of solids under pressure has developed only recently, but it has, however, been very rapid as indicated by the large number of important articles reviewed in this paper. The results obtained have led, in numerous cases, to an increase in our knowledge concerning magnetic materials and, more particularly, of the fundamental mechanisms which are responsible for their principal magnetic properties. These results allow, in some cases, a comparison between theories generally conceived for a static system for which the volume is constant and the experimental results obtained at constant pressure. However, in general, to make an effective theoretical analysis of the magnetic behaviour of a solid, it is necessary for the investigator to have at his disposal the behaviour of a large number of parameters capable of being modified under the effect of pressure. Indeed, as the results reviewed above indicate, many analyses have been frustrated by the lack of supplementary experimental information. With the present availability of high pressure apparatus and experience in its use, it is hoped that in the future investigators will address themselves to a greater variety of high pressure experiments and to greater detail in these experiments. As examples, one can cite the following types of experiments which are sorely needed in order to help understand the properties of magnetic materials.

1. The variation with pressure of different crystallographic parameters, and of crystallographic and magnetic structures determined by X-ray and neutron diffraction.
2. The variation with pressure of the magnetic moments of magnetic solids, especially at low temperatures. This work should be extended to materials where the magnetic moments can appear, or disappear, under the action of pressure, for example in cerium and certain types of dilute alloys.
3. The pressure variation of the magnetocrystalline anisotropy constants of magnetic materials.

These experiments should be performed with single crystal samples, wherever possible, so as to give the most meaningful results.

Finally, a continuing effort should be made to develop methods for measurements at static pressures up to several hundred kilobars, as well as methods utilizing pressures produced by shock waves.

Acknowledgements

The authors wish to thank Dr J. B. Goodenough, I. S. Jacobs, D. Landau, N. Menyuk and R. Pauthenet for their help in part of this work. Thanks are due to H. Bartholin, F. Chaissé and R. Georges from the High Pressure Laboratory at the L.E.P.M. for many discussions and to Miss G. Meneroud for her help in typing and translating the manuscript. One of the authors (D. Bloch) thanks the " Direction des Recherches et Moyens d'Essais " (D.R.M.E.) for financial support. The other author (A. S. Pavlovic) expresses his appreciation to West Virginia University for financial aid and to Professors Néel and Pauthenet for extending the hospitality of L.E.P.M. to him.

References

Adams, L. H. and Green, J. W. (1931). *Phil. Mag.* **12**, 367.
Anderson. C. T. (1931). *J. Am. chem. Soc.* **53**, 476.

Anderson, P. W. (1963). *In* " Solid State Physics ", ed. by F. Seitz and D. Turnbull, Vol. 14, p. 99. Academic Press, New York.
Argyle, B. E., Miyata, N. and Schultz, T. D. (1967). *Phys. Rev.* **160**, 413.
Artman, J. O., Murphy, J. C. and Foner, S. (1965). *Phys. Rev.* **138**, A192.
Assayag, G. and Bizette, H. (1954). *C.r. hebd. Seanc. Acad. Sci., Paris* **239**, 238.
Astrov, D. N., Novikova, S. I. and Orlova, M. P. (1960). *Soviet Phys. JETP* **37**, 851.
Astrov, D. N., Kytin, G. A. and Orlova, M. P. (1961). *Fizika tverd. Tela* **4**, 1055 (in Russian).
Austin, I. G. and Mishra, P. K. (1967). *Phil. Mag.* **15**, 529.
Azumi, K. and Goldman, J. E. (1954). *Phys. Rev.* **93**, 630.
Bacon, G. E. (1961). *Acta Cryst.* **14**, 823.
Barson, F., Legvold, S. and Spedding, F. H. (1957). *Phys. Rev.* **105**, 418.
Bartholin, H. and Bloch, D. (1967). *C.r. hebd. Seanc. Acad. Sci., Paris* **264**, 1135.
Bartholin, H., Bloch, D. and Georges, R. E. (1967). *C.r. hebd. Seanc. Acad. Sci., Paris* **264**, 360.
Bartholin, H. and Bloch, D. (1968). *Physics Chem. Solids* **29**, 1063.
Bean, C. P. and Rodbell, D. S. (1962). *Phys. Rev.* **126**, 104.
Belov, K. P., Nikitin, S. A. and Ped'ko, A. V. (1964). *Soviet Phys. JETP* **18**, 20.
Benedek, G. B. and Kushida, T. (1960). *Phys. Rev.* **118**, 46.
Bertaut, E. F. (1954). *J. phys. Rad.* **15**, 775.
Bertaut, E. F. and Forrat, F. (1956a). *C.r. hebd. Seanc. Acad. Sci., Paris* **242**, 382.
Bertaut, E. F. and Forrat, F. (1956b). *C.r. hebd. Seanc. Acad. Sci., Paris* **244**, 96.
Bertaut, E. F., Roult, G., Aleonard, R., Pauthenet, R., Chevreton, M. and Jansen, R. (1966). *J. phys. Rad.* **25**, 582.
Birch, F. (1939). *Rev. scient. Instrum.* **10**, 137.
Birch, F. (1952). *J. geophys. Res.* **57**, 227.
Birss, R. R. and Hegarty, B. C. (1967). *J. scient. Instrum.* **44**, 627.
Bloch, D. and Pauthenet, R. (1962). *C.r. hebd. Seanc. Acad. Sci., Paris* **254**, 1222.
Bloch, D. and Pauthenet, R. (1965a). *In* " Proceedings of the International Conference on Magnetism " (Nottingham). A1964, p. 255. Institute of Physics, London.
Bloch, D. and Pauthenet, R. (1965b). *J. appl. Phys.* **36**, 1229.
Bloch, D. (1966a). *Ann. Phys.* **1**, 93.
Bloch, D. (1966b). *Physics chem. Solids* **27**, 887.
Bloch, D., Chaissé, F. and Pauthenet, R. (1966). *C.r. hebd. Seanc. Acad. Sci., Paris* **262**, 404.
Bloch, D., Chaissé, F. and Pauthenet, R. (1967). *In* "Magnetismus." VEB Deutscher Verlag für Gründstaff Industrie, Leipzig.
Bloch, D. and Chaissé, F. (1967). *J. appl. Phys.* **38**, 409.
Bloch, D., Charbit, P. and Georges, R. (1968). *C.r. hebd. Seanc. Acad. Sci., Paris* **266**, 430.
Bloch, D. and Georges, R. (1968). *Phys. Rev. Lett.* **20**, 1240.
Bouchaud, J. P., Fruchard, R., Pauthenet, R., Guillot, M., Bartholin, H. and Chaissé, F. (1966). *J. appl. Phys.* **37**, 971.
Boyd, E. L. (1966). *Phys. Rev.* **145**, 174.
Bozorth, R. M. (1954). *Phys. Rev.* **96**, 311.
Bridgman, P. W. (1958). " The Physics of High Pressure." G. Bell and Sons, London.
Brockhouse, B. N. (1953). *J. Chem. Phys.* **21**, 961.
Brugger, R. M., Bennion, R. B. and Worlton, T. G. (1967). *Phys. Lett.* **24A**, 714.

Bundy, F. P. (1961). *J. appl. Phys.* **32**, 483.
Bundy, F. P., Hibbard, W. R., Jr. and Strong, H. M. (1961). " Progress in Very High Pressure Research." John Wiley and Sons, New York.
Busch, G., Junod, P., Morris, R. G., Muheim, J. and Stutius, W. (1964). *Phys. Lett.* **11**, 9.
Busch, G., Schwob, P. and Vogt, O. (1966a). *Phys. Lett.* **20**,602.
Busch, G. Natterer E. and Neukomm H. R. (1966b). *Phys. Lett.* **23**, 190.
Cable, J. W., Wollan, E. W., Koehler, W. C. and Wilkinson, M. K. (1961). *J. appl. Phys.* **32**, 49S.
Chenevard, P. (1921). *C.r. hebd. Seanc. Acad. Sci., Paris* **172**, 320.
Chevreton, M., Bertaut, E. F. and Jellinek, F. (1963). *Acta Cryst.* **16**, 431.
Child, H. R., Koehler, W. C., Wollan, E. O. and Cable, J. W. (1965). *Phys. Rev.* **138**, A1655.
Ciszewski, R. (1963). *Physics solid St.* **3**, 1999.
Clendenen, R. L. and Drickamer, H. G. (1966). *J. chem. Phys.* **44**, 4223.
Coleman, W. E. and Pavlovic, A. S. (1964). *Phys. Rev.* **135**, A426.
Coles, B. A., Orton, J. W. and Owen, J. (1960). *Phys. Rev. Lett.* **4**, 116.
Corliss, L. M., Elliott, N. and Hastings, J. M. (1956). *Phys. Rev.* **104**, 924.
Corliss, L. M., Hastings, J. M. and Weiss, R. J. (1959). *Phys. Rev. Lett.* **3**, 211.
Coston, C. J., Ingalls, R. and Drickamer, H. G. (1966). *Phys. Rev.* **145**, 409.
Darnell, F. J. (1963a). *Phys. Rev.* **130**, 1825.
Darnell, F. J. (1963b). *Phys. Rev.* **132**, 1098.
Darnell, F. J. and Moore, E. P. (1963). *J. appl. Phys.* **34**, 1337.
de Blois, R. W. (1962). *In* " Proceedings of the International Conference on High Magnetic Fields " (M.I.T., 1967), ed. by H. Kolm, B. Lax, F. Bitter and R. Mills, p. 568. Technology Press, Cambridge, U.S.A.
de Blois, R. W. and Rodbell, D. S. (1963). *Phys. Rev.* **130**, 1347.
De Gennes, P. G. (1958). *C.r. hebd. Seanc. Acad. Sci., Paris* **247**, 1836.
De Gennes, P. G. (1962). *J. phys. Rad.* **23**, 510.
de Graaf, A. M. and Xavier, R. M. (1965). *Phys. Lett.* **18**, 225.
Doerner, W. A. and Flippen, R. B. (1965). *Phys. Rev.* **137**, A926.
Dzialoshinskii, I. E. (1958). *Physics chem. Solids* **4**, 241.
Ebert, H. and Kussman, A. (1937). *Phys. Z.* **38**, 437.
Ebert, H. and Kussman, A. (1939). *Phys. Z.* **39**, 598.
Elliott, R. J. (1965). *In* " Magnetism ", ed. by G. T. Rado and H. Suhl, Vol. IIA, p. 385. Academic Press, New York.
Enz, U. (1961). *J. appl. Phys.* **32**, 522.
Euler, F. and Bruce, J. (1965). *Acta Cryst.* **19**, 971.
Fallot, M. and Hocart, R. (1939). *Rev. scient. Instrum.* **8**, 498.
Fawcett, E. and White, G. K. (1967). *J. appl. Phys.* **38**, 1321.
Foex, M. (1948). *C.r. hebd. Seanc. Acad. Sci., Paris* **227**, 193.
Foiles, C. L. and Tomizuka, C. T. (1965). *J. appl. Phys.* **36**, 3839.
Fontaine, R. and Pauthenet, R. (1962). *C.r. hebd. Seanc. Acad. Sci., Paris* **254**, 650.
Franse, J. J. M., Winkel, R., Veen, R. J. and de Vries, G. (1967). *Physica's Grav.* **33**, 475.
Frisbie, F. C. (1905). *Phys. Rev.* **18**, 432.
Fujimori, H. (1966). *J. phys. Soc. Japan* **21**, 1860.
Fujiwara, H., Iwasaki, T., Tokumaga, T. and Tatsumoto, E. (1966). *J. phys. Soc. Japan* **21**, 2729.
Fujiwara, H., Tsukiji, N., Yamate, N. and Tatsumoto, E. (1967). *J. phys. Soc. Japan* **23**, 1176.

Gardner, W. E., Smith, T. F., Howlett, B. W., Chu, C. W. and Sweedler, A. (1968). *Phys. Rev.* **166**, 577.
Geller, S. and Gilleo, M. A. (1957). *Physics Chem. Solids* **3**, 30.
Georges, R. (1969). *C.r. hebd. Seanc. Acad. Sci., Paris* **268**, 16.
Gerstein, B. C., Griffel, M., Jennings, L. D., Miller, R. E., Skochdopole, R. E. and Spedding, F. H. (1957). *J. chem. Phys.* **27**, 394.
Gibbons, D. F. (1959). *Phys. Rev.* **115**, 1194.
Goodenough, J. B. and Kafalas, J. A. (1967). *Phys. Rev.* **157**, 389.
Graham, C. D., Jr. (1965). *In* " Proceedings of the International Conference on Magnetism " (Nottingham, 1964), p. 740. Institute of Physics, London.
Grazhdankina, N. P. (1957). *J. exp. theor. Phys.* **33**, 1524.
Grazhdankina, N. P., Gaidukov, L. G., Rodionov, K. P., Oleinik, M. I. and Shchpanov, V. A. (1961). *Soviet Phys JETP.* **13**, 297.
Grazhdankina, N. P. and Rodionov, K. P. (1963). *Soviet Phys. JETP* **16**, 1429.
Grazhdankina, N. P. (1965). *Soviet Phys. JETP* **21**, 840.
Grazhdankina, N. P. and Bersenev, Yu. S. (1966). *J. exp. theor. Phys.* **51**, 1052; also *Soviet Phys. JETP* **24**, 702 (1967).
Grazhdankina, N. P. (1967). *Soviet Phys. JETP* **25**, 258.
Greenwald, S. (1953). *Acta Cryst.* **6**, 396.
Greenwald, S. W. (1956). *Nature, Lond.* **177**, 286.
Griffel, M., Skochdopole, R. E. and Spedding, F. H. (1954). *Phys. Rev.* **93**, 657.
Griffel, M., Skochdopole, R. E. and Spedding, F. H. (1956). *J. chem. Phys.* **25**, 75.
Guillaud, C. (1949). *Ann. Phys.* **4**, 671.
Guillaud, C. and Greveaux, H. (1950). *C.r. hebd. Seanc. Acad. Sci., Paris* **230**, 1458.
Guillot, M. and Pauthenet, R. (1966). *C.r. hebd. Seanc. Acad. Sci., Paris* **263**, 527.
Gugan, D. (1958). *Proc. phys. Soc. Lond.* **72**, 1013.
Gugan, D. and Rowlands, G. (1958). *Proc. phys. Soc. Lond.* **72**, 207.
Hägg, G. and Sucksdorff, I. (1933). *Z. phys. Chem.* B**22**, 444.
Hanneman, R. G. and Strong, H. M. (1965). *J. appl. Phys.* **36**, 523.
Hanneman, R. G. and Strong, H. M. (1966). *J. appl. Phys.* **37**, 672.
Haraldsen, H. (1941). *Z. anorg. allg. Chem.* **246**, 195.
Hastings, J. M., Elliott, N. and Corliss, L. M. (1959). *Phys. Rev.* **115**, 13.
Hasuo, M. (1964). *J. Sci. Hiroshima Univ.* Ser. A-II, **28**, 71.
Heller, P. and Benedek, G. B. (1962). *Phys. Rev. Lett.* **8**, 428.
Herpin, A., Meriel, P. and Meyer, A. P. (1958). *C.r. hebd. Seanc. Acad. Sci., Paris* **246**, 3178.
Herpin, A., Meriel, P. and Villain, J. (1959). *C.r. hebd. Seanc. Acad. Sci., Paris* **249**, 1334.
Herpin, A. (1961). *J. phys. Rad.* **22**, 337.
Herpin, A. (1962). *J. phys. Rad.* **23**, 453.
Hirahara, E. and Murakami, M. (1958). *Physics chem. Solids* **7**, 281.
Hirone, T., Kaneko, T. and Kondo, K. (1963). *J. phys. Soc. Japan* **18**, 65.
Hirone, T., Kaneko, T. and Kondo, K. (1965). *In* " Physics of Solids at High Pressures ", ed. by C. T. Tomizuka and R. M. Emrick, p. 298. Academic Press, New York.
Howe, L. and Myers, H. P. (1957). *Phil. Mag.* **2**, 554.
Huber, E. E., Jr. and Ridgley, D. H. (1964). *Phys. Rev.* **135**, A1033.
Ido, H., Kaneko, T. and Kamigaki, K. (1967). *J. phys. Soc. Japan* **22**, 1418.
Jacobs, I. S. and Lawrence, P. E. (1965). *J. appl. Phys.* **35**, 996.
Janusz, T. P. (1960). Technical Report 150, Laboratory for Insulation Research. M.I.T., Cambridge, U.S.A.

Jayaraman, A. and Sherwood, R. C. (1964a). *Phys. Rev. Lett.* **12**, 22.
Jayaraman, A. and Sherwood, R. C. (1964b). *Phys. Rev.* **134**, A691.
Jellinek, F. (1957). *Acta Cryst.* **10**, 620.
Jennings, L. D., Stanton, R. M. and Spedding, F. H. (1957). *J. chem. Phys.* **27**, 909.
Jones, E. D. and Morosin, B. (1967). *Phys. Rev.* **160**, 451.
Kamigaichi, T., Masumoto, K. and Hihara, T. (1965). *J. Sci. Hiroshima Univ.* Ser. A-II, **29**, 53.
Kamigaichi, T., Okamoto, T., Iwata, N. and Tatsumoto, E. (1966). *J. phys. Soc. Japan* **21**, 2730.
Kaminow, I. P. and Jones, R. V. (1960). Science Report 5, AFCRL-181, Gordon Mackay Lab. Harvard Univ., Cambridge, U.S.A.
Kaminow, I. P. and Jones, R. V. (1961). *Phys. Rev.* **123**, 1122.
Kaneko, T. (1960). *J. phys. Soc. Japan* **15**, 2247.
Kasuya, T. (1956). *Prog. theor. Phys., Osaka* **16**, 45.
Kawai, N. and Ono, F. (1966). *Phys. Lett.* **21**, 279.
Kawai, N., Sakakihara, M., Marizumi, A. and Sawaoka, A. (1967a). *J. phys. Soc. Japan* **23**, 475.
Kawai, N., Sawaoka, A. and Kaji, G. (1967b). *J. phys. Soc. Japan* **23**, 896.
Kawai, N. and Sawaoka, A. (1967). *Rev. sci. Instrum.* **38**, 1770.
Kawai, N. and Sawaoka, A. (1968). *Physics Chem. Solids* **29**, 575.
King, E. G. (1957). *J. Amer. chem. Soc.* **79**, 2399.
Kittel, C. (1948). *Rev. Mod. Phys.* **21**, 541.
Kittel, C. (1960). *Phys. Rev.* **120**, 335.
Klitzing, K. H. and Gielessen, I. (1958). *Z. Phys.* **150**, 409.
Koehler, W. C. (1965). *J. appl. Phys.* **36**, 1078.
Kondorskii, E. I. and Sedov, V. L. (1960). *Soviet Phys. JETP* **11**, 561.
Kondorskii, E. I., Vinokurova, L. V. (1965). *In* " Proceedings of the International Conference on Magnetism " (Nottingham, 1964), p. 260. Institute of Physics, London.
Kornetzki, M. (1934). *Z. Phys.* **87**, 560.
Kornetzki, M. (1935). *Z. Phys.* **98**, 289.
Kornetzki, M. (1943). *Z. Phys.* **44**, 296.
Kouvel, J. S. and Wilson, R. A. (1961). *J. appl. Phys.* **32**, 435.
Kouvel, J. S. (1963). *In* " Solids Under Pressure ", ed. by W. Paul and D. M. Warschauer, p. 277. McGraw-Hill, New York.
Kouvel, J. S. and Hartelius, C. C. (1964a). General Electric Res-Lab. Rept no. 64-gc-0283.
Kouvel, J. S. and Hartelius, C. C. (1964b). *J. appl. Phys.* **35**, 940.
Kouvel, J. S. and Hartelius, C. C. (1964c). *Physics Chem. Solids* **25**, 1357.
Kouvel, J. S. (1966). *J. appl. Phys.* **37**, 1257.
Kume, S., Endo, S., Koizumi, M., Okazi, C. and Hirota, E. (1966). *Rev. Sci. Instrum.* **37**, 289.
Kussman, A. and Jessen, K. (1962). *J. phys. Soc. Japan* **17**, Suppl. B1, 136.
Landry, P. and Stevenson, R. (1963). *Can. J. Phys.* **41**, 1273.
Landry, P. (1966). *Phys. Rev.* **156**, 578.
Laurens, J. A. J. and Alberts, L. (1964). *Sol. Stat. Comm.* **2**, 141.
Lazarev, B. and Kan, L. (1944). *Zh. Eksp. teor. Fiz.* **14**, 499.
Lee, E. W. (1964). *Proc. Phys. Soc. Lond.* **84**, 693.
Léger, J. M., Susse, C., Epain, R. and Vodar, B. (1966a). *Sol. Stat. Comm.* **4**, 197.
Léger, J. M., Susse, C. and Vodar, B. (1966b). *Sol. Stat. Comm.* **4**, 503.
Léger, J. M., Susse, C. and Vodar, B. (1967a). *Sol. Stat. Comm.* **5**, 755.

Léger, J. M., Susse, C. and Vodar, B. (1967b). *C.r. hebd. Seanc. Acad. Sci., Paris* **265**, 892C.

Lewis, G. K., Jr. and Drickamer, H. G. (1966). *J. chem. Phys.* **45**, 224.

Lines, M. E. and Jones, E. D. (1965). *Phys. Rev.* **139**, A1313.

Litster, J. D. and Benedek, G. B. (1966). *J. appl. Phys.* **37**, 1320.

Litvin, D. F. and Ponyatovsky, E. G. (1964). *Soviet Phys. Dokl.* **9**, 388.

Liu, S. H. (1961). *Phys. Rev.* **123**, 470.

Livshitz, L. D. and Genshaft, Y. S. (1964). *Soviet Phys. JETP* **19**, 560.

Livshitz, L. D. and Genshaft, Y. S. (1965). *Soviet Phys. JETP* **21**, 701.

McWhan, D. B. and Stevens, A. L. (1965). *Phys. Rev.* **139**, A682.

McWhan, D. B., Souers, P. C. and Jura, G. (1966a). *Phys. Rev.* **143**, 385.

McWhan, D. B., Corenzwit, E. and Stevens, A. L. (1966b). *J. appl. Phys.* **37**, 1355.

McWhan, D. B. (1966). *J. chem. Phys.* **44**, 3528.

McWhan, D. B. and Stevens, A. L. (1967). *Phys. Rev.* **154**, 438.

McWhan, D. B. and Rice, T. M. (1967). *Phys. Rev. Lett.* **19**, 846.

Matthias, B. T., Clogston, A. M., Williams, H. J., Corenzwit, E. and Sherwood, R. C. (1961). *Phys. Rev. Lett.* **7**, 7.

Menyuk, N., Kafalas, J. A., Dwight, K. and Goodenough, J. B. To be published.

Meyer, J. P. and Taglang, P. (1956). *J. Phys. Rad.* **17**, 457.

Millar, R. W. (1928). *J. Am. chem. Soc.* **50**, 1875.

Millar, R. W. (1929). *J. Am. chem. Soc.* **51**, 215.

Milstein, F. and Robinson, L. B. (1967a). *Phys. Rev.* **159**, 466.

Milstein, F. and Robinson, L. B. (1967b). *Phys. Rev. Lett.* **18**, 308.

Milton, J. E. and Scott, T. A. (1967). *Phys. Rev.* **160**, 387.

Mitsui, T. and Tomizuka, C. T. (1965). *Phys. Rev.* **137**, A564.

Miwa, H. (1965). *Proc. Phys. Soc., Lond.* **85**, 1197.

Montfort, C. E., III and Swenson, C. A. (1965). *Physics Chem. Solids* **26**, 623.

Morin, F. J. (1950). *Phys. Rev.* **78**, 819.

Moriya, T. (1960). *Phys. Rev.* **120**, 91.

Moruzzi, V. L. and Teaney, D. T. (1963). *Sol. Stat. Comm.* **1**, 127.

Nagamiya, T., Yosida, K. and Kubo, R. (1955). *Adv. Phys.* **4**, 1.

Nagamiya, T. (1962). *J. appl. Phys.* **33**, 1029.

Nagaoka, H. and Honda, K. (1898). *Phil. Mag.* **46**, 261.

Narath, A. (1964). *Phys. Rev.* **136**, A766; *Phys. Rev. Lett.* **13**, 12.

Narath, A. (1965). *Phys. Rev.* **139**, A1221.

Narath, A. and Schirber, J. E. (1966). *J. appl. Phys.* **37**, 1124.

Néel, L. (1932). *Ann. Phys.* **17**, 64.

Néel, L. (1938). *C.r. hebd. Seanc. Acad. Sci., Paris* **206**, 49.

Néel, L. (1948). *Ann. Phys.* **3**, 137.

Néel, L. (1951). *J. phys. Rad.* **12**, 258.

Néel, L. and Pauthenet, R. (1952). *C.r. hebd. Seanc. Acad. Sci., Paris* **234**, 2172.

Néel, L. (1954). *C.r. hebd. Seanc. Acad. Sci., Paris* **239**, 1954.

Okamoto, T., Iwata, N., Ishida, S. and Tatsumoto, E. (1966). *J. phys. Soc. Japan* **21**, 2727.

Okamoto. T., Fujii, H., Taurui, M., Fujiwara, H. and Tatsumoto, E. (1967a). *J. phys. Soc. Japan* **22**, 337.

Okamoto. T., Fujii, H., Hidaka, Y. and Tatsumoto, E. (1967b). *J. phys. Soc. Japan* **23**, 1174.

Ozawa K. and Anzai S. (1966). *Physics Solid St.* **17**, 697.

Ozawa, K., Anzai, S. and Hamaguchi. Y. (1966). *Phys. Lett.* **20**, 132.

Patrick, L, (1954). *Phys. Rev.* **93**, 384.

Paul, W. and Warschauer, D. M. (1963). *In* " Solids Under Pressure." McGraw-Hill, New York.
Pauthenet, R. (1952). *Ann. Phys.* **7**, 710.
Pauthenet, R. (1958). *Ann. Phys.* **3**, 424.
Ponyatovskii, E. G., Kut-Sar, A. R. and Dubovka, G. T. (1967). *Soviet Phys. Crystallogr.* **12**, 63.
Rebouillat, J. P. and Veyssié, J. J. (1964). *C.r. hebd. Seanc. Acad. Sci., Paris* **259**, 4239.
Rhyne, J. J. and Legvold, S. (1965). *Phys. Rev.* **140**, A2143.
Roberts, B. W. (1956). *Phys. Rev.* **104**, 607.
Robinson, L. B., Milstein, F. and Jayaraman, A. (1964). *Phys. Rev.* **134**, A187.
Robinson, L. B., Tan, S. and Sterrett, K. F. (1966). *Phys. Rev.* **141**, 548.
Rocher, Y. A. (1962). *Phil. Mag.* Suppl. **11**, 233.
Rodbell, D. S. (1961). *In* " Progress in Very High Pressure Research ", ed. by F. P. Bundy, W. R. Hibbard, Jr. and H. M. Strong, p. 283. John Wiley and Sons, New York.
Rodbell, D. S., Osika, R. M. and Lawrence, P. E. (1965). *J. appl. Phys.* **36**, 666.
Rooymans, C. J. M. (1965). *Sol. Stat. Comm.* **3**, 421.
Roth, W. L. (1958). *Phys. Rev.* **110**, 1333.
Rudermann M. A. and Kittel. C. (1954). *Phys. Rev.* **96**, 99.
Samara, G. A. and Giardini, A. A. (1965). *In* " Physics of Solids at High Pressures ", ed. by C. T. Tomizuka and R. M. Emrick, p. 308. Academic Press, New York.
Sawaoka, A. and Miyahara, S. (1965). *J. phys. Soc. Japan* **20**, 2087.
Sawaoka, A., Miyahara, S. and Minomura, S. (1966). *J. phys. Soc. Japan* **21**, 1017.
Sawaoka, A. and Kawai, N. (1967). *Phys. Lett.* **24A**, 503.
Sawaoka, A., Kawai, N. and Kikuchi, S. (1967). *Physics Solid St.* **24**, K83.
Schwob, P. and Vögt, O. (1967). *Phys. Lett.* **24A**, 242.
Shirane, G. and Takei, W. J. (1962). *J. phys. Soc. Japan* **17**, Suppl. B-III, 35.
Shull, C. G. and Smart, J. S. (1949). *Phys. Rev.* **76**, 1256.
Shull, C. G., Strauser, W. A. and Wollan, E. O. (1951). *Phys. Rev.* **83**, 333.
Smart, J. S. (1964). *In* " Magnetism ", ed. by G. T. Rado and H. Suhl, Vol. III, Chapter 2. Academic Press, New York.
Smart, J. S. (1966). " Effective Field Theories of Magnetism." Saunders Co., Philadelphia.
Smidt, F. A. and Daane, A. H. (1963). *Physics Chem. Solids* **24**, 361.
Smit, J. (1966). *J. appl. Phys.* **37**, 1455.
Smith, F. A., Bradley, C. C. and Bacon, G. E. (1966). *Physics Chem. Solids* **27**, 925.
Sokolova, G. K., Demchuk, K. M., Rodionov, K. P. and Samokhalov, A. A. (1966). *Soviet Phys. JETP* **22**, 317.
Souers, P. C. and Jura, G. (1964). *Science, N.Y.* **145**, 575.
Srivastava, C. V. and Stevenson, R. (1968). *Can. J. Phys.* **46**, 2703.
Stager, R. A. and Drickamer, H. G. (1964). *Phys. Rev.* **133**, A830.
Starr, C., Bitter, F. and Kaufman, A. R. (1940). *Phys. Rev.* **58**, 977.
Steinberger, R. L. (1933). *Physics* **4**, 153.
Stevenson, R. and Robinson, M. C. (1965). *Can. J. Phys.* **43**, 1744.
Stevenson, R. (1966). *Can. J. Phys.* **44**, 281.
Stoelinga, J. H. M., Gersdorf, R. and de Vries, G. (1965). *Physics* **31**, 349.

Swenson, C. A., Legvold, S., Good, R. and Spedding, F. H. (1960). Ames (Iowa), Laboratory Report 15–191, p. 49.
Swenson, C. A. (1960). *In* " Solid State Physics ", ed. by F. Seitz and D. Turnbull, Vol. II, p. 41. Academic Press, New York.
Syono, Y. and Ishikawa, Y. (1967). *Phys. Rev. Lett.* **19**, 747.
Tatsumoto, E., Fujiwara, H., Tange, H. and Kato, Y. (1962a). *Phys. Rev.* **128**, 2179.
Tatsumoto, E., Kamigaichi, T., Fujiwara, H., Kato, Y. and Tange, H. (1962b). *J. phys. Soc. Japan* **17**, 592.
Tatsumoto, E., Fujiwara, H., Fujii, H., Iwata, N. and Okamoto, T. (1968). *J. appl. Phys.* **39**, 894.
Todd, S. S. and Bonnickson, K. R. (1951). *J. Am. chem. Soc.* **73**, 3894.
Tomlinson, H. (1887). *Proc. R. Soc. Lond.* **42**, 224.
Uchida, E., Kondoh, H. and Fukuoka, N. (1957). *J. phys. Soc. Japan* **11**, 27.
Umebayashi, H., Frazer, B. C., Shirane, G. and Daniels, W. B. (1966). *Phys. Rev. Lett.* **22**, 407.
Umebayashi, H., Shirane, G., Frazer, B. C. and Daniels, W. B. (1968). *Phys. Rev.* **165**, 688; also *J. phys. Soc. Japan* **24**, 368.
van Laar, B. (1965). *Phys. Rev.* **138**, A584.
Veerman, J. and Rathenau, G. W. (1965). *In* " Proceedings of the International Conference on Magnetism " (Nottingham), p. 737. Institute of Physics, London.
Vinokurova, L. I. and Kondorskii, E. I. (1964). *Soviet Phys. JETP* **19**, 777.
Vinokurova, L. I. and Kondorskii, E. I. (1965). *Soviet Phys. JETP* **21**, 283.
Waldron, R. D. (1955). *Phys. Rev.* **99**, 1727.
Wayne, R. C. (1967). *Bull. Am. phys. Soc.* **13**, 1036.
Wayne, R. C. and Anderson, D. H. (1967). *Phys. Rev.* **155**, 496.
Wazzan, A. R., Vitt, R. S. and Robinson, L. B. (1967). *Phys. Rev.* **159**, 400.
Weil, L. (1950). *C.r. hebd. Seanc. Acad. Sci., Paris* **231**, 122.
Weiss, P. (1907). *J. de Phys.* (4)**6**, 661.
Weissmuth, A. (1882). *Sber. Akad. Wiss. Wien* **86**, 539.
Werner, K. (1959). *Ann. Phys. Leipzig* **7**, 403.
Werner, S. A., Arrott, A. and Kendrick, H. (1967). *J. appl. Phys.* **38**, 1243.
Werner, S. A., Arrott, A. and Atoji, M. (1968). *J. appl. Phys.* **39**, 671.
Williams, G. and Pavlovic, A. S. (1968). *J. appl. Phys.* **39**, 571.
Willis, B. T. M. and Rooksby, H. P. (1954). *Proc. phys. Soc. Lond.* **67**B, 290.
Wisniewski, R. (1962). *Acta Phys. pol.* **22**, 159.
Worlton, T. G., Bennion, R. B. and Brugger, R. M. (1967). *Phys. Lett.* **24**A, 653.
Xavier, R. M. (1967). *Phys. Lett.* **25**A, 244.
Yeh, C. S. (1925). *Proc. Am. Acad. Arts Sci.* **60**, 503.
Yosida, K. (1957). *Phys. Rev.* **106**, 893.
Yuzuri, M., Wanatabe, H., Nagazaki, S., Maeda, S. and Hirone, T. (1957). *J. phys. Soc. Japan* **12**, 385.
Zakharov, A. I., Kadomtseva, A. M., Levitin, R. Z. and Ponyatovskii, E. G. (1964). *Soviet Phys. JETP* **19**, 1348.

CHAPTER 3

The Effects of Hydrostatic Pressure on Ferroelectric Properties†

GEORGE A. SAMARA

Sandia Laboratories, Albuquerque, New Mexico, U.S.A.

I. INTRODUCTION

Rochelle salt, the sodium-potassium salt of tartaric acid, ($NaKC_4H_4O_6 \cdot 4H_2O$) is the oldest known ferroelectric material. Its unusual dielectric properties were first recognized by Pockels in 1894, but it was Valasek's (1921) investigation of this material which revealed the phenomenon later to be called ferroelectricity. Valasek observed

† This work was supported by the U.S. Atomic Energy Commission.

that the electric polarization was not determined uniquely by the applied field but rather exhibited a hysteresis. He pointed out the analogy (purely phenomenological) to ferromagnetic behaviour, but it was Mueller (1935) who first used the name ferroelectricity to describe the phenomenon. Up to about 1955 only a few ferroelectric crystals were known and there was a conviction that the phenomenon was restricted to a few special classes of materials. Since then, however, a very large number of new ferroelectrics have been discovered, and it now appears that ferroelectricity is one of the most common co-operative phenomena. Ferroelectrics now represent a very important class of materials with a wide range of technical applications. Comprehensive reviews of the subject of ferroelectricity can be found in the books by Megaw (1957), Jona and Shirane (1962) and Fatuzzo and Merz (1967), and in the review paper by Känzig (1957).

The phenomenon of ferroelectricity is very structure-sensitive and, as will become clear later, ferroelectric properties are among the most pressure-sensitive of all physical properties. This reflects to a large extent the very intimate relationship between the onset of ferroelectric ordering and the accompanying spontaneous strain in the lattice, and it emphasizes the role of pressure as an important variable in the study of ferroelectric properties.

There has been a considerable number of studies on the effects of hydrostatic pressure on the equilibrium properties of a number of isolated ferroelectrics, but the subject does not appear to have been treated in a systematic manner. Accordingly, the present chapter represents an attempt to bring together the knowledge gained from these studies and to assess their contribution to the understanding of the ferroelectric phenomenon. It should be noted at the start that although ferroelectrics have certain properties in common, they are an extremely varied group—almost each one exhibiting a different mechanism for the onset of spontaneous polarization. Ferroelectrics also usually possess complicated chemical and crystallographic structures so that the phenomenon is reasonably well understood on an atomistic scale only in the simplest cases.

The present state of the theory is such that we have a fairly good understanding of the transition in two types of ferroelectrics: (1) perovskite ferroelectrics of the barium–titanate type, and (2) hydrogen-bonded ferroelectrics of the potassium dihydrogen–phosphate type. In the case of the pervoskites there is strong evidence that the ferroelectric transition is associated with an instability of one of the optical phonon modes of the lattice. On lowering the temperature of the crystal in the high temperature nonpolar (or paraelectric) phase, the frequency of the relevant mode decreases and finally becomes unstable at a critical temperature, and simultaneously the crystal transforms to a polar (or ferroelectric) phase. In the case of the hydrogen-bonded ferroelectrics, the motion of the proton in the bond plays a key role. The transition to the ferroelectric state is believed to be triggered by a cooperative ordering of the protons to which the displacements of the other ions are strongly coupled.

Much of our discussion will deal with the above two types of ferroelectrics with emphasis on the interpretation of the pressure results, in terms of model theories. We shall also discuss, but in less detail, other ferroelectrics where the pressure results turn out to be especially revealing and helpful in developing a better understanding of the behaviour of these materials. In particular, we shall discuss the effects of pressure on the properties of (1) Rochelle salt, triglycine sulphate and related ferroelectrics, (2) potassium nitrate and sodium nitrite, and (3) a number of miscellaneous ferroelectrics. Adequate references to the majority of the papers covering the subject will be given, but in the limited space available the review is necessarily not exhaustive or complete.

Because of the strong interdependence between the lattice strain and the spontaneous polarization, widely different results on ferroelectric properties are observed by the application of one-dimensional and, two-dimensional stresses and hydrostatic pressure. Therefore, it is extremely important to know exactly the nature of the applied stress. Results obtained in quasihydrostatic apparatus are very difficult, if not impossible, to interpret. In this paper the effects of hydrostatic pressure are dealt with exclusively. The study of the effects of one-dimensional stress on the dielectric properties of ferroelectrics is important, especially in connection with their piezoelectric response, although this aspect will not be dealt with here.

The chapter is organized as follows. Section II presents a brief general review of dielectric properties, some of the terms and relationships to be used later in the chapter are defined here. Section III gives a definition of ferroelectricity and summarizes some of the general characteristic properties of ferroelectrics. The phenomenological theory of ferroelectric transitions is discussed in Section IV. This theory has been very successful in correlating and interpreting ferroelectric properties and is particularly useful in analysing pressure effects. In the following four sections the dielectric properties of various ferroelectrics will be presented and discussed with emphasis on the interpretation of the effects of pressure on these properties. Finally, Section IX gives a brief overall summary.

Throughout the chapter it will often be convenient to use the abbreviations FE for ferroelectric and PE for paraelectric. As was already mentioned, the paraelectric phase refers to the phase above the transition, or Curie, point. Other abbreviations will be defined and used in various sections. No discussion will be given of apparatus and experimental techniques. The reader is referred to original referenced articles for details.

II. Dielectric Properties—General Theoretical Considerations

When a dielectric (insulator) slab is placed in a static electric field, it acquires a surface charge. The polarization so induced arises from the alignment of electric dipoles and the displacement of positive and negative charges in the dielectric. For an isotropic, linear dielectric, the polarization **P** is proportional and parallel to the applied field **E**. The electric flux density, or electric displacement, **D** is defined by (in cgs units)

$$\mathbf{D} = \mathbf{E} + 4\pi\mathbf{P}$$
$$= \epsilon\mathbf{E}, \tag{1}$$

where ϵ is the static dielectric constant of the medium. From eqn (1) it follows that

$$\epsilon = 1 + 4\pi\chi \tag{2}$$

where, by definition, χ $(=P/E)$ is the dielectric susceptibility. For isotropic dielectrics ϵ and χ are scalar quantities which are dependent on the molecular properties of the dielectric.

One of the important considerations in the theoretical treatment of dielectric properties is the relationship between the dielectric constant and the polarizability. Directly related to this is the question of the value of the internal field at a given point in the lattice. The polarizability α of an atom (or molecule) is defined by

$$\boldsymbol{\mu} = \alpha\mathbf{F} \tag{3}$$

where $\boldsymbol{\mu}$ is the electric dipole moment and **F** is the local, or effective, field acting on the given atom. The polarization **P** is defined as the net dipole moment per unit volume and is given by

$$\mathbf{P} = \sum_i N_i \boldsymbol{\mu}_i = \sum_i N_i \alpha_i \mathbf{F}_i \tag{4}$$

where N_i is the number of dipoles per unit volume.

The local field **F** at a given lattice site i is generally written as

$$\mathbf{F}_i = \mathbf{E} + \mathbf{E}_{\text{int}} = \mathbf{E} + 4\pi \sum_j \phi_{ij} \mathbf{P}_j \quad (5)$$

where **E** is the applied field, and $\mathbf{E}_{\text{int}}$ is the internal field acting on the ion i due to the other ions j. It is usually expressed as a power series in odd powers of the polarization. For small field measurements (which is the case of most interest to us) only the first power term in the polarization need be considered as indicated in eqn (5). ϕ_{ij} is the internal field coefficient which is a dimensionless quantity that depends on the position of the ion in the lattice. For diagonal cubic crystals, that is crystals in which *all* ions have cubic environment, the Lorentz internal field is applicable (Fröhlich, 1949) and $\phi_{ij} = 1/3$. For ferroelectrics, even in their paraelectric phase, this is never the case, and the field at each site has to be computed. This is a rather difficult procedure for the complicated lattice structures of ferroelectrics, but it has been done for the perovskite lattice (e.g. $BaTiO_3$) and the results are summarized in the article by Slater (1950). It is found that the Ti ions and those oxygen ions which are in the same line with them (the line being parallel to the applied field) exert on each other fields several times the strength of the ordinary Lorentz field. It is the large field at the Ti site that Slater finds operating to lead to a large enhancement of the effect of the displacement of the Ti ion and thus produces ferroelectricity in $BaTiO_3$.

Once the internal field at each site is known, an expression for the dielectric constant in terms of the individual polarizabilities of the ions can be obtained; however, polarizabilities are also difficult to determine. The total polarizability α can be expressed as the sum of various contributions,

$$\alpha = \alpha_{op} + \alpha_{ir} + \alpha_d. \quad (6)$$

These contributions are not completely independent of one another, but they exhibit different frequency responses. The optical, or high frequency contribution, α_{op}, arises from displacements of the electronic charge distributions of the ions relative to their nuclei. The infrared contribution, α_{ir}, arises from the combined effects of the displacements of the ions and the resulting displacement of the electronic charge distributions. α_d is the dipolar contribution associated with permanent dipoles which have a frequency response in the d.c. to millimetre waves range. Other possible contributions to α, such as space charge effects, are neglected.

From the above discussion it is seen that experimental results are difficult to treat microscopically. It is often instructive, however, to use a macroscopic treatment. Equation (5) becomes

$$\mathbf{F} = \mathbf{E} + 4\pi\phi\mathbf{P}. \tag{7}$$

This macroscopic local field combined with eqns (2) and (4) leads to the following relationship for the dielectric constant

$$\frac{\epsilon - 1}{1 + \phi(\epsilon - 1)} = 4\pi\left(\frac{\alpha}{V}\right) \tag{8}$$

where α is the total polarizability of a macroscopic element of the material of volume V. In the Lorentz field approximation, $\phi = 1/3$, and eqn (8) yields the well-known Clausius–Mossotti relationship which is applicable in its macroscopic form to cubic or isotropic materials (Fröhlich, 1949; Bosman and Havinga, 1963).

For anisotropic dielectrics, ϵ, χ and α are tensors, and eqn (1) must be written as

$$\mathbf{D} = \boldsymbol{\epsilon} \cdot \mathbf{E} \tag{9}$$

with components

$$D_i = \sum_{j=1}^{3} \epsilon_{ij} E_j, \qquad (i = 1, 2, 3). \tag{10}$$

All other quantities follow similarly.

Until now only the static case has been considered. When a dielectric is subjected to an alternating field, both **D** and **P** will vary periodically with time. In general, however, **D** and **P** cannot follow the field instantaneously, and there will always be inertial and energy dissipation effects (losses) which cause a lag in phase between **E** and the response of the material. Thus, if

$$\mathbf{E} = \mathbf{E}_0 \cos \omega t \tag{11}$$

then

$$\mathbf{D} = \mathbf{D}_0 \cos (\omega t - \delta), \tag{12}$$

where δ is the loss angle.† It is independent of $\mathbf{E}_0$ but generally depends on frequency.

In the presence of dielectric losses and relaxation effects, ϵ and χ are complex quantities composed of charging (real) and loss (imaginary) components. Thus

$$\epsilon^* = \epsilon' - i\epsilon'' \tag{13a}$$

and

$$\chi^* = \chi' - i\chi''. \tag{13b}$$

† In an ideal dielectric no energy losses should occur, and the current should lead the applied voltage by exactly $\phi = 90°$. When losses are present $\phi < 90°$. δ is a measure of this difference, that is $\phi + \delta = 90°$. The current is related to the displacement through Maxwell's equations.

The loss angle is given by

$$\tan \delta = \epsilon''/\epsilon'. \tag{14}$$

It is simply related to the Q-factor of the dielectric by

$$Q = 1/\tan \delta, \tag{15}$$

and is obtained rather directly from experiment. ϵ' and ϵ'' are related, at any given frequency, by the Kramers–Krönig dispersion relations (see Fröhlich, 1949), but this point will not be pursued here. Throughout the remainder of this review the real part of the dielectric constant will mostly be considered, and for convenience the superscript will be deleted and ϵ will be written for ϵ'.

III. Definition of a Ferroelectric and Its General Characteristic Properties

Of the thirty-two possible crystal classes (that is point groups) eleven are centrosymmetric and thus cannot exhibit polar properties, and the remaining twenty-one lack a centre of symmetry and thus can possess one or more polar axes. Of these, twenty classes are *piezoelectric* (the one exception being cubic class 432). Piezoelectric crystals have the property that the application of mechanical stress induces polarization, and, conversely, the application of an electric field produces mechanical deformation. Ten of the twenty piezoelectric classes have a unique polar axis and thus are spontaneously polarized, that is polarized in the absence of an electric field. Crystals belonging to these ten classes are called *pyroelectric*. The intrinsic polarization of pyroelectric crystals is often difficult to detect experimentally because of the neutralization of the charges on the crystal surfaces by free charges from the atmosphere and by conduction within the crystal. However, since the polarization is a function of temperature, it is often possible to observe the spontaneous moment in these crystals by changing the temperature. *Ferroelectric* crystals belong to the pyroelectric family, but they also exhibit the additional property that the direction of the spontaneous polarization can be reversed by the application of an electric field. Thus we have the following simple definition for a ferroelectric crystal,

> "A ferroelectric crystal is a crystal which possesses *reversible* spontaneous polarization as exhibited by a dielectric hysteresis loop."

The key feature in this definition is the reversibility of the polarization, and the most characteristic property of a ferroelectric is the hysteresis loop. It should be pointed out, however, that there are ferroelectrics (e.g. $LiNbO_3$) in which it is very difficult to reverse the polarization at applied fields below the breakdown strength. In a virgin macroscopic crystal it is quite unlikely that the direction of the polarization will be the same throughout. Rather, the crystal will consist of domains, which are macroscopic regions of homogeneous polarization which differ only in the direction of the polarization vector.

On the basis of the domain concept, polarization reversal takes place by way of processes involving nucleation of new domains and domain-wall motion, and the occurrence of a dielectric hysteresis loop can be explained as follows (Fig. 1). Starting with a crystal with

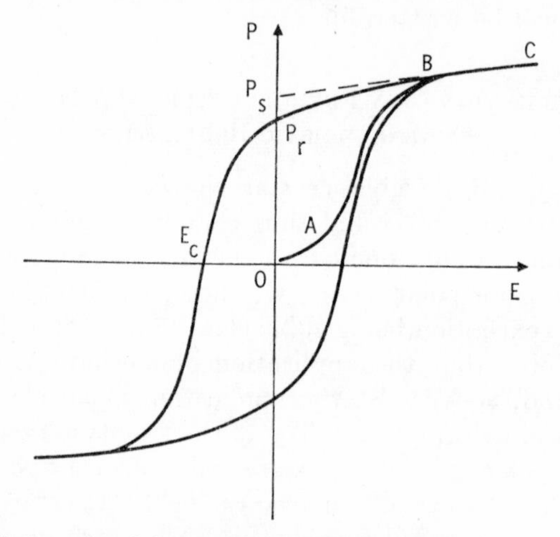

FIG. 1. Schematic ferroelectric hysteresis loop showing the spontaneous polarization P_s, the remanent polarization P_r, and the coercive field E_c.

zero net polarization, the application of a small d.c. field **E** will produce only a linear field dependence of polarization **P** (*OA* of Fig. 1) as is observed for normal dielectric crystals. For larger fields, domains whose polarization vectors are not along the direction of the field will rotate to the field direction and the polarization increases rapidly (*AB*). When all the domains are aligned in the direction of the applied field, the polarization saturates and the crystal becomes a single domain. Further increase in **E** produces a linear increase in **P** resulting from " normal " polarization effects (*BC*).

Extrapolation of the linear part of the P *versus* E curve (*BC*) to zero applied field yields the value of the spontaneous polarization P_s. This value of P is that which existed in each domain in the starting crystal. When the applied field is reduced to zero the crystal retains a remanent polarization P_r. A field in the opposite direction must be applied in order to reduce this polarization to zero. This defines the coercive field E_c, that is, the field required to make $P=0$ again.

The definitions of the dielectric constant and susceptibility given in Section II are adequate for normal dielectrics for which the relationship between D (or P) and E is linear up to relatively large fields. However, in the case of ferroelectrics ϵ and χ must be defined more precisely owing to the more complicated relationship between D (or P) and E (Fig. 1). For the present purposes we are going to define ϵ and χ as the slopes of the D *versus* E and P *versus* E curves, respectively, at the origin, that is the initial values,

$$\epsilon=(\partial D/\partial E)_{E=0} \text{ and } \chi=(\partial P/\partial E)_{E=0}. \tag{16}$$

They are determined from measurements made at very low a.c. fields so as not to reverse any domains.

There is another important consideration. In ferroelectric crystals, whereas it is true that every unit cell has a permanent dipole moment, this cannot, except in the simplest cases, be associated with any specific ions in the unit cell. Rather, all that can be said is that given a charge distribution $\rho(\mathbf{r})$, the dipole moment is

$$\mu=\iiint \rho(\mathbf{r}).\mathbf{r}\, dx\, dy\, dz \neq 0 \tag{17}$$

and is independent of the choice of origin.

Although ferroelectric crystals are a widely varied group, they possess a number of general characteristic properties, of which the following are mentioned:

1. The hysteresis loop disappears at a certain temperature, the Curie point T_c, above which the crystal behaves as a normal dielectric. It should be noted, however, that in some crystals melting or chemical decomposition may occur before the Curie point is reached. Examples where such behaviour occurs will be discussed later.
2. At the Curie point a ferroelectric crystal transforms to a phase of higher symmetry. This higher temperature phase is usually nonpolar.
3. The polar crystal structure of a ferroelectric can be derived from the high temperature paraelectric structure by a slight distortion of the crystal lattice. As will be discussed later, this is the important reason behind the success of the phenomenological theory of ferroelectricity which assumes that the same free energy function is applicable for both the FE and PE phases.

4. Ferroelectrics generally have a high dielectric constant (or susceptibility) which rises to a peak value at the Curie point.
5. Above the Curie point the dielectric constant ϵ of a ferroelectric (measured along the polar axis) usually obeys the Curie–Weiss law $\epsilon = C/(T - T_0)$, where C and T_0 are the Curie–Weiss constant and Curie–Weiss temperature, respectively.

Finally, we should mention that there are substances which, on cooling, undergo at a certain critical temperature a transition from a nonpolar to an antipolar state. In this state the crystal has a superlattice consisting of arrays of antiparallel dipoles. If, in a given crystal, the coupling energy between these arrays is comparable to that of the polar case, then the crystal is said to be *antiferroelectric*. An antiferroelectric crystal can usually be made ferroelectric by the application of a sufficiently large electric field, and it exhibits a large dielectric anomaly at its transition. Except for a brief discussion of the properties of $PbZrO_3$, $PbHfO_3$, and $NaNO_2$, antiferroelectrics will not be considered in this paper.

IV. Thermodynamic Theory of Ferroelectric Transitions

A. General Theory

It is now well established that many of the physical properties of ferroelectrics, especially those in the transition region, can be successfully correlated and interpreted in terms of the thermodynamic (or phenomenological) theory which was developed in some detail by Mueller (1940), Cady (1946), and Devonshire (1954). Although a thermodynamic theory gives a purely macroscopic picture and consequently does not describe the physical mechanism responsible for the ferroelectric properties of a given material, it has the distinct advantage of being independent of any particular microscopic model and, thus, leads to quite general conclusions. It, of course, treats only equilibrium properties and does not apply to non-equilibrium (transient) properties such as ferroelectric switching.

The internal energy U of a crystal can be expressed as a function of temperature and entropy as well as the mechanical (X) and electrical (E) stresses, or the mechanical (x) and electrical (P) strains. Two of the four variables X, E, x and P can be chosen as independent variables and the other two as dependent variables. The thermodynamic theory of ferroelectric transitions postulates that the free energy of the crystal (for example the Helmholtz free energy A, $A = U - TS$) can be expanded in powers and products of the components of strain (or stress) and polarization, and it is assumed that the power series converges after a finite number of terms. The theory also assumes that the same free energy function can be used to describe both the FE and PE phases. This is justified on the basis that the structure of the FE phase can be usually derived from that of the PE phase by a slight distortion of the lattice.

In the most general case, the free energy of the strained and polarized crystal can be written as (Jona and Shirane, 1962):

$$\begin{aligned} A(x, P, T) - A(0, 0, T) = {} & \tfrac{1}{2} \sum_{1}^{3} \gamma_{ij}^{x} P_i P_j + \tfrac{1}{3} \sum_{1}^{3} \omega_{ijk}^{x} P_i P_j P_k \\ & + \tfrac{1}{4} \sum_{1}^{3} \xi_{ijkl}^{x} P_i P_j P_k P_l + \tfrac{1}{5} \sum_{1}^{3} \psi_{ijklm}^{x} P_i P_j P_k P_l P_m \\ & + \tfrac{1}{6} \sum_{1}^{3} \zeta_{ijklmn}^{x} P_i P_j P_k P_l P_m P_n + \tfrac{1}{2} \sum_{1}^{3} c_{ijkl}^{P} x_{ij} x_{kl} \\ & + \sum_{1}^{3} a_{ijk} x_{ij} P_k + \tfrac{1}{2} \sum_{1}^{3} q_{ijkl} x_{ij} P_k P_l + \ \ldots \end{aligned} \tag{18}$$

where the summations are carried out for all the x, y and z (1 to 3) coordinates. $A(0, 0, T)$ is the free energy of the unstrained and unpolarized crystal and can be set equal to zero. The effect of temperature is incorporated by assuming that the coefficients of the power series are functions of temperature. The coefficients c, a and q are the tensors of the elastic, piezoelectric, and electrostrictive constants, respectively. The meaning of some of the other coefficients will become apparent later. The superscripts x and P imply conditions of constant strain and polarization, respectively.

For the present purposes it is more useful to express the free energy in terms of stress and polarization, and this can be obtained from eqn (18) by the use of the transformation (Forsbergh, 1954)

$$A(\mathbf{X}, \mathbf{P}) = A(\mathbf{x}, \mathbf{P}) + \mathbf{X} \cdot \mathbf{x}. \tag{19}$$

Considering, for simplicity, the case where the PE phase is non-piezoelectric and including only even powers of P (that is A is an even function), the result is

$$\begin{aligned} A(X, P, T) = {} & \tfrac{1}{2} \sum_{1}^{3} s_{ijkl} X_{ij} X_{kl} + \tfrac{1}{2} \sum_{1}^{3} q_{ijkl} X_{ij} P_k P_l \\ & + \tfrac{1}{2} \sum_{1}^{3} \gamma_{ij} P_i P_j + \tfrac{1}{4} \sum_{1}^{3} \xi_{ijkl} P_i P_j P_k P_l \\ & + \tfrac{1}{6} \sum_{1}^{3} \zeta_{ijklmn} P_i P_j P_k P_l P_m P_n + \ \ldots \end{aligned} \tag{20}$$

where the coefficients s are tensors of the elastic compliances.

In a ferroelectric crystal the polarization (in any given phase) is directed along only one crystallographic axis, say the z-axis, and therefore we can write

$$P_1 = P_2 = 0; \; P_3 = P. \tag{21}$$

Furthermore, when the applied stress is hydrostatic we have

$$X_{11} = X_{22} = X_{33} = p, \tag{22a}$$

and

$$X_{ij} = 0 \qquad (i \neq j). \tag{22b}$$

Under the above conditions eqn (20) reduces to the form

$$A(p, P, T) = \Lambda p^2 + \tfrac{1}{2}\Omega p P^2 + A(0, P, T) \tag{23}$$

where Λ and Ω are functions of the elastic and electrostrictive constants, and

$$A(0, P, T) = \tfrac{1}{2}\gamma P^2 + \tfrac{1}{4}\xi P^4 + \tfrac{1}{6}\zeta P^6 + \ldots \tag{24}$$

is the free energy of the unstressed crystal. The coefficients γ, ξ and ζ (especially γ) are functions of temperature.

The signs and magnitudes of the coefficients γ, ξ, and ζ determine the nature of the transition and the behaviour of the dielectric properties in the immediate vicinity of the Curie point. It is therefore instructive to examine, first, the behaviour of the unstressed crystal when it undergoes a ferroelectric transition. In the PE phase the coefficients γ and ζ are found to be positive for all known ferroelectrics whereas ξ can be either positive or negative. Therefore consider two cases: (1) γ, ξ, and ζ are all positive, and (2) γ and ζ are positive and ξ is negative.

Case (1). Since stable states are those for which the free energy is a minimum, the fact that all the coefficients in eqn (24) are positive implies that the only stable state is that corresponding to $P=0$, that is a non-ferroelectric state. If we assume that ξ and ζ are independent of temperature and that γ changes from positive to negative with decreasing temperature (as is observed experimentally), then it is seen that as soon as γ becomes negative, the free energy function will develop a maximum for $P=0$ and a pair of minima (at $\pm P$) for a non-vanishing value of P. Thus the stable state of the crystal will correspond to $P=0$ for $\gamma>0$ and to $P\neq 0$ for $\gamma<0$, with the crystal undergoing a ferroelectric transition at $\gamma=0$. A schematic representation is shown in Fig. 2. In this case the change in γ, and therefore in P, with temperature is continuous and the transition is of second order (Fig. 2).

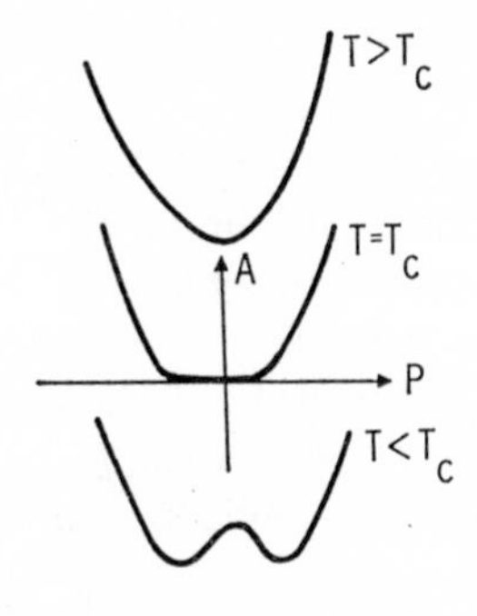

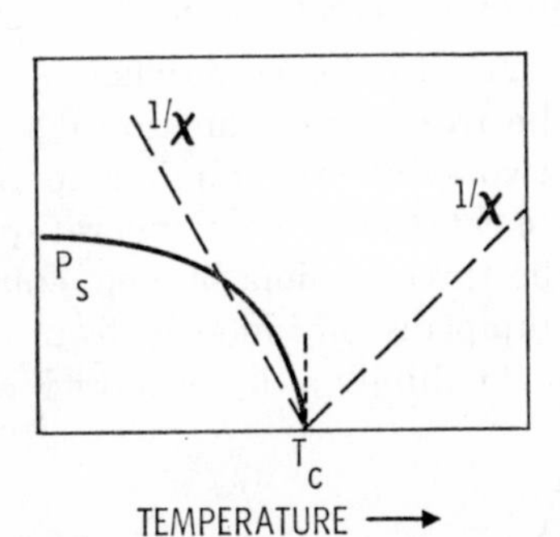

FIG. 2. Schematic temperature variation of the free energy (A) *versus* polarization (P) and of the spontaneous polarization and reciprocal susceptibility for a second-order ferroelectric transition.

From eqn (24) and the defining relations

$$(\partial A/\partial P)=E, \tag{25}$$

$$(\partial E/\partial P)=1/\chi, \tag{26}$$

and

$$\epsilon=1+4\pi\chi\simeq 4\pi\chi \text{ (for } \epsilon \gg 1), \tag{27}$$

it is easily shown that in the PE phase

$$(1/\chi)_{T>T_c}=(\partial E/\partial P)_{P=0}=\gamma=(4\pi/C)(T-T_0). \tag{28}$$

The last equality follows from the experimentally observed Curie–Weiss temperature dependence of ϵ, that is $\epsilon=C/(T-T_0)$. At the transition $T_c(=T_0)$, $\gamma=0$ and $\chi=\infty$. Thus the coefficient γ in eqn (24) represents the reciprocal of the dielectric susceptibility.

In the ferroelectric phase, and for temperatures near the transition (that is where P is small) it is often a good approximation to neglect in eqn (24) terms in P of order higher than 4. Remembering that $P=P_s$ (=spontaneous polarization) at $E=0$ and assuming that ξ is independent of temperature, eqns (24–26) yield

$$P_s^2=-(\gamma/\xi)\propto(T-T_0) \tag{29}$$

and

$$(1/\chi)_{T<T_c}=\gamma+3\xi P_s^2=-2\gamma. \tag{30}$$

Thus, under the above approximations, P_s^2 is a linear function of T near T_0 and the slope of the $(1/\chi)$ *versus* T curve in the FE phase is -2 times† that in the PE phase (Fig. 2). These conclusions are fairly

† Equation (30) is strictly applicable for the isothermal case. Near T_0 in the FE phase, P is a rapidly varying function of T and E, and the measured susceptibility is the adiabatic one. For comparison between experiment and theory, it is therefore necessary to make the appropriate correction (see for example, Jona and Shirane, 1962, p. 38).

well obeyed by ferroelectrics which exhibit second-order transition, for example, triglycine sulphate.

Case (2). In this case (where γ and ξ are >0 and $\xi<0$) it is possible for the free energy function eqn (24) to have one minimum for $P=0$ and two symmetric minima for $P\neq 0$, all at the same value of $\gamma>0$, that is at the same temperature. At some particular temperature, T_c, the three minima become equal, and the stable state of the crystal will jump discontinuously from one with $P=0$ to one with $P=\pm P_s$. Thus P exhibits a discontinuity at T_c (Fig. 3). It is evident that this

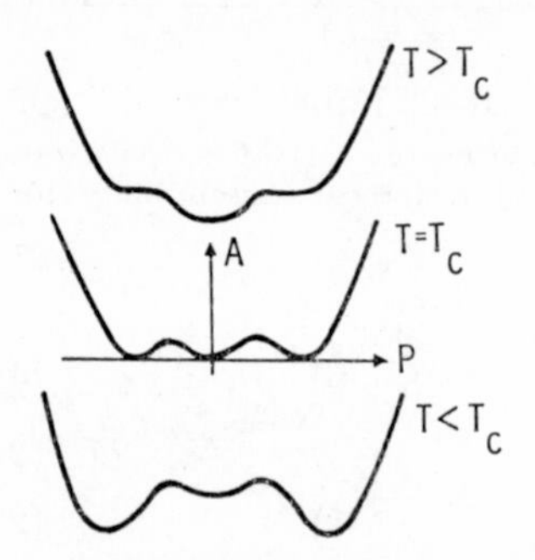

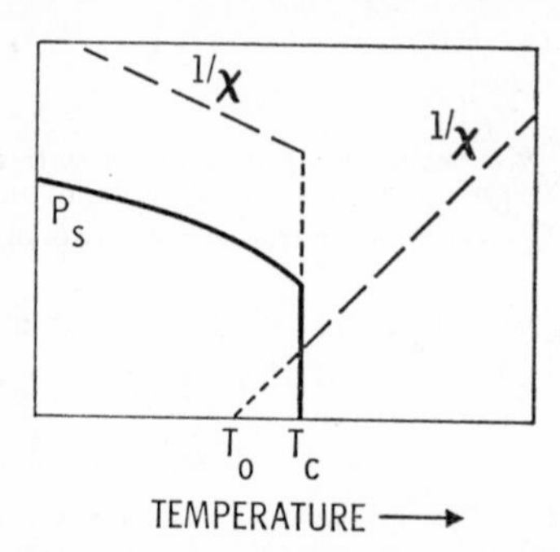

FIG. 3. Schematic temperature variation of the free energy (A) *versus* polarization (P) and of the spontaneous polarization and reciprocal susceptibility for a first-order ferroelectric transition.

transition is first order and should thus be accompanied by a latent heat and discontinuities in the entropy and volume.

From the fact that *at the transition* the free energy has the same value for both $P=0$ and $P=P_s$, it is seen that

$$A(0, T_c)=\tfrac{1}{2}\gamma P_s^2+\tfrac{1}{4}\xi P_s^4+\tfrac{1}{6}\zeta P_s^6=0. \tag{31}$$

Also noting that $E=0$ for $P=P_s$, it follows that

$$E=(\partial A/\partial T)=\gamma P_s+\xi P_s^3+\zeta P_s^5=0. \tag{32}$$

Equations (31) and (32) yield

$$P_s^2=-\tfrac{3}{4}(\xi/\zeta) \tag{33}$$

and

$$\gamma=\tfrac{3}{16}(\xi^2/\zeta), \tag{34}$$

where these are the values of P_s and γ (i.e. $1/\chi$) at the transition, T_c. Both P_s and γ change discontinuously at T_c. γ is positive, and its zero value (found by extrapolation) occurs at $T_0(<T_c)$ as shown in Fig. 3. The difference (T_c-T_0) is found to vary from a fraction of a degree to several tens of degrees for various ferroelectrics. $BaTiO_3$

and $NaNO_2$ are examples of ferroelectrics exhibiting first-order transitions.

B. EFFECTS OF PRESSURE

The effects of pressure enter explicitly in the first two terms of the free energy expansion given by eqn (23). In addition, the coefficients γ, ξ and ζ may vary with pressure, and this variation can lead to changes in the nature of phase transitions. In fact, as will be seen later, there is experimental evidence that the first-order transition in barium titanate tends towards one of second order with increasing pressure.

When pressure is a variable eqn (23) must be used in deriving the desired relationships for the dielectric properties. Thus, for example, corresponding to eqn (28) we have

$$(1/\chi)_{T>T_c^p} = (\gamma + \Omega p) = 4\pi\left(\frac{T - T_0^p}{C^p}\right) \tag{35}$$

where the last equality follows from the usually observed Curie–Weiss temperature dependence of the susceptiblity at a given pressure in the PE phase. The superscript p emphasizes that the quantities T_0 and C are functions of pressure. If it is assumed that T_0 varies linearly with pressure (i.e. $T_0^p = T_0^0 + k'p$) and further that the pressure dependence of C is negligible (which is often a good assumption), then it is seen from eqn (35) that

$$\gamma = 4\pi\left(\frac{T - T_0^0}{C}\right) \quad \text{and} \quad \Omega = 4\pi\left(\frac{k'}{C}\right) \tag{36}$$

where $k' = \mathrm{d}T_0/\mathrm{d}p$. In such a case eqn (35) predicts that, at constant temperature, $1/\chi$ varies linearly with pressure.

Similarly, the relationship corresponding to eqn (29) becomes

$$P_s^2 = -\frac{(\gamma + \Omega p)}{\xi} - \frac{\zeta}{\xi^3}(\gamma + \Omega p)^2, \tag{37}$$

where it has been assumed that the P^6 term in the free energy expansion is much smaller than the P^4 term, and higher order terms are neglected. Other relationships follow in a similar manner.

From the thermodynamic point of view, the most pronounced effect of pressure on ferroelectric properties is that of the displacement of the Curie point. As is well known, for a second-order transition the shift of the Curie point with hydrostatic pressure obeys the Ehrenfest relationship

$$\frac{dT_c}{dp} = \frac{(\beta^+ - \beta^-)T_c}{(c_p^+ - c_p^-)\rho}. \tag{38}$$

Here β is the volume expansion coefficient, c_p is the specific heat at constant pressure, and ρ is the density. The + and − designate values above and below T_c, respectively.

For a first-order transition, on the other hand, the Clausius–Clapeyron equation holds with

$$\frac{dT_c}{dp} = \frac{T_c(\Delta V)}{Q} = \frac{\Delta V}{\Delta S} \tag{39}$$

where ΔV and ΔS are the discontinuous volume and entropy changes at the transition, and Q is the latent heat. As we shall see in the following sections, eqns (38) and (39) hold well for many ferroelectrics.

V. Perovskite Ferroelectrics of the $BaTiO_3$-Type

Many oxides belonging to the perovskite family (after the mineral pervoskite–$CaTiO_3$) exhibit ferroelectric and antiferroelectric properties. These oxides, and in particular $BaTiO_3$, have been the subject of extensive experimental and theoretical investigation (see for example, Jona and Shirane, 1962). This is partly due to their use in a variety of practical applications and partly to the simplicity of their crystal structure (compared with other ferroelectrics) which makes them attractive from the theoretical point of view. The oxides have the general formula ABO_3, where A is a mono-, di-, or trivalent metal and B is a penta-, tetra-, or trivalent metal, respectively.

The spatial arrangement of the atoms in the ideal pervoskite structure, which is the high temperature PE phase of these oxides, is shown in Fig. 4. The structure is cubic with the B atom octahedrally surrounded by O atoms. Below the Curie point there occurs a small distortion of this structure to one of lower symmetry with an accompanying displacement of the ions from their original symmetric positions. The magnitude of the dipole moments, and thus the measured polarization, can generally be accounted for on the basis of the observed ionic displacements. One thus speaks of pervoskites as being " displacive-type " ferroelectrics as opposed, say, to " order-disorder type " involving the ordering of molecular dipoles, which are disordered above the Curie point.

The effects of pressure on the dielectric properties of a number of these oxides have been investigated. $BaTiO_3$ and $SrTiO_3$ have

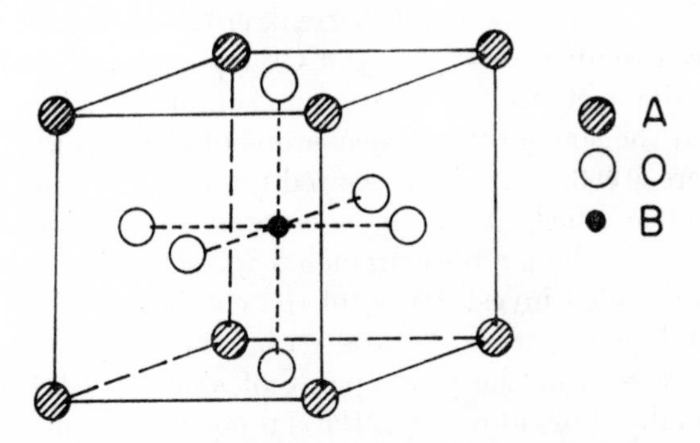

FIG. 4. The cubic perovskite structure ABO_3.

received the most attention, and we shall discuss the pressure results on these two compounds in some detail. Results on other important perovskites will be discussed only briefly.

A. DIELECTRIC PROPERTIES AND THEIR PRESSURE DEPENDENCE

1. *Barium Titanate—BaTiO$_3$*

Above its Curie point of ~120°C, $BaTiO_3$ crystallizes in the ideal cubic perovskite structure (space group Pm3m) and is paraelectric. Below 120°C, it transforms to a tetragonal ferroelectric phase (space group P4/mmm and $c/a = 1{\cdot}01$) with spontaneous polarization along the fourfold c-axis. At ~5°C there is a second transformation to an orthorhombic ferroelectric phase (space group Cmm2), and the polar axis is now parallel to a face diagonal, that is to one of the original cubic $\langle 110 \rangle$ directions. Below −90°C, a third transition to a phase with rhombohedral symmetry (space group R3m) occurs. This phase is also ferroelectric with the spontaneous polarization along a body diagonal, that is one of the original cubic $\langle 111 \rangle$ directions. Each of the transitions is accompanied by discontinuous changes in the dielectric and several other physical properties (see for example, Jona and Shirane, 1962).

$BeTiO_3$ has been the ferroelectric by far most widely investigated under pressure. The effects of hydrostatic pressure on its dielectric constant, polarization and transition temperatures have been studied.

Merz (1950), in one of the earliest pressure investigations on ferroelectrics, reported the effects of pressure to 3·5 kb on the Curie point and on the tetragonal–orthorhombic transition temperatures of single-crystal samples. Klimowski and Pietrzak (1960) measured the pressure (to 1·5 kb) and temperature (20°–160°C) dependences of the dielectric constant of single-crystal and ceramic samples. Klimowski (1962) later extended the single-crystal measurements to 12·5 kb. Leonidova and co-workers (1963, 1966) investigated the effect of pressure to 10 kb on the dielectric constant of single-crystal samples, and Minomura *et al.* (1964) made similar measurements to ~13 kb. Samara (1966) carried out a detailed investigation of the combined effects of pressure (to 25 kb) and temperature (20–150°C) on the polarization, dielectric constant and loss, and the Curie point of single-crystal and ceramic samples. Finally, Polandov *et al.* (1967) reported the effects of pressure to 10 kb on the spontaneous polarization of a single crystal.

Although there are quantitative differences among the various investigations, the qualitative features are similar, the predominant effect being a large decrease in the Curie point with increasing pressure. Some of the experimental results will be reviewed and discussed below. The author's work (Samara, 1966) has been the most extensive so far, and the illustrations will be taken from that work.

a. *Dielectric constant*

Figure 5 shows typical results on the effects of pressure on the temperature dependence of the dielectric constant, ϵ, of a single crystal and a polycrystalline (ceramic) sample. These results show the large characteristic anomaly in ϵ at the Curie point, T_c. For both samples, T_c decreases linearly with pressure with slopes $(\mathrm{d}T_c/\mathrm{d}p) = -5{\cdot}5$°K/kb for the crystal and $-5{\cdot}1$°K/kb for the ceramic. However, there is a striking difference between the two samples. Whereas for the crystal the value of ϵ in the immediate vicinity of the transition increases by ~60–70% in 15 kb, it decreases by over 50% over the same pressure range for the ceramic. This difference may be partly related to the pressure dependence of ϵ along the a-axis, ϵ_a, which apparently has not been measured. In the ceramic, of course, it is an average value of ϵ that is measured. There are also anisotropy field effects associated with the fact that although the applied pressure is hydrostatic, the stress experienced by the individual crystallites in the ceramic may be highly anisotropic.

Figure 6 shows the dielectric loss, tan δ, as a function of temperature for the same sample and conditions as in Fig. 5(a). A large discon-

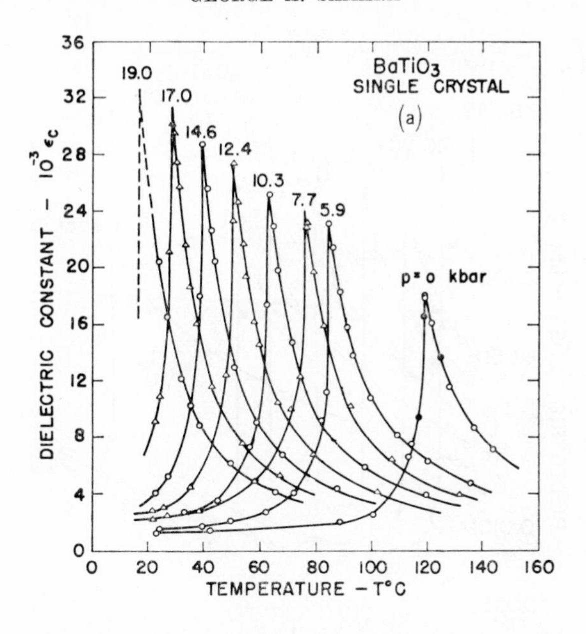

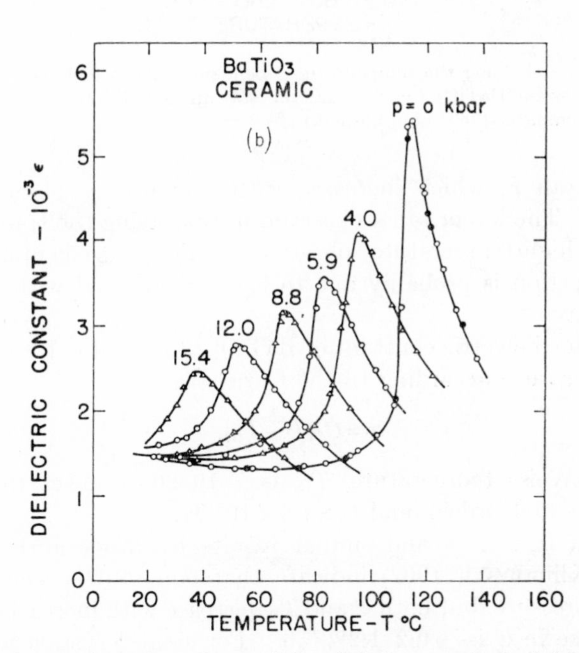

FIG. 5. The temperature dependence of the dielectric constant of (a) single-crystal and (b) ceramic $BaTiO_3$ at various hydrostatic pressures. Reproduced from Samara (1966) by permission of the American Institute of Physics.

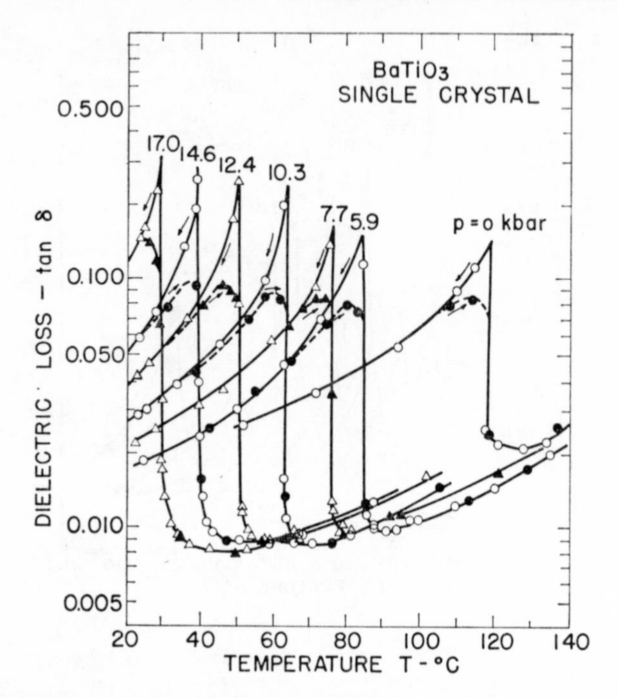

FIG. 6. Isobars showing the temperature dependence of the dielectric loss, tan δ, of single-crystal $BaTiO_3$ for the same crystal and conditions shown in Fig. 5(a). (After Samara, 1966; unpublished.)

tinuity in tan δ, which increases with pressure, is observed at the transition. The larger tan δ observed in traversing the transition from the PE (or high temperature) phase to the FE phase as opposed to the reverse direction is probably due to losses associated with domains in the FE phase.

At 1 b, the dielectric constant of $BaTiO_3$ in the cubic PE phase varies with temperature according to the Curie–Weiss law

$$\epsilon = C/(T - T_0). \tag{40}$$

The Curie–Weiss temperature, T_0, is $\sim$10–20°K lower than T_c (the transition is first order), and $C \simeq 1{\cdot}5 \times 10^5$°K.

The data in Fig. 5 and similar results by other authors (see for example, Klimowski, 1962) indicate that eqn (40) is well obeyed at any given pressure, but both C and T_0 decrease with increasing pressure. The decrease in C is $\sim$0·2–1·2%/kb. For single crystals T_0 decreases at a lower rate than does T_c, while for ceramic specimens the reverse is true. This is related to the fact that ϵ_{max} rises sharply with pressure in the case of single crystals, but decreases for ceramics. We can easily see this by remembering that ϵ_{max} occurs at $T = T_c$ and writing eqn (40) as

$$\epsilon_{max} = C/(T_c - T_0).$$

Thus a smaller ΔT ($\equiv T_c - T_0$) is associated with a higher ϵ_{max} and *vice versa*. (The small change in C with pressure is neglected in this argument.) Some typical results are shown in Fig. 7. For the single crystal, $(T_c - T_0)$ extrapolates to zero at ~23 kb, and presumably (see Section IV) the transition should become second order at this and higher pressures.

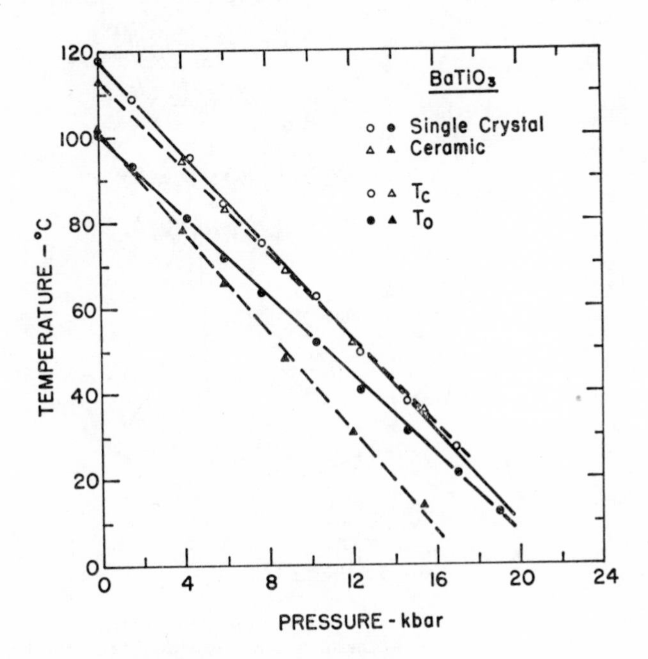

Fig. 7. The effects of hydrostatic pressure on the Curie point, T_c, and Curie–Weiss temperature, T_0, of single-crystal and ceramic $BaTiO_3$ samples. (After Samara, 1966; unpublished.)

The decrease of T_c with pressure indicates that the FE→PE transition in $BaTiO_3$ can be induced by pressure at constant temperature. That this is indeed so is shown in Fig. 8, where isotherms of ϵ_c *versus* pressure are given for two single crystals. In analogy to the Curie point, T_c, there is a transition pressure p_c taken as the pressure corresponding to the peak value of ϵ. Above p_c, that is in the PE phase, ϵ decreases monotonically with increasing pressure, as it does with increasing temperature above T_c. In fact, in this phase the results can be well represented by (see Fig. 8)

$$\epsilon = C^*/(p - p_0) \tag{41}$$

that is, a Curie–Weiss-like relationship expressed in terms of pressure. Here C^* and p_0 are constants at any given temperature, corresponding to C and T_0, respectively. At room temperature, $C^* \simeq 2{\cdot}8 \times 10^4$ kb and $p_0 \simeq 18$ kb, the latter being ~ 1–2 kb lower than p_c. Values of p_c, p_0, and C^* at different temperatures are given in Table II.

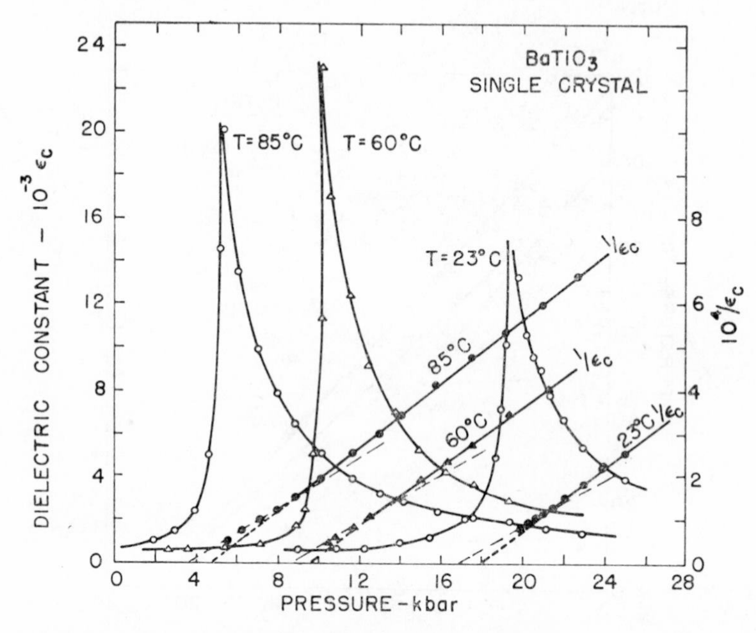

FIG. 8. Isotherms showing the effect of pressure on the dielectric constant of single-crystal $BaTiO_3$. The 23°C isotherm is for one crystal, and the 60 and 85°C isotherms are for another crystal. Reproduced from Samara (1966) by permission of the American Institute of Physics.

The above results are consistent with the predictions of the thermodynamic theory outlined in Section IV. On the basis of this theory, Goswami (1966) calculated the effect of hydrostatic pressure on the dielectric constant and Curie point of single crystal $BaTiO_3$ using known atmospheric pressure values for the various coefficients in the free energy expansion, eqn (20). He assumed that these coefficients are independent of pressure, and that the pressure effect enters only through the term involving the electrostrictive coefficients, that is the second term on the right-hand side of eqns (20) and (23). Values of ϵ in both the tetragonal and cubic phases are evaluated through the use of the defining relationships, eqns (25–27), in eqn (20), and the variation of T_c with pressure is obtained by imposing on the saturation polarization, P_s, the condition that, in the cubic phase, P_s ceases to be real. Goswami's calculations show that: (1) at 20°C there is a first-order FE–PE transition at 14·6 kb, (2) in the cubic phase ϵ obeys eqn (41) with $C^* = 2 \times 10^4$ kb at 150°C, and (3) T_c decreases linearly with pressure with slope $\mathrm{d}T_c/\mathrm{d}p = -6{\cdot}7$°K/kb. In view of the assumptions made, these results are in quite satisfactory agreement with the experimental results in Fig. 8 and Tables I and II.

Although it appears to be a reasonable first approximation, Goswami's assumption that the coefficients of the free energy expansion are independent of pressure is not strictly valid. Thus, for example, the results discussed above show that the Curie constant C (which is inversely proportional to the coefficient γ in eqn (20)) decreases significantly with increasing pressure. There is also evidence (Leonidova and Volk, 1966) that the coefficient ξ is also pressure dependent. Recently, Goswami and Cross (1968) explained the decrease in C in the framework of the thermodynamic theory by assuming a weak temperature dependence for the electrostrictive constants.

b. *Polarization*

The changes of the hysteresis loop of $BaTiO_3$ with pressure at constant temperature closely resemble the more familiar changes observed with increasing temperature at 1 b. Small decreases in polarization and coercive field occur up to the transition pressure where the polarization decreases rapidly. Figure 9(a) shows the pressure dependence of the polarization, P, for two crystals.

Since the Curie point of $BaTiO_3$ decreases with pressure, it may be asked whether or not the pressure dependence of P_s is the result of a simple displacement of the P_s *versus* T curve along the T-axis. Figure 9(b) shows isobars of P_s *versus* the reduced temperature $(T - T_c)$, that is P_s measured at a given separation from the Curie point. These results clearly show that the major part of the measured change of P_s with pressure at constant T is caused by the shift of T_c. There is only a small decrease in the magnitude of the dipole moment with pressure. This is perhaps not too surprising in view of the fact that the dipole moment is largely associated with the Ti–O displacement along the c-axis, and the compressibility of the crystal along this direction is only $\sim 4 \times 10^{-4}$ kb^{-1} (Kabalkina *et al.*, 1962).

Polandov *et al.* (1967) measured the effect of pressure to 10 kb on the

TABLE I. Values of the spontaneous polarization (P_s), Curie point (T_c) and its pressure coefficient, Curie–Weiss temperature (T_0), Curie–Weiss constant (C) for various ferroelectrics. For detailed references see the text

Ferroelectric		P_s (μC/cm^2),	at T (°C)		T_c (°C)	dT_c/dp (°C/kb)	T_0 (°C)	C (°C)
$BaTiO_3$		26	(20)		~120	−4·5 to −6·7	~105	(1·3 to 3) × 10^5
$BaTiO_3$	(ceramic)	< 20	(20)		~115	−4·0 to −5·1	~100	(1·2 to 1·5) × 10^5
$PbTiO_3$		> 80	(20)		490	—	—	1·1 × 10^5
$SrTiO_3$		—			—	—	−237	8·3 × 10^4
$KTaO_3$		—			−260	—	—	6 × 10^4
$Ba_{0·75}Sr_{0·25}TiO_3$	(ceramic)	—			55	−4·4[a]	—	—
$Ba_{0·05}Sr_{0·95}TiO_3$	(ceramic)	—			−232	−7·3[b]	—	—
$BaTi_{0·9}Zr_{0·1}O_3$	(ceramic)	—			90	−3·8[c]	—	—
$PbZrO_3$	(ceramic)	AFE			230	+4·4	—	1·6 × 10^5
$PbHfO_3$	(ceramic)	AFE		(i)	203	+5·5	110	1·65 × 10^5
				(ii)	160	~+6·0	—	—
KH_2PO_4		4·7	(−173)		−151	−4·5[d]	−151	3·3 × 10^3
					−153	−5·6[e]	—	—
KD_2PO_4		5·5	(−100)		−60	−2·6[d]	—	—
					−65	−3·9[f]	−65	4·0 × 10^3
$LiH_3(SeO_3)_2$		15·0	(20)		147	−6·0	144	4·1 × 10^4
$LiD_3(SeO_3)_2$		—			173	−5·9	159	2·9 × 10^4
TGS		2·8	(20)		49	+2·6	49	3·2 × 10^3
TGSe		3·2	(0)		22	+3·7	22	4·0 × 10^3
TGFB		3·2	(20)		73	+2·5	73	2·5 × 10^3
Rochelle salt		0·25	(5)	(i)	24	+11·0	24	2·29 × 10^3
				(ii)	−18	+3·6	−18	−1·18 × 10^3
Rochelle salt	(deuterated)	0·37	(5)	(i)	35	+10·2	35	1·6 × 10^3
				(ii)	−22	+3·3	−22	−1·1 × 10^3
KNO_3		6–8	(120)		~125	+22·0	41	6 × 10^3
$NaNO_2$		~7·0	(100)		165	+4·9	160	5 × 10^3
SbSI		25·0	(−10)		20	−37·0	16	2·33 × 10^5

[a] Richard (1961). [b] Hegenbarth and Frenzel (1967). [c] Polandov and Mylov (1964). [d] Umebayashi *et al.* (1967). [e] Hegenbarth and Ullwer (1967). [f] Samara (1967); Sample ~90% deuterated.

TABLE II. The ferroelectric transition pressure, p_c, for various ferroelectrics at the indicated temperature, T. Also given are the quantities p_0 and C^* defined by the equation, $\epsilon = C^*/(p - p_0)$, which describes the pressure dependence of the dielectric constant in the paraelectric phase

Ferroelectric	T (°C)	p_c (kb)	p_0 (kb)	C^* (kb)	Reference
$BaTiO_3$ (crystal No. 6)	23	19·6	18·0	$2{\cdot}77 \times 10^4$	Samara (1966)
$BaTiO_3$ (crystal No. 7)	60	10·4	9·6	$2{\cdot}90 \times 10^4$	Samara (1966)
	85	5·5	4·9	$2{\cdot}75 \times 10^4$	
$SrTiO_3$	20	—	—	$1{\cdot}25 \times 10^4$	Bosman and Havinga (1963)
	20	—	−40·0	$1{\cdot}26 \times 10^4$	Samara and Giardini (1965)
	20	—	—	$1{\cdot}25 \times 10^4$	Moreno and Gränicher (1964)
	−250 to −190	—	—	$1{\cdot}3 \times 10^4$	Hegenbarth and Frenzel (1967)
$KTaO_3$	20	—	−51·0	$1{\cdot}23 \times 10^4$	Wemple *et al.* (1966)
LHS	23	20·2	19·2	$4{\cdot}4 \times 10^3$	Samara (1968a)
TGS	55	2·3	2·3	$1{\cdot}17 \times 10^3$	Jona and Shirane (1960)
RS (lower PE phase)	22	10·5	10·5	$3{\cdot}96 \times 10^2$	Samara (1965b)
DRS (lower PE phase)	45	18·0	18·0	$1{\cdot}57 \times 10^2$	Samara (1968b)
DRS (upper PE phase)	45	1·0	1·0	$-1{\cdot}65 \times 10^2$	Samara (1968b)

magnitude of the jump in P_s at the onset of the transition. Measurements were carried out at fixed pressures as the crystal was cooled from the PE phase. They found that the square of the jump in P_s decreased linearly with pressure and extrapolated to zero at ~ 17 kb. A $\Delta P_s = 0$ at T_c of course implies that the transition is of second order.

The decrease in P_s at T_c observed by Polandov *et al.* is consistent with the results shown in Fig. 9(b), but there are quantitative differences. As will be discussed later, this decrease is also consistent with the tendency of the transition to change from first to second order with increasing pressure.

From the pressure variation of P_s at the transition and the known pressure dependences of T_c and the Curie constant C, it is possible to estimate the effects of pressure on the transition entropy and latent heat. This can be seen in the following way.

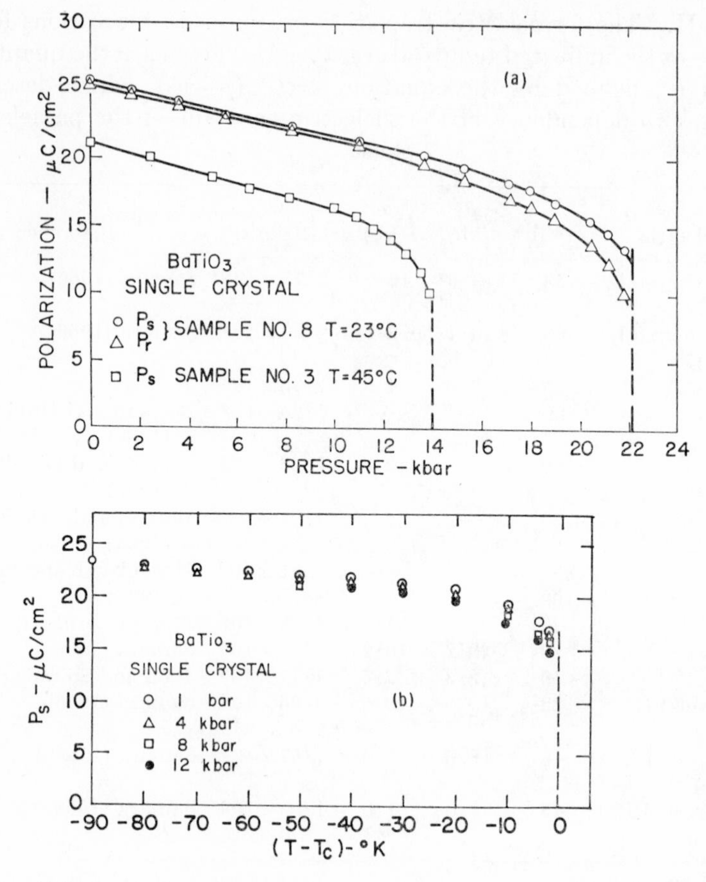

FIG. 9. (a) Pressure dependence of the spontaneous (P_s) and remanent (P_r) polarizations of single-crystal $BaTiO_3$. (b) Isobars of the spontaneous polarization of single-crystal $BaTiO_3$ as a function of the reduced temperature ($T-T_c$). Reproduced from Samara (1966) by permission of the American Institute of Physics.

The entropy is defined by (refer to eqn 23).

$$S = -(\partial A/\partial T)_P. \tag{42}$$

At any given pressure, and assuming that Λ, ξ, and ζ are independent of temperature, eqns (23) and (35) yield

$$\Delta S = S_0 - S = \frac{2\pi}{Cp} P_s^2 \tag{43}$$

where S_0 and S are the entropies of the PE and FE phases, respectively, and P_s is the jump in polarization at T_c. The latent heat Q is given by

$$Q = T_c \Delta S = \frac{2\pi}{Cp} T_c P_s^2 \tag{44}$$

from which it follows that

$$\frac{d \ln Q}{dp} = \frac{d \ln T_c}{dp} + 2\frac{d \ln P_s}{dp} - \frac{d \ln C_p}{dp}. \tag{45}$$

Since values of T_c, P_s, and C_p and their pressure derivatives vary considerably from one crystal to another, the best estimate of d ln Q/dp is obtained if all these quantities are measured on the same crystal. This has been done by Samara (1966; sample No. 6) whose data yield

$$\frac{d \ln Q}{dp} = [-1{\cdot}3 - 2(1{\cdot}2) - (-0{\cdot}7)] \times 10^{-2}/\text{kb}$$

$$= -3 \times 10^{-2}/\text{kb}$$

as compared with a value of -5×10^{-2}/kb calculated by Polandov *et al.* Most of the difference between the two values arises from the larger value of dlnP_s/dp measured by the latter authors. At 1 b $Q = 50 \pm 5$ cal/mole for $BaTiO_3$ (Jona and Shirane, 1962).

c. *Transition temperatures*

The decrease of the Curie point and the tetragonal-orthorhombic transition temperature (see below) with pressure are to be expected from a consideration of the lattice strains which occur at these transitions. Starting with the cubic phase, the onset of spontaneous polarization at the Curie point results in an elongation of the unit cell along the polar c-axis and contractions along the a-axes leading to a small increase ($\sim 1\%$) in unit cell volume. Since pressure favours the smaller volume, a stabilization of the cubic phase, or a decrease in T_c, is expected and is observed. Similarly, the volume of the orthorhombic phase is larger than that of the tetragonal phase, and thus pressure favours the latter.

Figure 10 gives a summary of the experimental data on the shift of T_c with pressure. There are large quantitative differences in the reported results, the initial slope dT_c/dp varying between ~ -4 and $-6{\cdot}7$°K/kb. The present author (Samara, 1966) observed a variation between $-4{\cdot}6$ and $-5{\cdot}9$°K/kb on eight different samples (six single crystals and two ceramics) from measurements performed in the same pressure apparatus under identical conditions. The causes of this variation are not known, but the presence of impurities and the domain

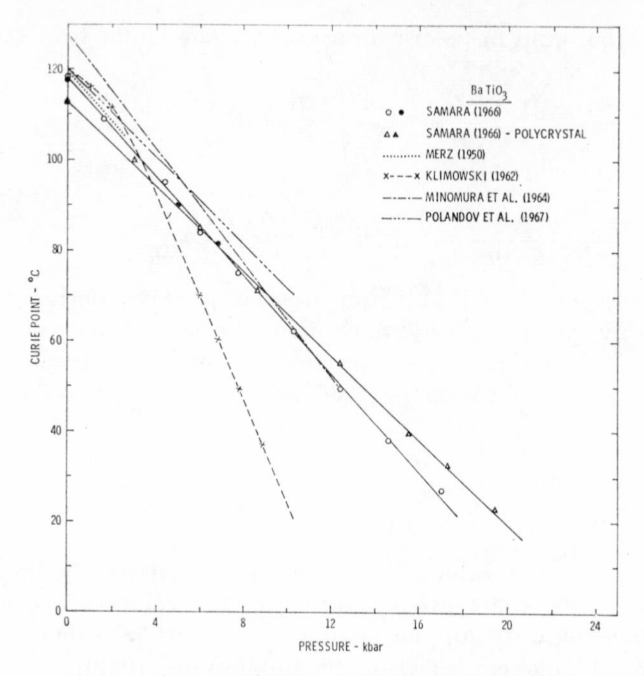

FIG. 10. Shift of the Curie point of $BaTiO_3$ with pressure according to various authors.

structure of samples must play a role. There is some evidence that (dT_c/dp) depends somewhat on whether the pressure is applied while the crystal is above or below the Curie point (Polandov *et al.*, 1967; Samara, 1966). The explanation lies in the effect of pressure on the domain structure. It has been postulated (Klimowski, 1962) that hydrostatic pressure causes plastic deformation at domain boundaries and causes *c*-domain crystals to change to mixed structures with 90° and 180° domains. Klimowski cited this as a possible explanation for the irreversible effects of pressure on ϵ which he observed in the tetragonal phase.

The effects of impurities have not been investigated systematically. Measurements on slightly Fe- and Pb-doped samples (Samara, 1966) did not establish any definite trends, possibly because of insufficient amounts of dopants in the samples used.

No explanation has been given for the non-linear pressure dependence of T_c reported by some authors for $p > 3$ kb (see Fig. 10). However, it may be re-emphasized here that, because of the intimate relation between the spontaneous polarization and the accompanying lattice strain, any deviations from hydrostatic conditions can influence the results significantly. Thus, for example, a component of stress in the c-direction superimposed on a hydrostatic stress will yield a larger dT_c/dp than would the hydrostatic stress alone.

For the tetragonal-orthorhombic transition near 5°C, Merz (1950) measured a decrease of the transition temperature of 3°K/kb up to 1·5 kb followed by an increase of 3°K/kb between 1·5 and 3 kb. Later measurements by Minomura *et al.* (1964) yielded a linear decrease of −2·8°K/kb up to 4·5 kb, and this result is substantiated by Samara's (1966) results. No measurements have been reported on the orthorhombic-rhombohedral transition near −90°C.

The three transitions in $BaTiO_3$ are of first order, and thus the shifts of the transition temperatures, T_t, with hydrostatic pressure can be expected to obey the Clausius–Clapeyron relationship (eqn 39). Approximate values of Q and ΔV at each of the three transitions have been given by Shirane and Takeda (1952). These are presented in Table III, where calculated and observed values of the pressure coefficients dT_t/dp are also compared.

Based on the above discussion, a tentative pressure–temperature phase diagram for $BaTiO_3$ is shown in Fig. 11. It is seen that the

TABLE III. Values of the transition temperatures, T_t, discontinuous volume change, ΔV, and latent heat, Q, for a number of ferroelectrics and antiferroelectrics exhibiting first-order phase transitions at T_t. Values of (dT_t/dp) calculated from the Clausius–Clapeyron eqn (39) are compared with the measured values

Ferro-electric	T_t (°C)	ΔV[a] (cm³/mole)	Q[a] (cal/mole)	$(dT_t/dp)_{calcd}$ (°C/kb)	$(dT_t/dp)^{b}_{exptl}$ (°C/kb)
$BaTiO_3$	120	−0·0374	50	−7·0	∼ −5·5
	5	−0·0084	22	−2·6	−2·8
	−90	+0·0036	8	+2·0	—
$PbTiO_3$	490	−0·162	1150	−2·6	< −8·0
$PbZrO_3$	230	+0·120	400	+3·8	+4·4
$PbHfO_3$	203	+0·085	—	—	+5·5
	160	+0·085	—	—	∼ +6·0
KH_2PO_4	−151	−0·022; −0·03	11	−5·9; −7·6	−4·5; −5·6
KD_2PO_4	−52	∼ −0·06	105	∼ −3·0	−2·6; −3·9
$NaNO_2$	165	+0·027	56	+5·0	+4·9
SbSI	20	−0·31	58	−39·0	−37·0

[a] Values of ΔV and Q were taken from Shirane and Takeda (1952) and Megaw (1957) for the pervoskites, from Reese (1968) for KH_2PO_4 and KD_2PO_4, from Gesi *et al.* (1965) for $NaNO_2$, and from Samara (1968c) for SbSI. [b] See the text for references to these values.

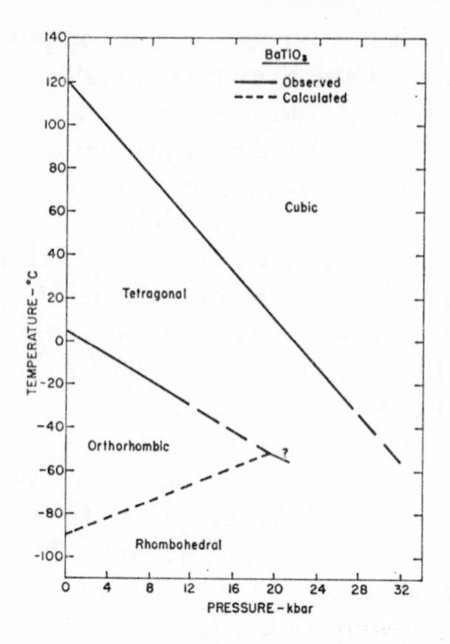

FIG. 11. Tentative temperature–pressure phase diagram for $BaTiO_3$.

range of stability of the orthorhombic phase decreases with increasing pressure and should vanish at a triple point at about 19 kb and −50°C.†

Finally, the single crystal data discussed above give strong indication that the first-order tetragonal–cubic transition in $BaTiO_3$ acquires, with increasing pressure, the characteristics of a continuous second-order transition. This point was recently discussed by Leonidova and Volk (1966) and Polandov *et al.* (1967). We recall that the experimental results discussed above show that the quantities $(T_c - T_0)$, the latent heat Q, and the jump in polarization at the transition decrease with increasing pressure and tend to zero. All these quantities should, of course, be zero for a second-order transition. In addition, the peak value of the dielectric constant, ϵ_{max}, increases rapidly with pressure. The simple thermodynamic considerations outlined in Section IV show that $\epsilon_{max} = \infty$ at a second-order transition, but this is never observed in real crystals because of thermal fluctuations, imperfections, domain effects, and so on. The experimental results shown in Figs 7–9 and more recent unpublished results by the author

† S. Minomura (private communication, October 1968) is presently investigating the effect of pressure on the orthorhombic–rhombohedral transition temperature. His preliminary results indicate that $dT_t/dp \simeq -(1\text{–}2)$°K/kb. If this negative shift turns out to be correct, then Shirane and Takeda's small positive value for ΔV must be in error.

show that the transition which is observed at ~ 20 kb and room temperature is still of first order exhibiting both temperature and pressure hysteresis.

This tendency of a first-order transition to acquire second-order characteristics with increasing pressure is in agreement with the theoretical considerations advanced by Ginzburg (1963) and Landau and Lifshitz (1962) which show that a first-order transition becomes second order at a certain critical temperature and pressure.

2. *Strontium Titanate—$SrTiO_3$*

The question of ferroelectricity in $SrTiO_3$ is still unresolved. Despite the fact that the material has been studied extensively, there is still considerable uncertainty as to its low temperature structure and dielectric properties. The crystal has the cubic perovskite structure down to 110°K at which temperature a small lattice distortion to tetragonal symmetry ($c/a \simeq 1{\cdot}00056$) occurs (Lytle, 1964). On further cooling, there is evidence of other small lattice distortions, but the nature of these is still in doubt. There are no dielectric anomalies associated with any of these distortions. The dielectric constant is well represented by a Curie–Weiss law (with $C = 8{\cdot}3 \times 10^4$°K and $T_0 = 36$°K) down to about 50°K. Below this temperature, ϵ deviates (falls below) from the Curie–Weiss law, a behaviour observed also for other pervoskites at low temperatures. This deviation is attributed to quantum mechanical effects (Barrett, 1952).

Hegenbarth (1964) reported that a maximum in the ϵ *versus* T curve of $SrTiO_3$ can be induced by the application of a d.c. bias field, and that the maximum shifts to higher temperatures with increasing field strength. The maximum occurs at 20°K for a field of 2 kV/cm and at 45°K for a field of 9 kV/cm. The maximum was explained in terms of the thermodynamic theory.

The effect of pressure, at and above room temperature, on the dielectric constant of $SrTiO_3$ has been reported by a number of authors as follows: Moreno and Gränicher (1964) to 2 kb, Bosman and Havinga (1963) to 5 kb, Samara and Giardini (1965) to 50 kb, and Samara (1966) to 14 kb. There is good agreement among the various results. One of the interesting features is that the relationship $\epsilon = C^*/(p - p_0)$, eqn (41), is well obeyed over a wide pressure range (Fig. 12). At 20°C, $C^* = 1{\cdot}26 \times 10^4$ kb and $p_0 = -40$ kb (Samara and Giardini, 1965). Other authors' values for C^*, $\equiv 1/[\partial(1/\epsilon)/\partial p]_T$, fall in the range $1{\cdot}20$–$1{\cdot}26 \times 10^4$ kb (Table II).

Recently, Hegenbarth and Frenzel (1967) reported the effect of

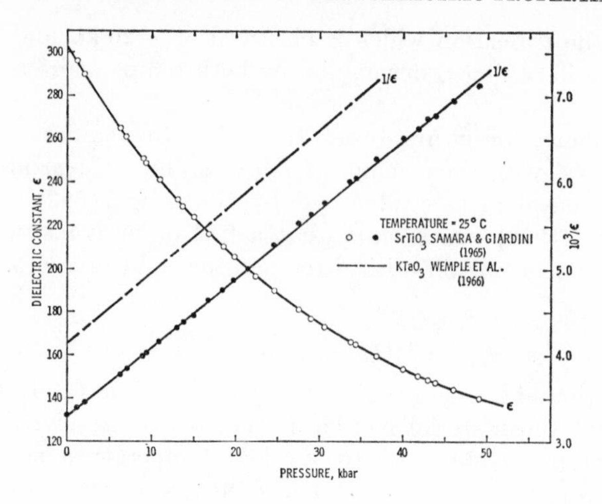

FIG. 12. The effects of pressure on the dielectric constants of single-crystal $SrTiO_3$ and $KTaO_3$.

pressure to $\sim 1{\cdot}5$ kb on ϵ of $SrTiO_3$ in the temperature range 20–80°K (Fig. 13a). Their results over this range yield $C^* = 1{\cdot}3 \times 10^4$ kb, in rather good agreement with the room temperature values cited above, thus indicating that the pressure coefficient of $1/\epsilon$ is not very temperature-dependent. The authors also investigated the effect of pressure on the field-induced maximum in the ϵ *versus* T curve. Their results, shown in Fig. 13(b), indicate that at a field of 4·1 kV/cm, T_{max} decreases linearly with pressure with slope $(dT_{max}/dp) = -16$°K/kb.

The pressure and temperature dependences of ϵ of $SrTiO_3$ will be discussed along with the $BaTiO_3$ results in subsection B below.

3. *Other Perovskites*

a. *Potassium tantalate—$KTaO_3$*

$KTaO_3$ has the ideal cubic perovskite structure down to ~ 13°K, at which temperature it undergoes a transition to the ferroelectric state (Hulm *et al.*, 1949). In the cubic PE phase ϵ obeys a Curie–Weiss law, with $C \simeq 6 \times 10^4$°K, down to ~ 50°K. Wemple *et al.* (1966) measured the effect of pressure on ϵ at room temperature and found that, as in the case of $BaTiO_3$ and $SrTiO_3$, the relationship

$$\epsilon = C^*/(p - p_0)$$

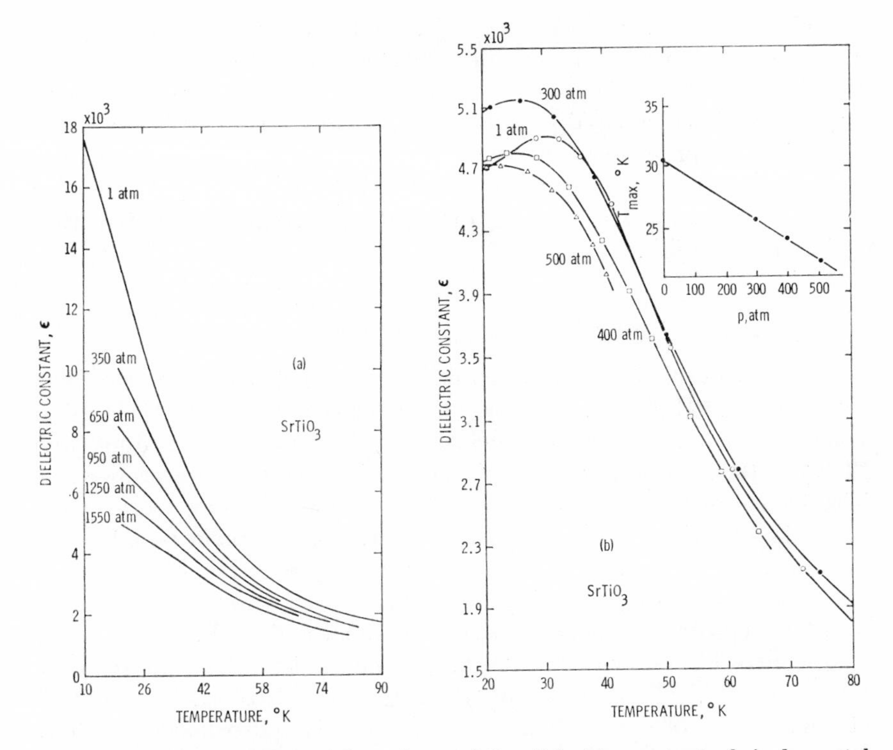

FIG. 13. (a) The temperature dependence of the dielectric constant of single-crystal $SrTiO_3$ at various pressures. (b) Isobars of the dielectric constant *versus* temperature for single-crystal $SrTiO_3$ measured at a biasing field of 4·1 kV/cm. Insert shows the shift of the peak with pressure. Reproduced from Hegenbarth and Frenzel (1967) by permission of the Iliffe Science and Technology Publications Ltd.

is well obeyed over a wide pressure range (Fig. 12). Their results yield $C^* = 1{\cdot}23 \times 10^4$ kb and $p_0 = -51$ kb, both values nearly the same as those for $SrTiO_3$ (see Table II).

b. *Lead titanate—$PbTiO_3$*

On lowering the temperature, $PbTiO_3$ transforms from the cubic PE phase to a tetragonal ($c/a = 1{\cdot}06$) FE phase at $T_c = 490$°C (Jona and Shirane, 1962). Accurate dielectric measurements near and above T_c are usually hampered by the large conductivity of the material at the high temperatures needed.

Kabalkina and Vereshchagin (1962) measured the effect of pressure to ~18 kb, on the room temperature lattice parameters of $PbTiO_3$, and observed a linear decrease of c/a. A linear extrapolation of their results suggested that the crystal becomes cubic at $p \simeq 27$ kb, and on this basis they predicted a linear decrease of the Curie point with pressure with slope $dT_c/dp \simeq -18°K/kb$. Samara (1965b) investigated the effects of pressure up to 28 kb on ϵ of a polycrystalline sample in the FE phase. No transition to the cubic PE phase was observed even at temperatures to $\sim 240°C$, but the results suggested that $-dT_c/dp < 8°K/kb$.

c. *Lead zirconate—$PbZrO_3$, and lead hafnate—$PbHfO_3$*

These two crystals have the cubic perovskite structure in their high temperature PE phases. But, unlike $BaTiO_3$ and $PbTiO_3$, they transform on cooling into antiferroelectric (AFE) phases (see Jona and Shirane, 1962, for a summary of the atmospheric pressure properties). The transition in $PbZrO_3$ occurs at 230°C and the low temperature phase is orthorhombic, but is often described in terms of pseudotetragonal axes a and c. $PbHfO_3$ transforms to a tetragonal phase at 203°C and, on further cooling, exhibits another transition at 160°C. The low temperature phase is also tetragonal and AFE.

In contrast again with $BaTiO_3$ and $PbTiO_3$, at the transitions to the "tetragonal" AFE phases in $PbZrO_3$ and $PbHfO_3$, the induced strains consist of contractions along the c-axis and elongations along the a-axes (that is, $c/a < 1{\cdot}0$) with a resulting decrease in unit cell volumes. Pressure would then be expected to favour the lower volume AFE phases, and the Curie or transition temperatures, T_c, should increase with increasing pressure. Indeed, this is found to be the case.

The effects of pressure on the Curie points were measured by Samara (1965b) for both $PbZrO_3$ and $PbHfO_3$ and by Rapoport (1966) for $PbZrO_3$. In the case of $PbHfO_3$ both transition temperatures increase with pressure. In Table III, experimental values of dT_c/dp are compared with values calculated using the Clausius–Clapeyron eqn (39)—the transitions are of first order.

d. *Mixed compounds*

There have been a number of measurements on the effects of pressure on the FE–PE transition temperatures of mixed perovskites such as $Ba_xSr_{1-x}TiO_3$ and $Ba(Ti_xZr_{1-x})O_3$. These results will not be discussed here, suffice it to say that in all cases the Curie points decrease with pressure by ~ 4–7°K/kb. A few values are listed in Table I.

B. THEORETICAL CONSIDERATIONS

Most of the model theories which have been advanced to explain the ferroelectric properties of the perovskites start with the treatment of the dielectric constant, ϵ, of the nonpolar cubic phase and its variation as conditions approach the transition temperature. As we have seen, in this phase ϵ increases with decreasing temperature according to the Curie–Weiss law $\epsilon = C/(T - T_0)$. In the early theories of Devonshire (1949) and Slater (1950), anharmonic restoring forces in the lattice potential are held responsible for this temperature dependence of ϵ, and more recent theories (Cochran, 1960; Anderson, 1960) have related it to the temperature dependence of a long wavelength transverse optic phonon mode of the lattice. In the nonpolar phase the frequency of this mode, ω_T, decreases with decreasing temperature, and $\omega_T \to 0$ as $T \to T_0$, whereby the lattice becomes unstable and a transition to the ferroelectric state occurs. In a strictly harmonic theory, of course, the polarizability and all lattice frequencies are rigorously temperature-independent.

As mentioned in Section II, one of the important considerations in the theoretical treatment of dielectric properties is the relationship between the macroscopic dielectric constant (and refractive index) and the microscopic polarizabilities of the individual ions. Directly related to this is the question of the value of the internal, or local, field at each lattice site. The first detailed treatment of this problem for the perovskite lattice, specifically $BaTiO_3$, was given by Slater (1950). He assumed a small temperature dependence for the total polarizability (see eqn (50) below) resulting from the displacement of the Ti ion in an anharmonic potential. The local field at each site was calculated under the assumption that each ion has an electronic (or optical) polarizability, but only the Ti ion has an ionic (or infrared) polarizability (cf Section II for the distinction between these polarizabilities). It was found that the local field at the Ti ion is much greater than the ordinary Lorentz field, and, according to Slater, it is this large field which leads to ferroelectricity in $BaTiO_3$. Triebwasser (1957) later extended the treatment by including the ionic polarizabilities of the O and Ba ions and calculated the local fields at the displaced positions of all the ions.

Ideally, a discussion of the temperature and pressure dependences of the dielectric constant should involve the evaluation of the effects of temperature and pressure on the individual polarizabilities. The pressure effects, at constant temperature, make it possible to separate the temperature dependence, at constant pressure, of each quantity into its pure volume-dependent and volume-independent (or pure temperature) contributions. The latter contributions are determined completely by anharmonicities.

Unfortunately, the determination of the individual polarizabilities and their temperature and pressure dependences is a complex problem which cannot be solved completely. This was recently discussed by Lawless and Gränicher (1967) who, on the basis of a number of simplifying assumptions, presented an approximate treatment for $BaTiO_3$ and $SrTiO_3$. We shall summarize their results below; however, before doing so we shall first find it instructive and useful to treat the results on a macroscopic scale. The overall results will then be related to the lattice dynamical theory.

1. *Polarizability and Internal Field*

a. *Macroscopic treatment*

From the discussion in Section II recall that the relationship between the dielectric constant ϵ and the total polarizability α of a macroscopic element of the crystal of volume V is

$$\frac{\epsilon-1}{1+\phi(\epsilon-1)}=4\pi\left(\frac{\alpha}{V}\right). \tag{46}$$

Differentiating eqn (46) with respect to temperature at constant pressure (assuming ϕ is independent of temperature) yields

$$\frac{1}{(\epsilon-1)[1+\phi(\epsilon-1)]}\left(\frac{\partial\epsilon}{\partial T}\right)_p=\beta\left(\frac{\partial\ln\alpha}{\partial\ln V}\right)_T-\beta+\left(\frac{\partial\ln\alpha}{\partial T}\right)_V \tag{47}$$

where $\beta=(\partial\ln V/\partial T)_p$ is the volume coefficient of thermal expansion. The first term on the right-hand side of eqn (47) represents the change in the polarizability of a given number of particles with change in volume. The second term $(-\beta)$ represents the contribution due to the decrease in the density of polarizable particles with increasing temperature. These two terms constitute the total contribution due to volume expansion. The third term is the change of the polarizability with temperature at constant volume, that is a pure temperature effect.

One would now like to know the relative contribution of each of these three terms to the temperature dependence of ϵ in the cubic phase. In eqn (47), the left-hand side can be evaluated from the measurement of ϵ *versus* T, provided that ϕ is known. Of the terms on the right-hand side, β can be evaluated from thermal expansion data, but there remain two unknowns: $(\partial \ln \alpha/\partial \ln V)_T$ and $(\partial \ln \alpha/\partial T)_V$. The first of these unknowns can be calculated from the measured pressure dependence of ϵ and the compressibility as follows. Differentiating eqn (46) with respect to pressure at constant temperature (assuming ϕ is independent of pressure) gives

$$\frac{1}{(\epsilon-1)[1+\phi(\epsilon-1)]}\left(\frac{\partial \epsilon}{\partial p}\right)_T = \left(\frac{\partial \ln \alpha}{\partial p}\right)_T + \kappa = -\kappa\left(\frac{\partial \ln \alpha}{\partial \ln V}\right)_T + \kappa \qquad (48)$$

where $\kappa = -(\partial \ln V/\partial p)_T$ is the volume compressibility of the material.

The main remaining question, then, concerns the appropriate value of the internal field coefficient ϕ to use in eqn (46). Since the perovskites are cubic in their PE phase, the *macroscopic* Clausius–Mossotti relationship (i.e. $\phi = 1/3$) should be applicable (see Section II). However, von Hippel *et al.* (1963), on the basis of semi-empirical arguments, have indicated that for the perovskites ϕ may be some two or three orders of magnitude smaller than the Lorentz factor of 1/3, and, in particular, $\phi \simeq 2{\cdot}8 \times 10^{-3}$ for $BaTiO_3$ and $4{\cdot}5 \times 10^{-4}$ for $SrTiO_3$. These values are arrived at by assuming for the polarizability a temperature dependence of the form

$$(\alpha/V) = c/T \qquad (49)$$

where c is a constant. Such a dependence can be easily shown to lead to a Curie–Weiss behaviour for the susceptibility with the result

$$\chi = (T_0/\phi)/(T - T_0). \qquad (50)$$

ϕ is then evaluated from $1/\chi$ *versus* T plots.

The applicability of such small values of ϕ for the perovskites seems somewhat doubtful. Slater (1950) had earlier argued against a temperature-dependence for α as given by eqn (49), and instead he, as did Devonshire (1949), assumed a slow decrease of (α/V) with temperature of the form

$$\frac{4\pi}{3}\left(\frac{\alpha}{V}\right) = 1 - c'\,(T - T_0). \qquad (51)$$

This linear temperature-dependence of α is obtained by introducing anharmonic terms into the potential energy of the lattice, and it also leads to a Curie–Weiss behaviour ($\phi = 1/3$) for χ.

In order to evaluate the various contributions to the temperature dependence of ϵ and to examine the effect of changing ϕ, in Table IV the various quantities in eqns (47) and (48) are tabulated for $BaTiO_3$ and $SrTiO_3$ for two different values of ϕ: (1) $\phi = 1/3$ and (2) ϕ as empirically evaluated by von Hippel and co-workers. An examination of the results

TABLE IV. The various contributions to the temperature dependence of the dielectric constants of $BaTiO_3$ and $SrTiO_3$ (defined by eqn (47)) calculated for two cases of the internal-field coefficient ϕ. Various quantities appearing in eqns (47) and (48) are also listed. (Data taken from Samara, 1966)

Compound	T (°C)	ϵ	$\frac{1}{K}\left(\frac{\partial \epsilon}{\partial p}\right)_T$[a] ($kb^{-1}$)	κ (kb^{-1})	$\left(\frac{\partial \ln \alpha}{\partial \ln V}\right)_T$	$\frac{1}{K}\left(\frac{\partial \epsilon}{\partial T}\right)_p$ (°K^{-1})	$\beta\left(\frac{\partial \ln \alpha}{\partial \ln V}\right)_T$ (°K^{-1})	$-\beta$ (°K^{-1})	$\left(\frac{\partial \ln \alpha}{\partial T}\right)_V$ (°K^{-1})
Case (i): $\phi = 1/3$									
$BaTiO_3$	150	2920	$-10 \cdot 8 \times 10^{-5}$	78×10^{-5}	$1 \cdot 13$	$-2 \cdot 0 \times 10^{-5}$	$5 \cdot 65 \times 10^{-5}$	$-5 \cdot 0 \times 10^{-5}$	$-2 \cdot 65 \times 10^{-5}$
$SrTiO_3$	25	307	$-24 \cdot 6 \times 10^{-5}$	57×20^{-5}	$1 \cdot 43$	$-3 \cdot 6 \times 10^{-5}$	$4 \cdot 30 \times 10^{-5}$	$-3 \cdot 0 \times 10^{-5}$	$-4 \cdot 90 \times 10^{-5}$
Case (ii): $\phi = 2 \cdot 8 \times 10^{-3}$ for $BaTiO_3$ and $4 \cdot 5 \times 10^{-4}$ for $SrTiO_3$									
$BaTiO_3$	150	2920	$-11 \cdot 4 \times 10^{-3}$	$0 \cdot 78 \times 10^{-3}$	$15 \cdot 6$	$-2 \cdot 10 \times 10^{-3}$	$0 \cdot 78 \times 10^{-3}$	$-0 \cdot 05 \times 10^{-3}$	$-2 \cdot 83 \times 10^{-3}$
$SrTiO_3$	25	307	$-22 \cdot 0 \times 10^{-3}$	$0 \cdot 57 \times 10^{-3}$	$39 \cdot 5$	$-3 \cdot 30 \times 10^{-3}$	$1 \cdot 19 \times 10^{-3}$	$-0 \cdot 03 \times 10^{-3}$	$-4 \cdot 46 \times 10^{-3}$

[a] $K = (\epsilon - 1)[1 + \phi(\epsilon - 1)]$.

in the table shows that the pure temperature contribution, that is $(\partial \ln \alpha/\partial T)_V$, is the predominant factor in determining the temperature dependence of the dielectric constant. This conclusion is not qualitatively affected by the value of the internal field coefficient chosen. The two volume contributions $\beta(\partial \ln \alpha/\partial \ln V)_T$ and $-\beta$ have opposite signs and for the case $\phi = 1/3$ are of comparable magnitude. For the case of small ϕ, the thermal expansion contribution, $-\beta$, is negligible. As may be expected *a priori*, the term $(\partial \ln \alpha/\partial \ln V)_T$ is positive, signifying that the ions become more polarizable as the volume they occupy in the crystal increases.

This analysis can be carried a little further. The total macroscopic polarizability of the material, α, can be separated into two components, as done in Section II,

$$\alpha = \alpha_{op} + \alpha_{ir}. \tag{52}$$

Since we are discussing only the nonpolar cubic phase, there are no permanent dipoles, and the dipolar contribution α_d (see eqn (6)) is zero.

With increasing pressure the ions in the lattice get closer together and, as a result, the elastic restoring forces increase. Therefore, α_{ir}, which arises from the ionic displacements and the resulting displacement of electronic charge distributions, can be expected to decrease. *A priori*, the pressure dependence of α_{op} (which arises from electronic displacements alone) is more difficult to predict since the amount of overlap between electronic wave functions is quite sensitive to changes in interatomic distance.

α_{op} is related to the high frequency dielectric constant n^2 (n = refractive index) by a relationship similar to eqn (46) with ϵ replaced by n^2 and α by α_{op}, and similarly for the temperature and pressure coefficients in eqns (47) and (48). Data on the pressure and temperature dependence of n are available only for $SrTiO_3$. Analysis of these results (see Samara, 1966) for the case $\phi = 1/3$ yields

$$(\partial \ln \alpha_{op}/\partial T)_V = -0{\cdot}48 \times 10^{-5}/°K$$

and

$$(\partial \ln \alpha_{ir}/\partial T)_V = -12{\cdot}1 \times 10^{-5}/°K.$$

In evaluating these contributions use was made of the relationship

$$\left(\frac{\partial \ln \alpha}{\partial T}\right)_V = \frac{\alpha_{op}}{\alpha}\left(\frac{\partial \ln \alpha_{op}}{\partial T}\right)_V + \frac{\alpha_{ir}}{\alpha}\left(\frac{\partial \ln \alpha_{ir}}{\partial T}\right)_V \tag{53}$$

which follows immediately from eqn (52).

The above considerations clearly show that the change in α_{ir} is the important factor in determining the temperature and pressure dependences of the dielectric constant of $SrTiO_3$; the change in α_{op} has a much smaller effect. The situation is undoubtedly similar for $BaTiO_3$ and other perovskites.

b. *Microscopic treatment*

Lawless and Gränicher (1967) presented an approximate microscopic treatment for cubic $BaTiO_3$ and $SrTiO_3$ based on a generalized internal-field theory involving the total polarizabilities (that is $\alpha_{op}+\alpha_{ir}$) of the Ti and Ba (or Sr) ions and only the electronic polarizability (α_{op}) of the O ions.

From the definition (see eqn (2)), the macroscopic dielectric constant is written as

$$\epsilon-1=4\pi P/E=4\pi \sum_i P_i/E \equiv \sum_i \theta_i \tag{54}$$

where P_i is the polarization at the lattice site i and $\theta_i=4\pi P_i/E$. The local field at this ith site is written as

$$F_i=E+\sum_j (4\pi/3+T_{ij})P_j. \tag{55}$$

Comparison with eqn (5) shows that $(4\pi/3+T_{ij})\equiv 4\pi\phi_{ij}$ in our notation. The T_{ij}'s are the internal field coefficients which are determined by the lattice geometry. They are given by Slater (1950) for the perovskite lattice. The polarizability α_i is defined by (cf. eqn (4))

$$F_i=(4\pi\tau/\alpha_i)P_i \tag{56}$$

where τ is the unit cell volume.

The model allows the independent determination, from experimental data, of the polarizabilities of two of the ions provided that the polarizability of the third ion is given (the polarizabilities of the three oxygen ions are equal, by symmetry). Lawless and Gränicher determined the temperature variation of the polarizabilities parametrically in α_O, that is at each temperature α_O was assumed known, and α_{Ba} (or α_{Sr}) and α_{Ti} were calculated. By taking limiting values of $\alpha_O(T)$, the allowed range of values for the other $\alpha(T)$'s were determined. The interesting feature of the results is that, although one cannot uniquely assign a value for any one of the α's, a unique and rather narrow range of values for that α can be assigned. However, one surprising and somewhat puzzling result is that $\alpha_{Ba}(T)$ and $\alpha_{Sr}(T)$ behave differently: α_{Ba} increases where α_{Sr} decreases with increasing temperature. The cause of this difference, if real, is not understood.

The determination of the explicit temperature and volume dependences of the individual polarizabilities from the calculated $\alpha(T)$

data is not possible. However, by making other simplifying assumptions and by comparing the various contributions to the temperature and pressure dependences of ϵ and n (refractive index), it is possible to draw some useful conclusions. In terms of the individual polarizabilities, the T and p dependences of ϵ are given by (cf. eqns (47) and (48))

$$\left(\frac{\partial \epsilon}{\partial T}\right)_p = -\frac{\beta}{\kappa}\left(\frac{\partial \epsilon}{\partial p}\right)_T + \frac{C'_{\mathrm{O}}}{\tau}\left[\left(\frac{\partial \alpha_{\mathrm{O}}}{\partial T}\right)_V + \frac{C'_{\mathrm{Ti}}}{C'_{\mathrm{O}}}\left(\frac{\partial \alpha_{\mathrm{Ti}}}{\partial T}\right)_V + \frac{C'_{\mathrm{Ba}}}{C'_{\mathrm{O}}}\left(\frac{\partial \alpha_{\mathrm{Ba}}}{\partial T}\right)_V\right] \quad (57)$$

and

$$\left(\frac{\partial \epsilon}{\partial p}\right)_T = \kappa \sum_j \alpha_j C'_j - \kappa C'_{\mathrm{O}}\left[\left(\frac{\partial \alpha_{\mathrm{O}}}{\partial V}\right)_T + \frac{C'_{\mathrm{Ti}}}{C'_{\mathrm{O}}}\left(\frac{\partial \alpha_{\mathrm{Ti}}}{\partial V}\right)_T + \frac{C'_{\mathrm{Ba}}}{C'_{\mathrm{O}}}\left(\frac{\partial \alpha_{\mathrm{Ba}}}{\partial V}\right)_T\right] \quad (58)$$

where

$$C'_j = \tau(\partial \epsilon/\partial \alpha_j) = \tau \sum_i (\partial \theta_i/\partial \alpha_j). \quad (59)$$

Lawless and Gränicher neglected the terms $(\partial \alpha_{\mathrm{Ba}}/\partial T)_V$ and $(\partial \alpha_{\mathrm{Ba}}/\partial V)_T$, and similarly for α_{Sr}, because the ratio $C'_{\mathrm{Ba}}/C'_{\mathrm{O}}$ was found to be small. Unfortunately, even then the explicit T and V dependences of α_{O} and α_{Ti} cannot be evaluated. Instead, the authors evaluated the quantities $[(\partial \alpha_{\mathrm{O}}/\partial T)_V + (C'_{\mathrm{Ti}}/C'_{\mathrm{O}})(\partial \alpha_{\mathrm{Ti}}/\partial T)_V]$ and $[(\partial \alpha_{\mathrm{O}}/\partial V)_T + (C'_{\mathrm{Ti}}/C'_{\mathrm{O}})(\partial \alpha_{\mathrm{Ti}}/\partial V)_T]$. The first of these was found to be negative and the second positive for both $BaTiO_3$ and $SrTiO_3$. In addition, particularly for $BaTiO_3$, the first quantity dominates $(\partial \epsilon/\partial T)_p$ because of the near cancellation of the pure volume dependent terms. These conclusions are the same as those arrived at earlier from the macroscopic treatment (see Table IV).

On the further assumption that $d\alpha/dT$ from the optical (refractive index) data can be associated solely with the oxygen polarizability, that is with $d\alpha_{\mathrm{O}}/dT$, Lawless and Gränicher argue that, in the case of $BaTiO_3$, the term $[(\partial \alpha_{\mathrm{O}}/\partial T)_V + (C'_{\mathrm{Ti}}/C'_{\mathrm{O}})(\partial \alpha_{\mathrm{Ti}}/\partial T)_V]$ is dominated by the contribution $(\partial \alpha_{\mathrm{Ti}}/\partial T)_V$, and thus implying that α_{O} exhibits a very weak or no explicit temperature dependence. This, it should be recalled, is the essence of the Slater model according to which it is the explicit T dependence of the ionic polarizability of the Ti ion (arising from anharmonicities in the potential energy) which is the dominant term in determining $(\partial \epsilon/\partial T)_p$.

Lawless and Gränicher find that the $SrTiO_3$ results are more difficult to interpret, because the cancellation of the pure volume-dependent contributions to $(\partial \epsilon/\partial T)_p$ is not nearly as complete as in the case of $BaTiO_3$. This feature is also reflected in the results given in Table IV.

2. *Lattice Dynamical Theory*

The conclusion that the explicit T-dependence of the ionic (infrared) polarizabilities is the predominant contribution to $(\partial\epsilon/\partial T)_p$, is important from the standpoint of the lattice dynamical theory of ferroelectricity first advanced by Cochran (1960) and Anderson (1960). In this theory the PE→FE transition in the perovskites is considered to be the result of an instability of the crystal lattice against a transverse optic mode of long wavelength (at $\mathbf{q}=0$) caused by the near cancellation of short- and long-range forces.

The situation can be illustrated simply for the case of a diatomic cubic crystal. For this case, the expressions for the frequencies of the transverse, ω_T, and longitudinal, ω_L, optic modes of zero wave vector ($\mathbf{q}$) are, according to Cochran,

$$\mu\omega_T^2 = R_0' - \frac{4\pi(\epsilon_\infty+2)(z'e)^2}{9v} \tag{60}$$

and

$$\mu\omega_L^2 = R_0' + \frac{8\pi(\epsilon_\infty+2)(z'e)^2}{9\epsilon_\infty v}. \tag{61}$$

Here μ is the reduced mass, $\epsilon_\infty(=n^2)$ is the optical dielectric constant, $z'e$ is the effective ionic charge, v is the unit cell volume, and R_0' is the restoring short-range force. It should be noted that the close cancellation of the forces which determine ω_T, does not imply any such cancellation for ω_L.

Since lattice vibrations are not completely harmonic, Cochran assumes that near $T=T_0$,

$$\omega_T^2 = \beta^*(T-T_0) \tag{62}$$

where β^* is a constant temperature coefficient and T_0 is the temperature at which $\omega_T=0$, and the lattice becomes unstable.

Of prime importance in the theory is the Lyddane–Sachs–Teller (LST) (1941) relationship which for a diatomic lattice is

$$\epsilon/\epsilon_\infty = (\omega_L/\omega_T)^2. \tag{63}$$

Hence, the Curie–Weiss behaviour of ϵ reflects the temperature dependence of ω_T, that is

$$(1/\epsilon) \propto \omega_T^2 \propto (T-T_0). \tag{64a}$$

Cochran generalized the derivation of the LST relationship to apply to the cubic perovskite structure with the result

$$\frac{\epsilon}{\epsilon_\infty} = \frac{(\omega_2\omega_3\omega_4)_L^2}{(\omega_2\omega_3\omega_4)_T^2}, \tag{65}$$

where the frequencies, ω_2, ω_3 and ω_4 represent the three infrared active vibrations of the lattice. The absorption frequencies $(\omega_3)_T$ and $(\omega_4)_T$ have been measured for both $BaTiO_3$ and $SrTiO_3$ and show no unusual temperature dependences (Spitzer *et al.*, 1962). Thus, according to the theory, $(\omega_2)_T$ exhibits the anomalous temperature behaviour, and

$$(1/\epsilon) \propto (\omega_2)_T^2 \propto (T - T_0). \tag{64b}$$

Experimental support for this T dependence of $(\omega_2)_T$ in $BaTiO_3$, $SrTiO_3$ and $KTaO_3$ has come from far-infrared dielectric dispersion (Barker and Tinkham, 1962; Spitzer *et al.*, 1962) and inelastic neutron scattering (Cowley, 1962; Shirane *et al.*, 1967a, b) studies.

In analogy with eqn (64b) the linear pressure dependence of $1/\epsilon$ (eqn 41) suggests that $(\omega_2)_T$ should vary with pressure as

$$(1/\epsilon) \propto (\omega_2)_T^2 \propto (p - p_0). \tag{66}$$

This also follows from eqn (64b) since T_0 varies linearly with pressure.

Although there are no data on the pressure dependence of $(\omega_2)_T$, some supporting evidence for the applicability of eqn (66) comes from the electrical conductivity measurements of Wemple *et al.* (1966; see also Di Domenico *et al.*, 1968). These authors measured the effect of pressure on the conductivity, σ, of several cubic perovskites and observed a large linear increase (similar to that for $1/\epsilon$ *versus* p). This change in σ could not be explained on the basis of repopulation effects or change in the effective mass and was thus attributed to a strongly pressure-dependent electron relaxation time. Lattice scattering is dominant in these materials and this scattering is said to be due mainly to the long wavelength transverse optic mode, that is the FE mode $(\omega_2)_T$. In the perovskites the electron mean free path is of the order of the lattice parameter and so electrons can be scattered by the local distortions of the unit cell produced by $(\omega_2)_T$.

Wemple and co-workers' (1966) data showed that the mobility, μ_e, can be represented by

$$\mu_e = \mathrm{f}(T)[(1/\epsilon) + B] \tag{67}$$

where $\mathrm{f}(T)$ is a temperature dependent factor, and B is a constant. Thus, they relate μ to $1/\epsilon$, and it follows that

$$\sigma \propto \mu_e \propto (\omega_2)_T^2. \tag{68}$$

VI. Hydrogen-Bonded Ferroelectrics of the KH_2PO_4-Type

This group of ferroelectrics consists of hydrogen (H)-bonded crystals in which the motion of the proton in the H-bond plays a key and reasonably well understood role in the onset of the ferroelectric transition. This is evidenced by a remarkably large isotope shift of the transition temperature observed on replacing the protons by deuterons. Thus, for example, the Curie point of KH_2PO_4, which is the most important member of this group, increases from 122 to 221°K on complete deuteration. The length of the H-bond turns out to be crucial in determining the dielectric properties of these substances and, as we shall see later, pressure is a particularly important variable in studying these properties.

In this section we shall discuss the effects of hydrostatic pressure on the ferroelectric properties of KH_2PO_4 and $LiH_3(SeO_3)_2$ and their deuterated analogues. Whereas KH_2PO_4 is the classic H-bonded ferroelectric whose properties are well known, $LiH_3(SeO_3)_2$ is a relatively new ferroelectric which exhibits features quite similar to those of KH_2PO_4. In particular, their properties are similarly affected by pressure.

A. Potassium Di-Hydrogen Phosphate

1. *Introductory Remarks*

Potassium di-hydrogen phosphate, KH_2PO_4 (abbreviated KDP), is one of the oldest known ferroelectrics. Its FE properties were discovered in 1935 by Busch and Scherrer. Since then much experimental and theoretical work has been reported on its atmospheric pressure properties, and this is adequately covered in the reviews and texts cited in the introduction.

At room temperature and pressure it crystallizes in a tetragonal non-centrosymmetric phase, space group $I\bar{4}2d$. At its Curie point ($T_c = 122$°K) it transforms to an FE phase with orthorhombic symmetry, space group Fdd2. The polar axis lies along the c-axis of the tetragonal phase. The spontaneous polarization and the dielectric constant exhibit the characteristic temperature dependence shown in Fig. 14. At 100°K, that is 22°K below the transition point, $P_s = 4{\cdot}7$ μC/cm^2, a value about one-fifth that of $BaTiO_3$ at 20°C. In the PE tetragonal phase the dielectric constant along the c-axis obeys, over a 50°K range above T_c, the Curie–Weiss law $\epsilon_c = C/(T - T_0)$ with $C = 3250$°K and $T_0 \simeq T_c$.

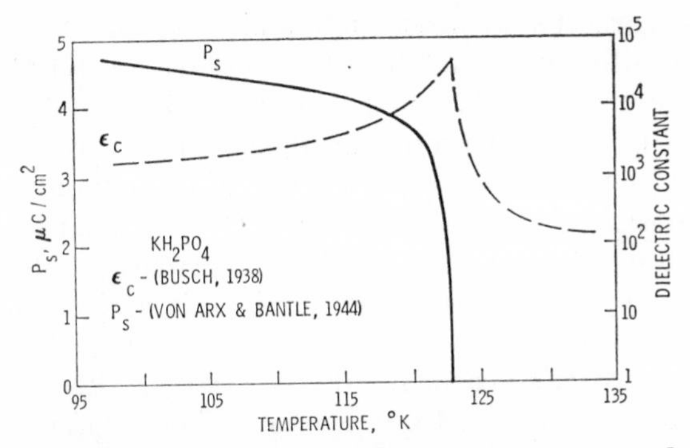

FIG. 14. The temperature dependence of the dielectric constant, ϵ_c, and the spontaneous polarization, P_s, of KH_2PO_4 at atmospheric pressure.

2. *Structural and Theoretical Considerations*

Considerable insight into the nature of the FE transition in KDP can be gained from a consideration of its crystal structure which is shown in Fig. 15. The most characteristic feature of this structure is that it consists of tetrahedral phosphate (PO_4) groups connected by a system of short O–H $\cdots$ O hydrogen bonds. K and P ions alternate each other along c at a distance $c/2$. The PO_4 tetrahedra are spaced $c/4$ apart along c and are connected in such a way that there is an H-bond between one " upper " oxygen of one tetrahedron and one " lower " oxygen of the neighbouring tetrahedron. The O–H $\cdots$ O bonds lie in planes very nearly perpendicular to the c-axis.

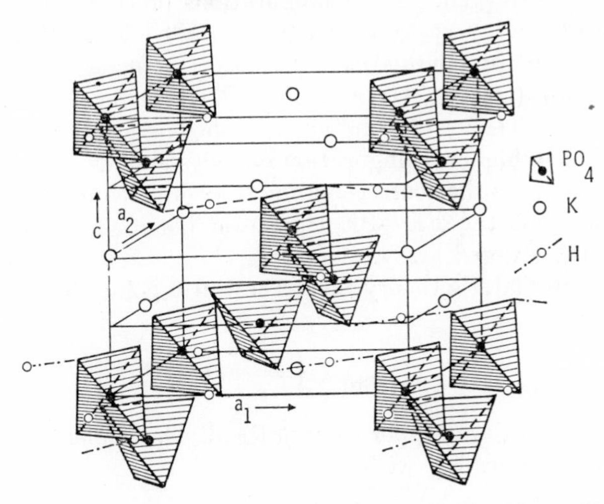

FIG. 15. The crystal structure of KH_2PO_4. (After West, 1930.)

One of the earliest and most successful molecular theories of ferroelectricity was that proposed by Slater (1941) for KDP. According to this theory the ferroelectric transition is triggered by an order–disorder transition of the proton arrangement in the H-bonds. It is assumed that each proton can occupy one of two symmetrically placed equilibrium positions along the O–H $\cdots$ O bond, that is a symmetric double-minimum potential well. The distribution of the protons between the two positions, or energy minima, is subject to two restricting conditions: (1) There is one and only one H in each bond, and (2) there are only two H's near any one PO_4 group. From (2) it follows that the predominant structural configuration consists of $(H_2PO_4)^-$ and K^+ ions, and that there are six possible configurations for each H_2PO_4 group as illustrated in Fig. 16. Of these, (a) and (b) correspond

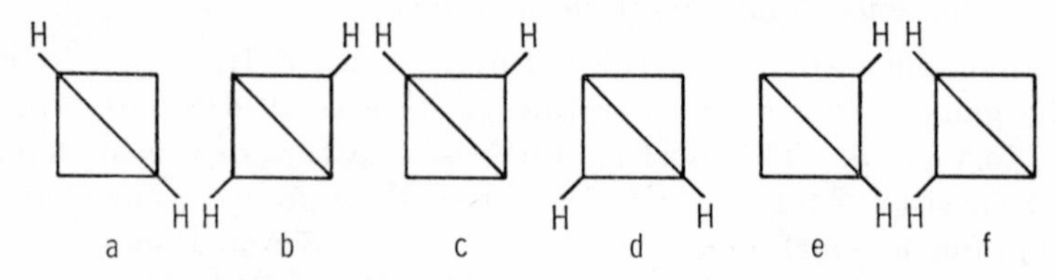

FIG. 16. Schematic representation of the six possible orientations of two protons around each PO_4 tetrahedron. The polar c-axis is perpendicular to the paper.

to the H's being nearest to both " upper " or both " lower " oxygens. In each of the remaining four, the H's are nearest to an " upper " and a " lower " oxygen. The configurations (a) and (b) are taken to be the lowest energy state with energy $\varepsilon = 0$. The other four configurations have an energy ε_0. All other configurations in which more than two (i.e. H_3PO_4 and $(H_4PO_4)^+$) or less than two (i.e. $(HPO_4)^{2-}$ and $(PO_4)^{3-}$) protons are attached to each PO_4 group are assumed to have higher configuration energies and are neglected. The FE state is taken as that corresponding to complete order of the proton arrangement, and the FE$\rightarrow$PE transition results from a change to disorder of the ordered array.

According to Slater's theory, near the Curie point the susceptibility χ is given by

$$\chi = \frac{1}{k \ln 2} \cdot \left(\frac{N\mu^2}{T - T_0} \right) \tag{69}$$

where N is the density of dipoles and μ the dipole moment per molecule. This is a Curie–Weiss law with

$$C/4\pi = N\mu^2/k \ln 2 \tag{70}$$

and $$T_0 = (C/4\pi)(\varepsilon_0/N\mu^2). \tag{71}$$

The transition entropy is given by:

$$\Delta S = \tfrac{1}{2}Nk \ln 2 = 0{\cdot}69 \text{ cal/mole} - {}^\circ\text{K}. \tag{72}$$

The proton ordering scheme proposed by Slater received beautiful confirmation years later from the neutron diffraction results of Bacon and Pease (1955), whose results show that in the FE phase the protons are indeed ordered and are asymmetrically placed between the oxygens. Further, the protons shift to mirror-image positions on reversing the polarization. In the PE phase the proton density is symmetrical and elongated along the O – – – O axes. This can result from the proton being symmetrically placed with a very high zero-point vibrational amplitude, or from the averaging of a situation in which there are two potential minima with the proton spending equal time in both of them.

The Slater theory can provide a good account for the observed entropy change at the transition and predicts a first-order transition. The order of the transition has been the subject of considerable controversy. Experimentally (see, for example, Fig. 14), the transition appears to be either diffuse first order, or very sharp second order; but it should be noted that crystal strains and defects can affect the sharpness of a first-order transition and make it look like second-order. Recently, very careful calorimetric studies by Reese (1968) on a single crystal of KDP have indicated that the transition is indeed first order with a latent heat of 11 cal/mole.

Agreement between theory and experiment can be improved by taking into account some of the proton configurations neglected by Slater. This is done, for example, in the treatments of Takagi (1948) and Silsbee *et al.* (1964). The inclusion of the high energy configurations can lead to either a first- or a second-order transition depending on the choice of the higher configuration energies. However, the most serious difficulty with the model is its inability to provide an explanation for the isotope effect.

In order to explain the large isotope shift in T_c, Pirenne (1949) proposed a model in which the kinetic energy of the proton plays an important role. This model was further developed, mostly by Blinc and co-workers (1960, 1966), into what is now known as the proton tunnelling model. According to this model, in the FE phase the protons are ordered in the double minimum potential wells along the H-bonds, whereas in the PE phase the proton distribution is symmetrical along the bonds. This results from a statistical averaging of a situation in which the protons tunnel across the energy barrier between the two potential minima. The isotope shift is believed to be a quantum effect caused by the lower tunnelling probability of the deuteron, that is a higher transition temperature.

Although this model can explain the isotope shift, it has been criticized (Tokunaga and Matsubara, 1966; Kobayashi, 1968) on the grounds that it treats only the proton system and does not incorporate satisfactorily the proton–lattice coupling. The evidence for strong proton–lattice coupling is quite ample. The displacements of the protons in the H-bonds are at right angles to the resulting polarization, and, therefore, they are not directly responsible for the polarization. Rather, in the FE phase the ordered arrangement of the protons exerts forces on the other ions which result in their displacement. In fact, the measured polarization can be well accounted for by the displacements of the K^+, P^{5+} and O^{2-} ions along the *c*-axis relative to their equilibrium positions in the PE phase (see, for example, Jona and Shirane, 1962). Further evidence comes from the fact that there is a latent heat (Reese, 1968) and a sudden volume change (Cook, 1967) at the transition, that is the transition is first order. These features emphasize that the transition in KDP is not pure order–disorder, but rather that it is a "mixed" type involving both an order–disorder arrangement of the protons and an accompanying deformation of the lattice. This point has been recently emphasized by the theoretical treatments of Tokunaga and Matsubara (1966) and Kobayashi (1968), both of which incorporate proton–lattice coupling as essential for a complete description of the FE transition in KDP. This will be considered in more detail later in the discussion of the experimental results.

3. *Results of Pressure Experiments*

Until very recently no results were available on the effects of pressure on the FE properties of KDP. This is not unrelated to the fact that its Curie temperature is relatively low, and thus the difficulties associated with *hydrostatic* pressure generation at low temperatures. However, during the last year or so, the pressure dependence of the Curie temperatures and dielectric properties of KDP and KD_2PO_4 were reported by three different groups.

Umebayashi *et al.* (1967) carried out high pressure (up to 2 kb) neutron diffraction experiments on both KDP and KD_2PO_4 in an apparatus using He gas as the pressure fluid. The Curie temperature and its pressure coefficient were determined from the temperature dependence of the neutron peak intensity of certain reflections (specifically (020) and (004)) at various pressures. A sharp increase in peak intensity is observed on cooling through the transition. The Curie temperatures of both compounds are found to decrease linearly with pressure with slopes given in Table I. One of the most interesting features of the results is the large isotope effect in the pressure coefficient of T_c. $|dT_c/dp|$ is found to be about 1·7 times smaller in the deuterated case, whereas T_c itself is about 1·7 times larger.

Hegenbarth and Ullwer (1967) investigated the effects of pressure to about 1·5 kb on the Curie temperature and dielectric constant of KDP. Helium gas was used as pressure fluid. The Curie point decreased linearly with pressure with slope $-5{\cdot}6$°K/kb. This is compared with Umebayashi and co-workers' value in Table I.

Samara (1967) reported the effects of hydrostatic pressure to 3 kb on the dielectric constant and Curie–Weiss temperature of KD_2PO_4. The measurements were made in an apparatus using low viscosity silicone oil as pressure transmitting fluid. In the PE phase the Curie–Weiss law was well obeyed over a relatively large temperature range. At 1 b the Curie constant C was 4040°K, and it decreased by $1{\cdot}5 \pm 0{\cdot}3$%/kb. The Curie–Weiss temperature T_0 decreased linearly with slope $dT_0/dp = -3{\cdot}9 \pm 0{\cdot}2$°K/kb (Table I). Within experimental error, the Curie point T_c (corresponding to ϵ_{max}) decreased at the same rate.

Although there are considerable quantitative differences in the reported values of dT_c/dp (see Table I), the important features of the pressure measurements so far are the decrease of the transition temperature and Curie constant with pressure and the isotope effect in (dT_c/dp). We shall discuss these, along with similar results on $LiH_3(SeO_3)_2$, later in this section.

No explanation has been given for the differences in the reported values of dT_c/dp given in Table I, but it is worth noting that the dielectric constant measurements of Hegenbarth and Ullwer and those of Samara yielded larger values than do the neutron diffraction measurements of Umebayashi and co-workers.

B. LITHIUM TRI-HYDROGEN SELENITE

1. *General Considerations and Comparison with KH_2PO_4*

Lithium tri-hydrogen selenite, $LiH_3(SeO_3)_2$ (abbreviated LHS), is a relatively new H-bonded ferroelectric whose properties have not been studied in detail, and for which studies pressure is an essential variable. Ferroelectricity in LHS was first reported by Pepinsky and Vedam

(1959). It is ferroelectric at room temperature and exhibits by far the largest spontaneous polarization yet observed for an H-bonded ferroelectric, namely 15 $\mu C/cm^2$ at 20°C. Compare this value with 5 $\mu C/cm^2$ for KDP and 0·25 $\mu C/cm^2$ for Rochelle salt. LHS also has the somewhat unusual property that the forces responsible for the ordering of the dipoles are strong enough for the material to remain ferroelectric up to its melting point of 110°C. Both small-field dielectric constant and specific heat measurement over the temperature range $-175°$ to 110°C failed to reveal any transitions below the melting point (Pepinsky and Vedam, 1959).

From the theoretical point of view, the most important properties of ferroelectrics are those observed in the vicinity of the Curie point. The study of these properties in LHS is not possible at atmospheric pressure because of melting. However, it has been found that a ferroelectric-to-paraelectric transition can be induced in LHS by the application of hydrostatic pressure (Samara and Anderson, 1966). The transition is made possible by the simultaneous raising of the melting point and lowering of the Curie point with pressure. A detailed study of the effects of hydrostatic pressure (to 26 kb) and temperature (20–140°C) on the ferroelectric properties of LHS and its deuterated analogue $LiD_3(SeO_3)_2$ (abbreviated LDS) were recently reported by the author (Samara, 1968a). It is found that deuteration raises the Curie point by about 6%.

Although the isotope shift in LHS is not as large as in KDP, there is a body of evidence suggesting that the mechanisms for the ferroelectric transitions in the two materials are similar. The most important evidence comes from a consideration of the crystal structure. At room temperature and pressure LHS has a monoclinic structure (space group Pn) with the ferroelectric axis nearly along the normal to the (001) plane. The structure of the PE phase is probably also monoclinic space group $P2_1/n$ (Samara, 1968a). Room temperature X-ray (Vedam *et al.*, 1960) and neutron (van den Hende and Boutin, 1963) diffraction studies at 1 b have revealed that the LHS structure consists of pyramidal selenite (SeO_3^{2-}) ions joined together by a system of O–H ··· O hydrogen bonds and Li ions. The H-bonds are fairly strong (that is tightly bound and of short O ··· O length) and lie nearly in the planes normal to the polar direction. Two of the H's are located in O–H ··· O bonds of length 2·52 and 2·57 Å while the third H forms an O–H group with an oxygen of one of the selenite groups. The neutron diffraction work suggested that the two H's in the H-bonds are ordered in double minimum potential wells between the two pairs of oxygen atoms. Some supporting evidence for this ordering of the H's in the FE phase has come from proton magnetic resonance (Blinc and Pintar, 1961; Gavrilova-Podol'skaya *et al.*, 1967) and infrared absorption (Khanna *et al.*, 1965) studies on LHS and $NaH_3(SeO_3)_2$.

It will be recalled that the presence of strong O–H ··· O bonds in planes normal to the polar direction and the ordering of the protons in the FE phase in LHS are features very similar to those observed in the structure of KDP. In addition, the transition entropies associated with the ferroelectric ordering of the two materials are comparable (Samara, 1968a). Also important is the fact that their ferroelectric properties are similarly affected by pressure.

2. *Results of Pressure Experiments*

Figure 17 shows, for LHS, typical isobars of the dielectric constant ϵ, measured along the polar axis, as a function of temperature. At 1 b both ϵ and the dielectric loss, tan δ, increase very rapidly for temperatures above 90°C because of the large increase in sample conductivity as conditions approach the melting point (110°C). At high pressures ϵ exhibits the large anomaly characteristic of a FE–PE transition. The

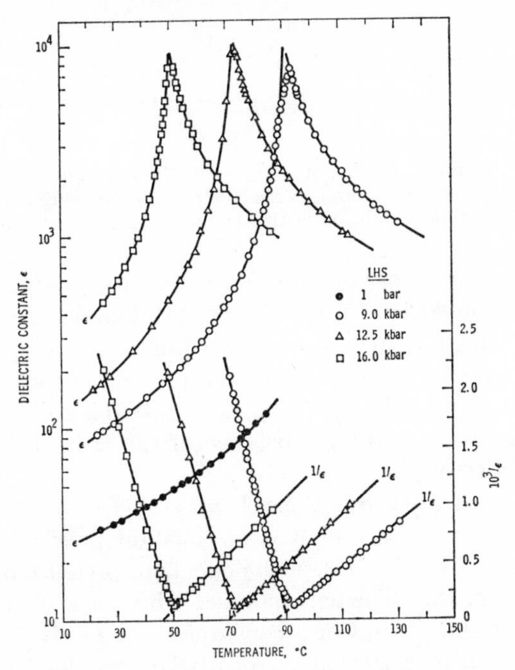

FIG. 17. The temperature dependence of the dielectric constant of $LiH_3(SeO_3)_2$ and its reciprocal measured along the polar axis at various hydrostatic pressures. (After Samara, unpublished results.)

main effect of pressure is to displace the transition to lower temperatures. LDS exhibits qualitatively similar behaviour.

The shifts of the Curie points, T_c, of LHS and LDS with pressure are summarized in Fig. 18 and Table I. Linear decreases in T_c are observed.

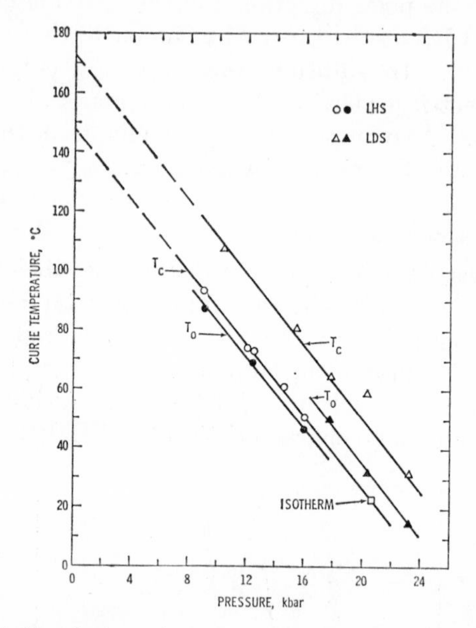

FIG. 18. Shifts of the Curie points, T_c, and Curie–Weiss temperatures, T_0, of $LiH_3(SeO_3)_2$ and $LiD_3(SeO_3)_2$ with hydrostatic pressure. Reproduced from Samara (1968a) by permission of the American Institute of Physics.

Extrapolation of the experimental results to zero pressure indicates that LHS and LDS would have Curie points of 147 °C and 172 °C, respectively, at 1 b if they did not melt before reaching these temperatures. Although much smaller than in KDP, the present isotope shift of ~6% (LDS sample thought to be over 90% deuterated) is still quite significant and again demonstrates the importance of the H-bond in determining the FE properties of these materials.

In the PE phases the Curie–Weiss law $\epsilon = C/(T - T_0)$ and the relationship $\epsilon = C^*/(p - p_0)$ are well obeyed at constant pressure and temperature, respectively (see Fig. 17). The effects of pressure on the T_0's are also shown in Fig. 18. The extrapolated 1 b values are 143 °C for LHS and 159 °C for LDS. The Curie constants, C, are given in Table I.

The spontaneous polarization, P_s, of LHS decreases with both increasing temperature and pressure. But, as for most ferroelectrics, the change of P_s with pressure at constant temperature is mostly associated

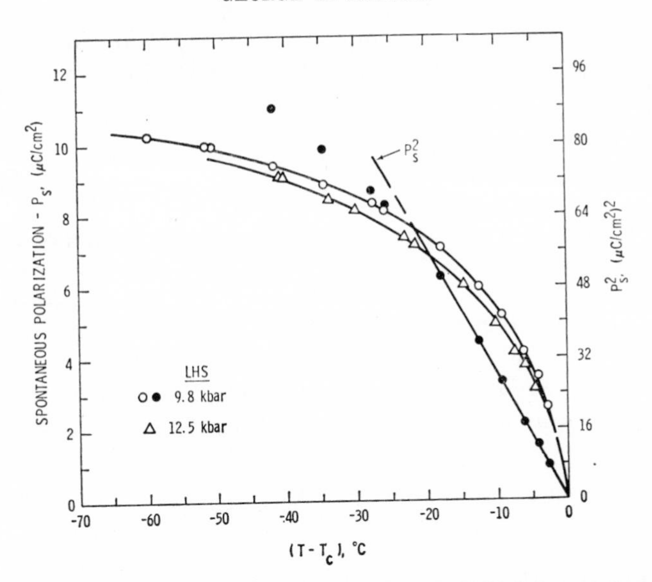

FIG. 19. Isobars of the spontaneous polarization of $LiH_3(SeO_3)_2$ as a function of the reduced temperature $(T-T_c)$. The linear temperature dependence of P_s^2 (shown only for the 9·8-kb isobar) in the neighbourhood of the transition is in agreement with the prediction of the thermodynamic theory. Reproduced from Samara (1968a) by permission of the American Institute of Physics.

with the shift of the Curie point. This is illustrated in Fig. 19 which shows two isobars of P_s *versus* the reduced temperature $(T-T_c)$. There is only a small intrinsic decrease of P_s with pressure amounting to $\simeq 1\%$/kb at temperatures far removed from T_c. Figure 19 also shows that P_s^2 varies linearly with temperature near T_c. This behaviour is predicted for a second-order transition by the thermodynamic theory of Section IV.

3. *Phenomenological Theory*

The pressure and temperature dependence of the spontaneous polarization and dielectric constant of LHS can be successfully correlated and explained in terms of the phenomenological theory described in Section IV (Samara, 1968a). The theory is particularly useful in this case because it makes possible the calculation of certain quantities which cannot be measured directly at 1b due to the melting, for example the changes in entropy and specific heat at the transition, both of which provide important information concerning the nature of the transition.

The entropy is defined by (see eqn (23))

$$S = -(\partial A/\partial T)_P. \tag{73}$$

At constant pressure, and assuming that ξ and ζ are independent of temperature (a reasonably good first approximation for many ferroelectrics), eqn (24) yields for the change in entropy

$$\Delta S = S_0 - S = \frac{2\pi}{C} P_s^2 \tag{74a}$$

where S_0 and S_2 are the entropies of the PE and FE phases, respectively, and P_s^2 is given by eqn (37). Using the saturation values of $P_s = 15$ μC/cm^2 and $C = 4{\cdot}1 \times 10^4$°K and the values 3·185 g/cm^3 for the density and 264 for the molecular weight, eqn (74a) yields $\Delta S = 0{\cdot}61$ cal/mole°K for LHS. This value is of the same magnitude as that for KDP. From eqn (74a) and the definition $c = T(\partial S/\partial T)$, the excess specific heat (at constant polarization) due to the temperature variation of the polarization is given by

$$\Delta c_P = c_P - c_{P_0} = -\frac{2\pi T}{C}\left(\frac{\partial P_2^s}{\partial T}\right). \tag{74b}$$

At 12·5 kb, the known experimental values of the parameters yield $\Delta c_P = 2{\cdot}36$ cal/mole°K.

C. EFFECT OF PRESSURE ON THE POTENTIAL FIELD OF THE PROTON AND SHIFT OF THE CURIE POINT

From the thermodynamic point of view, the decrease of the Curie point of KDP (and probably LHS) with pressure can be accounted for on the basis of the sign of the volume change (ΔV) and latent heat (Q) of the transition. In fact, use of the Clausius–Clapeyron equation, $(\mathrm{d}T_c/\mathrm{d}p) = T_c \Delta V/Q$, with the observed values of ΔV and Q yields values of $\mathrm{d}T_c/\mathrm{d}p$ comparable to those observed experimentally for both KDP and KD_2PO_4 (see Table III). However, if we accept the ordering of the protons in the potential minima as the triggering mechanism for the ferroelectric transition (and this ordering in turn leads to the deformation of the lattice), then at least part of the explanation for the decrease of the Curie points of KDP and LHS must be sought in the effect of pressure on the potential field of the proton in the O–H ··· O bond.

Analytically, the simplest form of the double-minimum potential well is provided by the one-dimensional double-harmonic oscillator whose wave equation is given by (see, for example, Merzbacher, 1961)

$$-\frac{\hbar^2}{2m}\frac{d^2\psi(x)}{\mathrm{d}x^2} + \tfrac{1}{2}K(|x| - a)^2\psi(x) = E\psi(x). \tag{75}$$

Here
$$V(x) = (K/2)(|x| - a)^2 \tag{76}$$
is the potential energy of the system.† Note that $V(x)$ is made up of two parabolic potentials with minima at $\pm a$. The parabolas are joined together at $x = 0$ with a potential barrier of height

$$V(0) = V_0 = \tfrac{1}{2}Ka^2.$$

As the parameter a is varied from 0 to ∞ two limiting cases obtain: (1) $a = 0$ where the system reduces to that of a single harmonic oscillator with non-degenerate eigenvalues, and (2) $a = \infty$ where the system reduces to two separate harmonic oscillators divided by an infinite barrier. Here the eigenvalues are doubly degenerate with the same energies as in (1). For intermediate values of a the degeneracy is broken, each energy level splitting into two levels and thus providing a qualitative explanation for the existence of two O–H stretching bands near 3000 cm^{-1} in KDP- and LHS-type ferroelectrics. If it is assumed that the main effect of pressure is to reduce a, then it is seen from eqn (76) that this lowers the energy barrier between the two potential minima which in turn should lead to a lower Curie point, as observed.

The actual situation in the H-bond is certainly more complicated than that in the case of the double-oscillator and more elaborate treatments are needed. The total energy of the H-bond is made up of at last four contributions (Coulson, 1957): (1) attractive energy, (2) repulsive energy, (3) charge delocalization energy—a quantum contribution accounting for the fluctuation of the charge density as the proton moves in the H-bond, and (4) dispersion energy. Detailed theoretical treatments satisfactorily accounting for all these contributions are very difficult and have not yet been done. However, a number of reasonably successful semi-empirical models, dealing mostly with the interpretation of infrared and Raman spectra of H-bonded crystals at atmospheric pressure, have been proposed. One such model is due to Lippincott and Schroeder (1955) and Reid (1959). In the model, the potential function for the motion of the proton in the H-bond is written as the sum of two diatomic molecule potentials, one (V_1) for O–H and the other (V_2) for H $\cdots$ O, plus an oxygen–oxygen interaction term (W). Thus

$$V(r, R) = V_1(r) + V_2(R - r) + W(R) \tag{77}$$

where V_1 and V_2 are of the form

$$V(r) = D\{1 - \exp\left[-n(r - r_0)^2/2r\right]\}. \tag{78}$$

† This potential is of course symmetric about $x = 0$, but some asymmetry can be easily introduced by adding a constant term on either side of $x = 0$.

In eqn (78) D is the bond dissociation energy, r is the interatomic distance, r_0 is the equilibrium interatomic distance, and $n = \kappa_0 r_0/D$, where κ_0 is the force constant. In the general case, the dissociation energy in V_2, that is D_2, is reduced from that in V_1 by an empirically determined factor F so that $D_2 = FD_1$. This accounts for the fact that the O–H bond is stronger than the H $\cdots$ O bond. In the limiting case of short and symmetrical H-bonds, $F = 1$, whereas for longer asymmetrical bonds $F < 1$.

The O $\cdots$ O interaction term W is considered a function of the O $\cdots$ O distance, R, only and is taken as the sum of a van der Waals repulsion term and an electrostatic attraction term, that is

$$W(R) = A \exp(-bR) + BR^{-m}. \tag{79}$$

The coefficients are evaluated empirically from the best fit of experimental data.

The overall potential function for the motion of the proton in the H-bond is then given by:

$$\begin{aligned} V(r, R) = D\{1 - \exp[-n(r - r_0)^2/2r]\} \\ + FD\{1 - \exp[-n(R - r - r_0)^2/2F(R - r)]\} \\ + A \exp(-bR) + BR^{-m}. \end{aligned} \tag{80}$$

Here D, n, and r_0 are experimental values.

Although it is hardly likely that such a simple potential as eqn (80) can completely describe the situation in ferroelectrics such as KDP and LHS, nevertheless its use leads to two pertinent results. With decreasing O $\cdots$ O distance R ($\equiv$increasing pressure), (1) the O–H distance r increases and (2) the height of the energy barrier between the two potential minima decreases. The various coefficients needed in eqn (80) are not known for KDP and LHS, but they are known for ice (Reid, 1959)—a classic H-bonded crystal—and the effects can be demonstrated by using the ice-model coefficients.

Calculations based on this model do indeed show a decrease in both the height of the potential barrier and the separation of the two minima with decreasing R. Figure 20 shows the sensitivity of the potential curve for changing values of R near $R \simeq 2{\cdot}5$ Å, which is close to the observed H-bond length in KDP-type ferroelectrics. Reid (1959) pointed out that for individual proton motion the potential curve is asymmetric ($F < 1$), and this motion is hindered by a sharply rising potential (dotted curve). Furthermore, this curve does not have a second minimum.

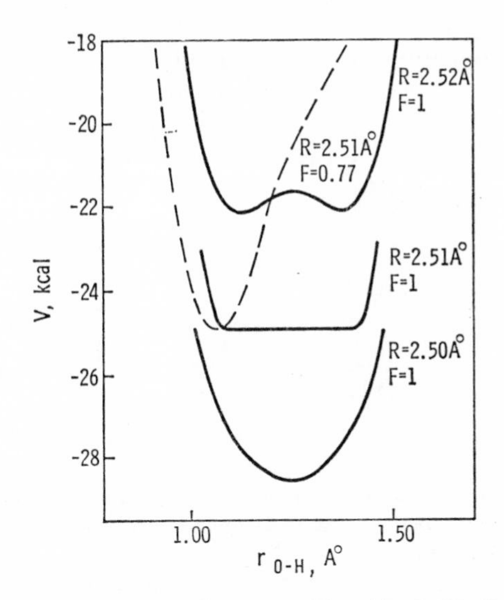

FIG. 20. Variation of the potential energy, V, with O–H distance, r, along the O–H ··· O bond. The extreme sensitivity of V to change in bond length, R, is shown. Reproduced from Reid (1959) by permission of the American Institute of Physics.

However, in the case of H-bonded ferroelectrics where a whole array of protons can be easily flipped from one ordered configuration to the opposite one by the application of a small coercive field, the potential must be nearly symmetric, that is $F \simeq 1$. In this case the potential curve is extremely sensitive to changes in R. Reid has also suggested that in the PE phase protons move collectively, since individual protons are again confined by the dotted curve.

The relation between r and R is obtained from the equilibrium condition $[(\partial V \partial r) = 0$ and is shown in Fig. 21. For the ferroelectric case of interest to us here, the results show that the separation of the two potential minima along the H-bond should decrease as R decreases. Particularly important, the results suggest that if R becomes sufficiently short, then the proton will move to the centre of the bond and the ferroelectric ordering should vanish. The critical distance for this to occur appears to be $R \simeq 2{\cdot}40$ Å. Ideally, the test of this prediction would require dielectric and crystallographic (both X-ray and neutron diffraction) measurements at high pressures and low temperatures. Unfortunately there are at present no data on the variation of R with pressure for any of the H-bonded ferroelectrics of interest.

On the above model, the decrease of the Curie points of KDP and LHS with pressure can result from either a thermal activation of the protons

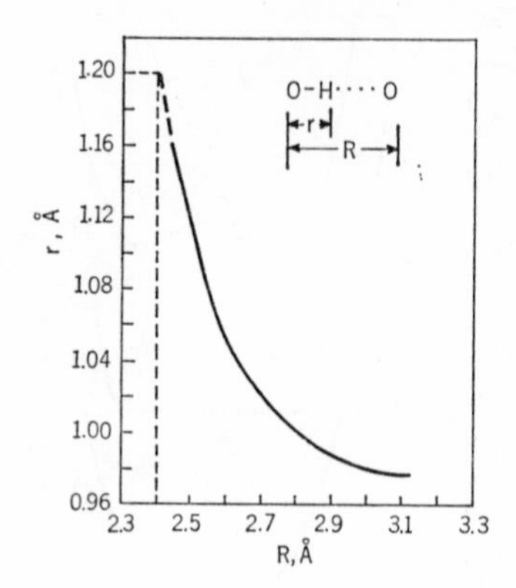

FIG. 21. The relation between the O–H distance, r, and the O ··· O bond length, R, for the O–H ··· O bond based on the model and parameters given by Lippincott and Schroeder (1955).

across a lower barrier or an increase in the tunnelling probability of the protons through the lower and narrower barrier, or both. At present, the experimental situation, including the pressure results, favours a tunnelling model for KDP, whereas the case of LHS is not as well understood. Due to its much higher Curie point (420°K *versus* 122°K), tunnelling should be less important in LHS than in KDP.

The above model can also provide a part of the explanation for the isotope shift. It has been observed (Ubbelohde and Gallagher, 1955) that deuteration causes a marked expansion (up to 0·04 Å) of the O–H ··· O bond length for short ($R < 2{\cdot}6$ Å) H-bonds with a negligibly small effect for long ($R \simeq 2{\cdot}65$–$2{\cdot}8$ Å) bonds, and this effect is then the opposite of the pressure effect. On the basis of the experimental results, therefore, the isotope shift should be larger for the shorter bonds, and this appears to be consistent with experiment. For example, the O–H ··· O bond lengths in KDP are 2·50 Å and those in LHS are 2·52 and 2·57 Å. KDP exhibits a much larger isotope shift.

At least two other factors are important in the explanation of the isotope effect. Both are quantum effects which are consequences of the larger mass of the deuteron. The larger mass leads to, (1) a lower zero-point energy and (2) a lower tunnelling probability through the barrier for the deuteron. Both effects lead to a higher Curie temperature, and these effects can be expected to become less important the higher the transition temperature. This is also consistent with the experimental observations—for example compare KDP and LHS.

D. MODEL CALCULATION OF THE EFFECTS OF PRESSURE ON THE CURIE POINT

The conclusions of the last section concerning the effects of pressure on the Curie point of KDP-type ferroelectrics are qualitatively consistent with the more elaborate theoretical treatments of the problem. It is then of interest to inquire as to whether it is possible to predict quantitatively the observed results from the theoretical models.

In the Blinc and Svetina (1966), Tokunaga and Matsubara (1966), and Kobayashi (1968) theories, the transition temperature is expressed in terms of, (1) the configuration energies of the protons around a given PO_4 group, (2) the kinetic energy of the protons, and (3) the proton–lattice interaction energy. Blinc and Svetina treat only the proton system, and the proton–lattice interaction is incorporated only indirectly *via* long-range proton–proton coupling. However, as pointed out earlier there is evidence for strong proton–lattice coupling, and this coupling must be included for a complete description of the phase transition in KDP-type ferroelectrics.

The theories of Tokunaga and Matsubara and Kobayashi attempt to incorporate the displacements of the ions along the c-axis as essential features. In particular, Kobayashi's model combines the proton tunnelling model with a lattice dynamical model similar to that proposed by Cochran (1961). In Kobayashi's model, the proton tunnelling mode couples strongly with the optical phonon mode of the [K–PO_4] lattice vibration along the c-axis. The frequency of one of the coupled modes tends to zero as the temperature approaches the Curie point from above. It is emphasized that in this case it is the ordering of the protons in the double-minimum potential well which makes the frequency of the coupled mode go to zero. This is in contrast to the case of the displacive type ferroelectrics of the $BaTiO_3$-type discussed in Section V, where anharmonic terms in the lattice vibrations cause the frequency of the transverse optic lattice vibration to tend to zero as $T \to T_c$. Below the Curie point, the coupled mode in KDP-type ferroelectrics is " frozen " out, causing large spontaneous polarization along the c-axis.

Kobayashi finds that the transition temperature of the coupled system, T_c, is given by

$$T_c = T_c^* + (NF^2/k\Omega_0^2), \tag{81}$$

where T_c^* is the transition temperature of the proton system without any coupling with lattice vibrations, N is the density of lattice sites, F is the coupling constant between the proton system and the optical mode of the lattice vibrations of the K and P ions along the c-axis, and Ω_0 is the optical mode frequency. In KDP and KD_2PO_4 the contribution of the last term in eqn (81) to T_c is estimated to be $\sim 10°K$ (Kobayashi, 1968), which is small compared to T_c. Thus, to a reasonable approximation, $T_c \simeq T_c^*$. This conclusion was also arrived at by Blinc and Svetina, and Tokunaga and Matsubara. It is then sufficient to examine the effect of pressure on T_c^*.

In terms of the Blinc and Svetina theory, T_c^* and its pressure dependence can be written as

$$T_c^* = T_c^*(\varepsilon_0, \Gamma, \gamma') \tag{82}$$

and
$$\left(\frac{dT_c^*}{dp}\right) = \left(\frac{\partial T_c^*}{\partial \varepsilon_0}\right)\frac{d\varepsilon_0}{dp} + \left(\frac{\partial T_c^*}{\partial \Gamma}\right)\frac{d\Gamma}{dp} + \left(\frac{\partial T_c^*}{\partial \gamma'}\right)\frac{d\gamma'}{dp}. \tag{83}$$

Here only the Slater configurations (that is configurations with only two H's near a PO_4 group) have been included. ε_0 is the Slater short-range configuration energy. Γ is the proton tunnelling matrix element which accounts for the finite overlap of the protonic wave functions between the two proton sites along the H-bond. γ' is the long-range dipole–dipole interaction energy. The partial derivatives in eqn (83) can be evaluated from the results given by Blinc and Svetina, but one is left with the pressure derivatives. The functional dependences of ε_0, Γ, and γ' involve a large number of parameters and their pressure derivatives cannot be evaluated without some assumptions and approximations.

On the basis of a number of simplifying assumptions, including the assumption that the principal effect of pressure can be reduced to the decrease of the width of the potential barrier, 2Δ, Blinc and Zeks (1968) have recently shown that eqn (83) reduces to

$$\frac{dT_c^*}{dp} = \left[\left(\frac{\partial T_c^*}{\partial \varepsilon_0}\right)\varepsilon_0 - \left(\frac{\partial T_c^*}{\partial \Gamma}\right)\frac{\Gamma\eta}{2} + \left(\frac{\partial T_c^*}{\partial \gamma'}\right)\gamma'\alpha\right]\frac{a}{2\Delta}\left(\frac{d \ln a}{dp}\right) \tag{84}$$

where $\eta = 2q^2 - 1$ with $q^2 = 2m\epsilon_0\Delta^2/\hbar^2$ and $\alpha = 1-(9/2)\ (2\Delta/a)$. In these relationships a is the lattice parameter of the tetragonal KDP unit cell, and ϵ_0 and m are the zero-point energy and mass of the proton (or deuteron). By substituting the appropriate values of the parameters in eqn (84), these authors calculated dT_c^*/dp and compared the results with experimental values. They obtained values of $-4{\cdot}70°K/kb$ for KDP and $-3{\cdot}36°K/kb$ for KD_2PO_4, both in surprisingly good agreement with the experimental values of dT_c/dp given in Table I. Interestingly enough, it turns out that it is the larger value of the tunnelling term $(\partial T_c^*/\partial\Gamma)(\Gamma\eta/2)$ in eqn (84) which is responsible for the larger dT_c^*/dp in KDP than in KD_2PO_4. Neglect of the tunnelling term would give the wrong sign of the isotope effect on dT_c^*/dp.

Using the same model, Blinc and Žekš also calculated the pressure dependences of the Curie–Weiss constant C and the spontaneous polarization P_s. For dln C/dp they find values of $-1{\cdot}3\%$/kb for KDP and $-0{\cdot}91\%$/kb for KD_2PO_4. The latter value is in reasonable agreement with the experimental result of $-1{\cdot}5 \pm 0{\cdot}3\%$/kb obtained by Samara (1967). The pressure dependence of C for KDP apparently has not yet been reported. For dln P_s/dp they estimate values of -1 and $-0{\cdot}6\%$/kb for KDP and KD_2PO_4, respectively. No experimental P_s *versus* p data are available at present.

Finally, it should be pointed out that in the case of KD_2PO_4, where T_c is relatively high, quantum effects are not as important as in the case of KDP, and one can use classical approximations such as the results of Slater's theory. From eqn (70) it is seen that $C \propto N\mu^2$. From the measured decrease of C with pressure, namely

$$d \ln C/dp = -1{\cdot}5\%/\text{kb},$$

and the fact that N increases by $0{\cdot}3\%$/kb (the volume compressibility is $d \ln V/dp = -0{\cdot}32\%$/kb for KD_2PO_4) one gets $d \ln \mu/dp = -0{\cdot}9\%$/kb. The maximum polarization can be written as $P_s = N\mu$, and the above results yield dln $P_s/dp = -0{\cdot}6\%$/kb.

From eqns (70) and (71) it is also seen that $T_0 \propto \varepsilon_0$, so that, in this approximation,

$$d \ln T_0/dp = d \ln \varepsilon_0/dp.$$

From the data in Table I one then gets $d \ln \varepsilon_0/dp \simeq -1{\cdot}5\%$/kb. The value of ε_0 is considered to be related (see for example, Silsbee *et al.*, 1964) to the off-centre distance of the proton along the H-bond. The decrease of ε_0 with pressure indicates that this distance decreases with pressure, a conclusion that we arrived at earlier from different considerations.

VII. Other Hydrogen-Bonded Ferroelectrics

In this section we discuss a number of important ferroelectrics in which the motion of protons in H-bonds is again believed to play an important role in the onset of spontaneous polarization. However, because of the complex chemical and crystallographic structures of these ferroelectrics, this role is much less understood than in the case of KDP-type ferroelectrics discussed in Section VI. The mechanisms for FE ordering in the present crystals are not necessarily similar, but, important from the point of view of our considerations, is the fact that they behave similarly under pressure, and this behaviour is different from that of KDP-type ferroelectrics. Whereas for the latter materials the Curie point and polarization decrease with increasing pressure, these quantities increase for the crystals discussed in this section. These crystals include tri-glycine sulphate and its isomorphs, Rochelle salt, and guanidinium aluminium sulphate hexa-hydrate.

A. TRI-GLYCINE SULPHATE AND ISOMORPHOUS CRYSTALS

1. *General Remarks*

Tri-glycine sulphate $(NH_2CH_2COOH)_3 . H_2SO_4$ (abbreviated TGS) and its two isomorphs, tri-glycine selenate (abbreviated TGSe) and tri-glycine fluoberyllate $(NH_2CH_2COOH)_3 . H_2BeF_4$ (abbreviated TGFB) form an important group of room temperature ferroelectrics whose properties have been thoroughly investigated (see Jona and Shirane, 1962; Wieder and Parkerson, 1966). Although from an atomistic point of view they are quite complex, their phenomenological behaviour is fairly simple and fits well the thermodynamic theory outlined in Section IV.

The crystal structure of these ferroelectrics is monoclinic both above and below their Curie point. In the PE phase the space group is $P2_1/m$, and, on cooling into the FE phase, the mirror plane disappears, the space group becoming $P2_1$. Spontaneous polarization occurs along the two-fold monoclinic b-axis, and the dielectric constant ϵ_b exhibits a large anomaly at the Curie point. The spontaneous polarizations are 3–4 $\mu C/cm^2$ at 0°C for all three crystals (Table I). Their Curie points are 22°C for TGSe, 49°C for TGS, and 73°C for TGFB, and the transitions are of second order. In the PE phase the Curie–Weiss law $\epsilon_b = C/(T - T_0)$ is well obeyed with $C \simeq 3200$, 4000, and 2500°K for TGS, TGSe, and TGFB, respectively.

2. *Effects of Pressure*

The effects of pressure on the spontaneous polarization, dielectric constant ϵ_b, and Curie point of each of the three crystals have been studied, with TGS having received the most attention. There is generally good agreement among the results reported by various authors; in all cases the main effect of pressure is to displace the Curie point to higher temperatures, and most of the changes in other properties are associated with this displacement. The shift of the Curie

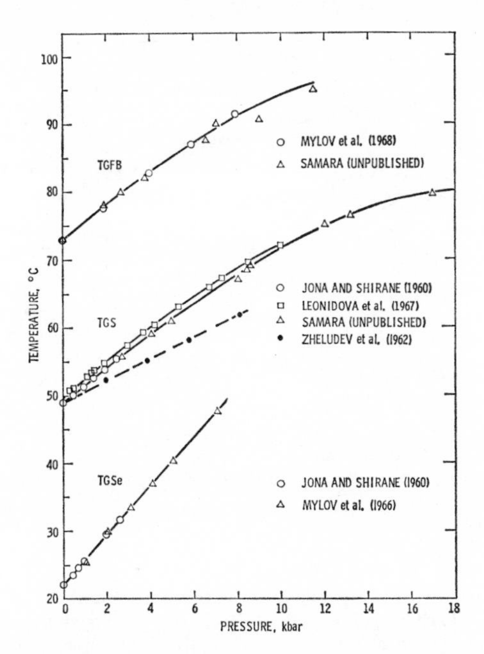

FIG. 22. The effects of hydrostatic pressure on the Curie points of tri-glycine sulphate (TGS), tri-glycine selenate (TGSe), and tri-glycine fluoberyllate (TGFB).

points with pressure are summarized in Fig. 22. At low pressures the shifts are linear with slopes $dT_c/dp = 2 \cdot 6$, $3 \cdot 7$ and $2 \cdot 5$°C/kb for TGS, TGSe and TGFB, respectively. Zheludev *et al.*'s (1962) value for TGS, 1·6°C/kb, is about 40% lower than those reported by others. The pressure dependence of the polarization reported by these authors is also anomalous.

Figure 23 shows the effects of temperature and pressure on the spontaneous polarization of TGS according to Jona and Shirane (1960). P_s is plotted *versus* the reduced temperature $(T - T_c)$ for different isobars. Similar results were obtained for TGSe. It is seen that the application of pressure results in a simple displacement of the P_s curve along the T-axis, with no apparent effect on the magnitude of the dipole moment.

From the results in Fig. 23 it is clear that at constant temperature P_s should increase with pressure, and this is found to be the case as shown by the data in Fig. 25(b). Unpublished results on TGS by the present author obtained from measurements to substantially higher pressures (see Fig. 31(b)) are in good agreement with those of Jona and Shirane. However, these results are in marked contrast with those

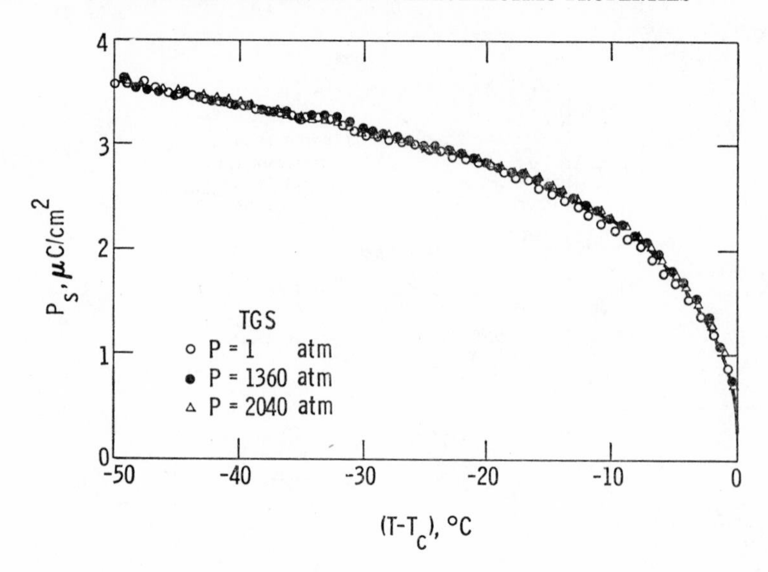

FIG. 23. Isobars of the spontaneous polarization as a function of the reduced temperature $(T-T_c)$ for tri-glycine sulphate. Reproduced from Jona and Shirane (1960) by permission of the American Institute of Physics.

reported by Zheludev and co-workers. These authors reported the effects of pressure to 25 kb on the hysteresis loop of TGS. At room temperature P_s was found to decrease very rapidly and continuously and vanish at ~20 kb, the loop becoming a straight line and thus implying a transition to the PE state. This finding is indeed anomalous and difficult to accommodate with the data shown in Figs 22 and 23.

Pressure also displaces the whole dielectric constant curve to higher temperatures for all three crystals. Figure 24 shows typical results on TGS plotted as $1/\epsilon_b$ *versus* T isobars. The Curie–Weiss law is well obeyed with C independent of pressure, to within the experimental accuracy. At constant temperature, ϵ_b in the PE phase varies as $\epsilon_b = C^*/(p_0 - p)$. Jona and Shirane find that for TGS $C^* = 1{\cdot}17 \times 10^3$ kb at 55°C, and a similar result was reported by Leonidova *et al.* (1967).

Mylov's *et al.*'s (1966, 1968) results on TGSe and TGFB are qualitatively similar to the above results on TGS. However, they observed a large decrease in the peak value of ϵ_b with increasing pressure. The decrease is ~60% in 5 kb for TGSe and ~50% in 8 kb for TGFB. As will be shown later, a similar result obtains for Rochelle salt. The effect is most probably due to partial clamping of the sample caused by the increased viscosity of the pressure transmitting fluid.

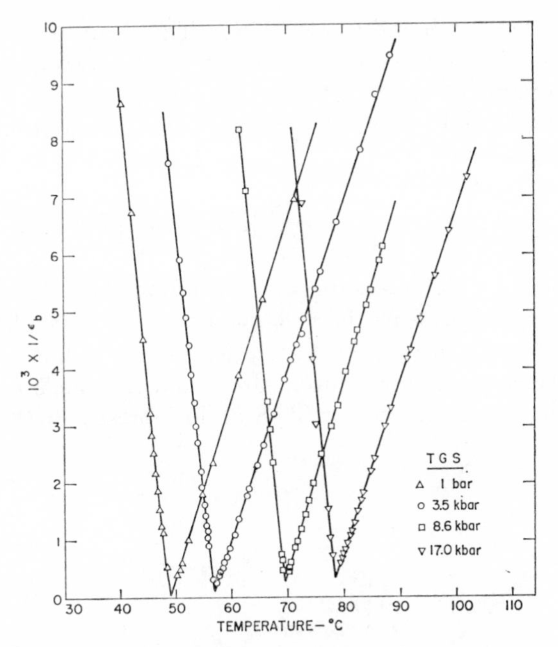

FIG. 24. Isobars of the reciprocal dielectric constant, $1/\epsilon_b$, *versus* temperature for triglycine sulphate. (After Samara, unpublished.)

3. *Discussion*

As mentioned earlier the dielectric properties of TGS and its isomorphs obey the predictions of the thermodynamic theory. For second-order transitions, to good approximation, one needs to consider terms in the free energy expansion only up to fourth power in the polarization. Equation (23) then becomes

$$A(p, P, T) = \Lambda p^2 + \tfrac{1}{2}(\gamma + \Omega p)P^2 + \tfrac{1}{4}\xi P^4. \tag{85}$$

This yields for the dielectric constant in the PE phase

$$\left(\frac{1}{\epsilon_b}\right)_{T>T_0^p} = \frac{1}{4\pi}(\gamma + \Omega p) = \frac{T - T_0^p}{C^p} = \frac{T - (T_0^0 + k'p)}{C^p} \tag{86}$$

where the last two equalities follow from the observed Curie–Weiss T-dependence of ϵ_b at constant pressure and the linear shift of T_0 $(=T_c)$ with pressure, at least at low pressures (Fig. 22). Since C is found to be nearly independent of pressure (hence the superscript p is dropped) eqn (86) indicates that $1/\epsilon_b$ is linear in p with slope $= k'/C$. Comparison with the observed behaviour, that is $\epsilon_b = C^*/(p_0 - p)$, shows that

$$k'/C = 1/C^*. \tag{87}$$

For TGS, $k' = 2{\cdot}6\,°K/kb$ and $C = 3200\,°K$. The theoretical slope is then $k'/C = 8{\cdot}1 \times 10^{-4}/kb$, in good agreement with Jona and Shirane's experimental value of $1/C^* = 8{\cdot}5 \times 10^{-4}/kb$.

From eqns (85) and (86) it also follows that the spontaneous polarization near T_c is given by

$$P_s^2 = -\frac{(\gamma + \Omega p)}{\xi} = \frac{4\pi}{C\xi}\,[(T_0^0 + k'p) - T\,]. \tag{88}$$

With C independent of pressure, and assuming that ξ is independent of temperature and pressure, eqn (88) shows that P_s^2 should be a linear function of temperature at constant pressure and a linear function of pressure at constant temperature. Jona and Shirane's data indicate that this behaviour obtains for TGS at low pressures (Fig. 25) with

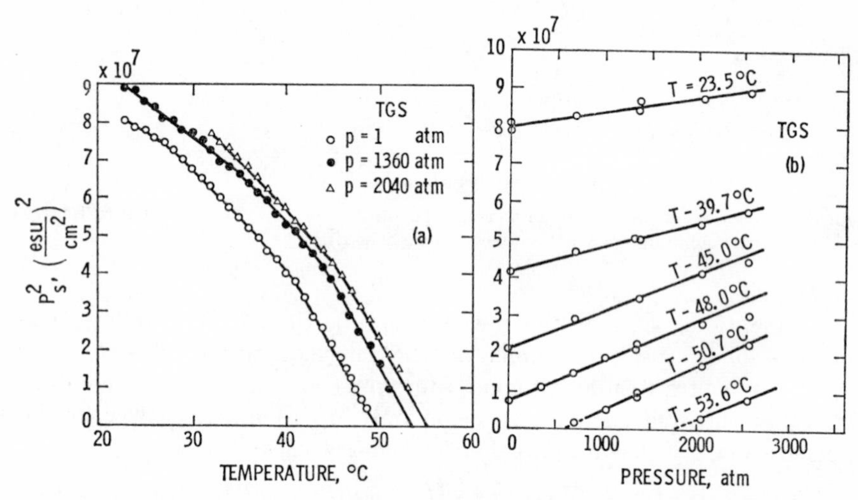

FIG. 25. Variation of the square of the spontaneous polarization of tri-glycine sulphate (a) with temperature at constant pressure and (b) with pressure at constant temperature. The linear behaviour near the transition is in agreement with the predictions of the thermodynamic theory as discussed in the text. Reproduced from Jona and Shirane (1960) by permission of the American Institute of Physics.

fairly good agreement between the slopes of the linear regions of this figure and the theoretical slopes calculated on the basis of the known values of the coefficients in eqn (88) and which are listed in Table I [$\xi = 8{\cdot}0 \times 10^{-10}\,(esu/cm^2)^{-2}$].

Leonidova *et al.*'s (1967) data to ~ 10 kb indicate that C increases slightly and ξ decreases slightly with increasing pressure and thus suggest that the product $C\xi$ in eqn (88) should remain nearly independent of pressure. The linear dependence of P_s^2 on p and T near the transition should then remain valid to fairly high pressures.

B. ROCHELLE SALT

1. *General Remarks*

Rochelle salt, $NaKC_4H_4O_6.4H_2O$ (abbreviated RS) is the oldest known ferroelectric and, as such, its properties have been thoroughly investigated. A detailed account of the experimental and theoretical results is given in the book by Jona and Shirane (1962).

RS is noted for exhibiting two Curie points, an upper (T_c^u) at 24°C and a lower (T_c^l) at −18°C. The crystal is ferroelectric only in the relatively small temperature region between them where the structure is monoclinic (space group $P2_1$). The structure in the paraelectric phase above 24°C is orthorhombic (space group $P2_12_12$). The low temperature phase may be antiferroelectric, but this has not been fully established. Spontaneous polarization occurs along the direction of the orthorhombic a-axis. The dielectric constant ϵ_a exhibits large anomalies at both Curie points (Fig. 26), and it obeys the Curie–

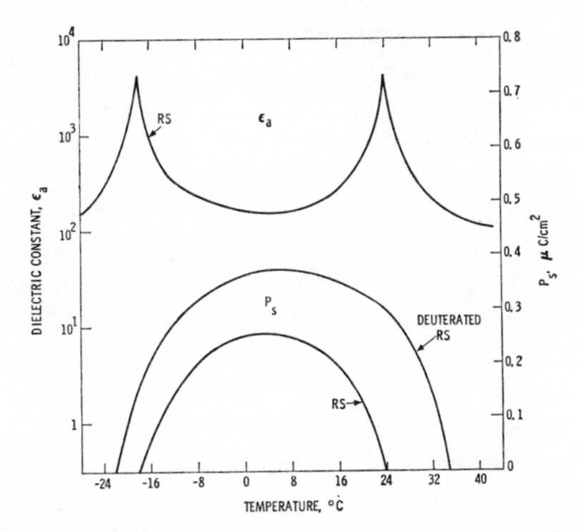

FIG. 26. Temperature dependence of the dielectric constant ϵ_a of Rochelle salt (after Hablützel, 1939) and of the spontaneous polarization P_s of Rochelle salt and deuterated Rochelle salt (after Mason, 1950) at 1 b.

Weiss law over narrow temperature ranges just above T_c^u and just below T_c^l with Curie constants $C = 2{\cdot}29 \times 10^3$°K and $-1{\cdot}18 \times 10^{-3}$°K, respectively. The two transitions are of second order, and the Curie temperatures T_0 coincide with the T_c's within experimental error.

The maximum value of the spontaneous polarization is 0·25 μC/cm^2 and occurs at $\sim$5°C at 1 b (Fig. 26).

There is strong evidence that the H-bond plays an important role in the FE behaviour of RS, but unfortunately due to the complexity of the chemical and crystallographic structure of the material, the exact nature of this role is not fully understood. The isotope shift in RS is not as spectacular as in KDP, but nevertheless it is quite significant. Only those protons in the $-$OH groups and in the water of crystallization are replaced by deuterons by the usual deuteration techniques (thus yielding $NaKC_4H_2D_2O_6.4D_2O$), but those are the ones believed to be responsible for the FE behaviour of RS (see Jona and Shirane, 1962). The Curie points of the deuterated salt are -22 and $+35$°C. Thus, deuteration increases the FE range by 36%, and it also increases the maximum spontaneous polarization by $\simeq$50% (see Fig. 26).

2. *Effects of Pressure*

Bancroft (1938) studied the effects of hydrostatic pressure, up to 10 kb, on the dielectric constant and Curie points of RS. Samara (1965a) extended these measurements to 20 kb and in addition investigated the effects of pressure on the spontaneous polarization and coercive field. The effects of pressure to 20 kb on the dielectric constant and Curie points of deuterated RS were recently reported by Samara (1968b).

Figure 27 shows isobars of ϵ_a *versus* T near the upper and lower Curie points. It is seen that the main effect of pressure is a large displacement of both Curie points to higher temperatures. The measurements near T_c^u could not be extended beyond 5 kb because of decomposition of RS at the elevated temperatures required. It is noteworthy that the peak value of ϵ_a at T_c^l decreases rapidly with increasing pressure. This behaviour has been attributed to a suppression of the piezoelectric deformation of the crystal caused by the increase in the viscosity of the pressure-transmitting fluid (*n*- and *iso*-pentane mixture) with increased pressure (Samara, 1965a).

Deuterated RS exhibits similar behaviour. Figure 28 shows two isotherms of ϵ_a *versus* pressure. For temperatures above 35°C (24°C for RS) two transitions are observed, the one occurring at the lower pressure corresponding to the upper Curie point. As we have seen earlier for several other ferroelectrics, in the nonpolar phases ϵ_a can be well expressed by $\epsilon_a = C^*/(p-p_0)$. Figure 28 shows $1/\epsilon_a$ *versus* p plots for the lower PE phase of deuterated RS. At 45°C,

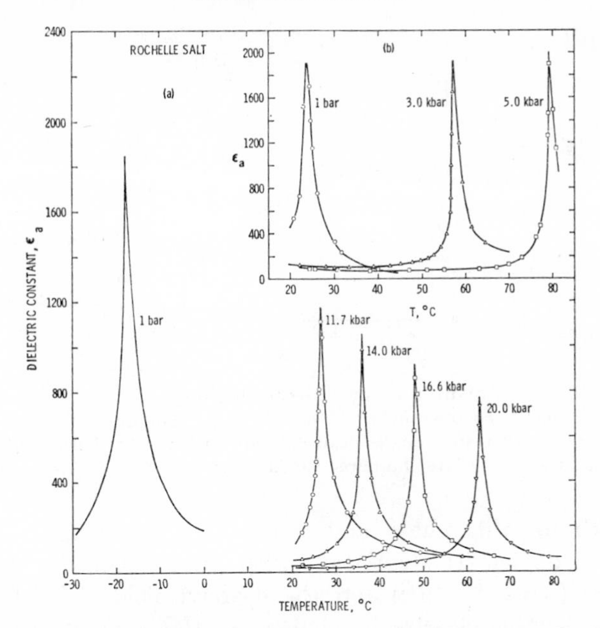

FIG. 27. Isobars of the dielectric constant ϵ_a *versus* temperature for Rochelle salt showing the behaviour (a) near the lower Curie point and (b) near the upper Curie point. (After Samara, 1965a.)

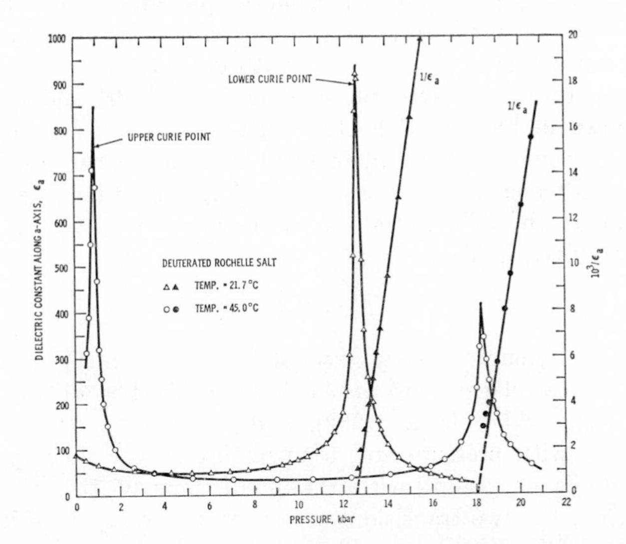

FIG. 28. Pressure dependence of the dielectric constant ϵ_a of deuterated Rochelle salt showing the linear dependence of $(1/\epsilon_a)$ *versus* pressure in the lower PE phase. Two transitions are observed at temperatures above 35°C. (After Samara, 1968b.)

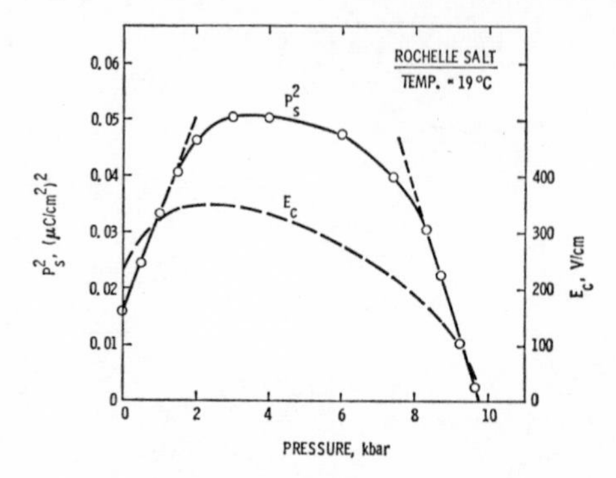

FIG. 29. The effects of hydrostatic pressure on the square of the spontaneous polarization and on the coercive field of Rochelle salt. The linear pressure dependence of P^2 near both transitions is in agreement with the predictions of the thermodynamic theory. (After Samara, 1965a.)

$C^* = 157$ kb for this phase and -165 kb for the upper PE phase.

Figure 29 depicts the variation with pressure of the spontaneous polarization (plotted at P_s^2) and the coercive field E_c of RS. These variations resemble closely the changes in these quantities caused by decreasing temperature at 1b. The peak value of P_s observed at 3·5 kb at room temperature is nearly the same ($\simeq 0{\cdot}25$ μC/cm^2) as that obtained at $\simeq 5$°C at atmospheric pressure. The linear dependence of P_s^2 on pressure (and also on temperature) in the vicinity of both Curie points is in agreement with eqn (88).

The temperature–pressure phase diagrams of RS and deuterated RS are shown in Fig. 30. The two salts behave quite similarly, the main feature being the large increase in the temperature range of the FE phase with increasing pressure. At 5 kb the FE range is about double that at 1b. The initial slopes for the shifts of T_c^u and T_c^l with pressure are given in Table I.

3. *Discussion*

The thermodynamic theory discussed in Section IV has proved successful in correlating the atmospheric properties of RS (see for example, Jona and Shirane, 1962), and, as pointed out above, the change of P_s with pressure and temperature in the vicinity of both Curie points is consistent with the predictions of this theory. In addition, since the two transitions are of second order, the shifts of the Curie points with pressure should obey Ehrenfest's relation, eqn (38). Although there is considerable disagreement in the reported thermal expansion and specific heat anomalies at the transitions, use of the

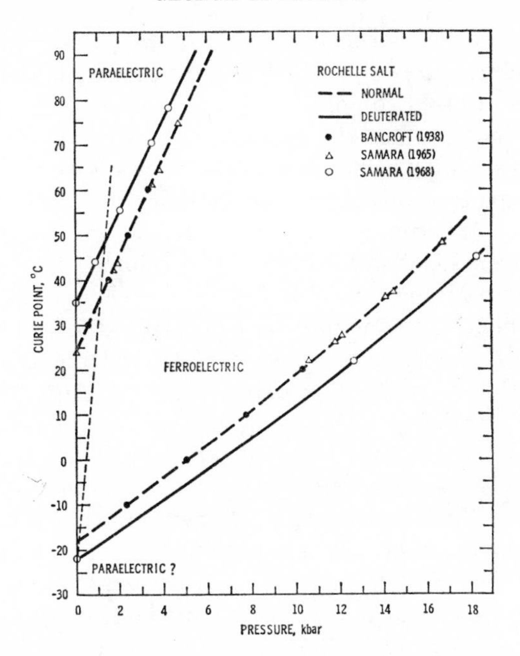

FIG. 30. Temperature–pressure phase diagram for normal and deuterated Rochelle salt. The dotted curve represents a line of constant volume.

available data in eqn (38) yields reasonably good agreement between calculated and observed values of dT_c/dp, at least for T_c^u (Samara, 1965a).

There are two results of the pressure experiment which are especially interesting. These are the large increase in the FE range of the material with increasing pressure and the relatively small effect of pressure on the maximum value of P_s. These results were discussed by Samara (1965) in terms of the model theories of Mason (1950), Devonshire (1957), and Mitsui (1958). These theories have been able to explain in a qualitative way the existence of two Curie points and several of the observed properties of RS. They are local field theories which are basically similar in that they assume that the ferroelectricity of RS is associated with certain hydrogen dipoles having two positions of equilibrium with different energies. The sign of the dipole changes when the hydrogen moves from one equilibrium position to the other. Mason's theory, based on the early structure determination by Beevers and Hughes (1941), considers the motion of the hydrogen in the $O_{(1)}-(H_2O)_{(10)}$ hydrogen bond (in the Beever and Hughes notation) as the FE dipole. Mitsui, on the basis of a more recent X-ray re-examination of the structure (Mazzi *et al.*, 1957) and of neutron diffraction studies (Frazer *et al.*, 1954), considers the hydroxyl group $(OH)_{(5)}$ as the rotatable dipole.

In order to account for the existence of two Curie points, Mason's theory assumes a dipole moment which varies with temperature due to thermal expansion. This assumption becomes difficult to accept in view of the experimental results (Fig. 30) which show that RS exhibits two transitions even at constant volume. Devonshire (1957) has shown that Mason's model does not give two Curie points at constant volume. By assuming that the two states of the hydrogen atom of the FE dipole have different statistical weights, he was able to overcome this difficulty in Mason's theory. But, the theory still fails to give a satisfactory explanation for the increase in the FE range of the material with increasing pressure, and to satisfactorily account for the observed effect, the theory would require a large increase in the dipole moment, μ. However, this would be in the wrong direction as μ can, in general, be expected to decrease with decreasing volume, and furthermore, the measured pressure dependence of the polarization does not indicate any large variation of μ with volume.

Mitsui's theory is quite similar to those of Mason and Devonshire, but the treatment is more rigorous. The internal field is described in terms of two parameters b and c which are defined by

$$b = N\mu^2(\beta' - \beta)/2\phi \tag{89}$$

and

$$c = (\beta' - \beta)/(\beta' + \beta) \tag{90}$$

where 2ϕ is the difference in potential energy between the two equilibrium positions of the dipoles, and the β's are parameters representing interaction between dipoles. β is of the order of $4\pi/3$, while $\beta' \gg 4\pi/3$ and includes both ionic and electronic polarization effects.

The values of b and c determine the state of polarization of the material. For a certain range of values of b and c the crystal is ferroelectric in a temperature range limited by two transition temperatures. Thus, in terms of this theory the increase in the FE range of RS with increasing pressure should be due to the effect of pressure on b and c. It can be argued that large changes in the β's would be needed to explain the experimental results, but such changes are difficult to justify (Samara, 1965a).

Thus, it can be concluded that although the above mentioned theories have had some success in explaining a number of experimental observations, they do not give a satisfactory explanation for the pressure effects.

C. GUANIDINIUM ALUMINIUM SULPHATE HEXA-HYDRATE

1. *General Remarks*

Guanidinium aluminium sulphate hexa-hydrate, $C(NH_2)_3.Al(SO_4)_2.6H_2O$ (abbreviated GASH), is another example (see also $LiH_3(SeO_3)_2$) of a room-temperature ferroelectric which does not exhibit a Curie point. It decomposes at ~200°C before undergoing any phase transitions. The polar phase is trigonal, space group C_{3v}^2 – P31m, and the polar direction is along the *c*-axis.

The FE properties of GASH were discovered and investigated at 1 b by Holden *et al.* (1956). The dielectric constants along both the *c*- and *a*-axes are quite small (6 and 5, respectively, at 20°C) and are practically independent of temperature up to ~100°C. Above this temperature dehydration of the crystal initiates. The spontaneous polarization is also small (0·35 μC/cm² at 20°C) and decreases nearly linearly with increasing temperature between ~ –60 and +100°C (Fig. 31(a)).

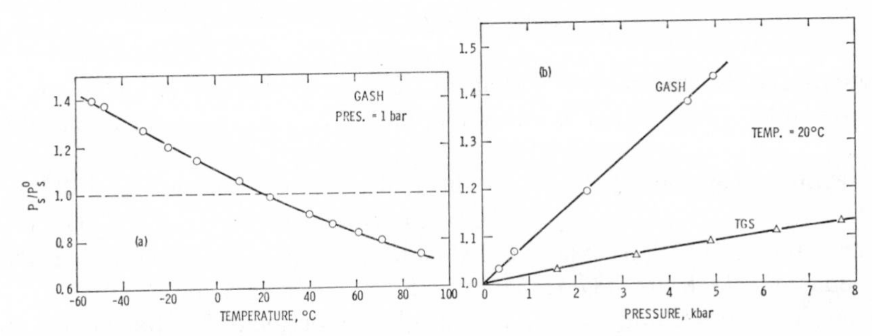

FIG. 31. (a) Fractional change of the spontaneous polarization of GASH *versus* temperature (P_s^0 is the value of P_s at 1 b and 20°C) based on the data given by Holden *et al.* (1956). (b) Fractional change of the spontaneous polarization *versus* pressure for GASH (after Merz, 1956) and TGS (after Samara, unpublished).

2. *Effects of Pressure*

Merz (1956) studied the effects of hydrostatic pressure to 5 kb, at room temperature, on the hysteresis loop of GASH. The changes of the loop with increasing pressure resemble closely the changes observed on lowering the temperature at 1b. Both P_s and the coercive field E_c increase linearly with pressure, the increase in P_s (Fig. 31(b)) being remarkably large with slope d ln $P_s/dp \simeq 0·09$ kb⁻¹.

These results indicate that the Curie point of GASH, if it exists at all, shifts to higher temperatures with increasing pressure. In the hope that high hydrostatic pressure may suppress dehydration of the crystal and thus make it possible to observe a Curie point transition, the present author (unpublished) measured the effects of pressure to 20 kb and temperature to $\sim 200°C$ on ϵ_c; however, no indication of a transition, or an approach to one, was observed. At higher temperatures decomposition still occurred.

As for other ferroelectrics, at least a part of the observed increase in P_s with pressure can probably be associated with an increase in T_c (hypothetical) and the resulting displacement of the P_s *versus* T curve to higher temperatures. However, there is one feature of the GASH results which deserves special attention, and that is the relatively large (compare TGS, Fig. 31(b)) and nearly linear increase in P_s observed on both increasing pressure and decreasing temperature. Merz (1956) examined this point and showed that, to a good approximation, the change in P_s (and also in E_c) is a function of the volume change alone regardless of whether the volume change is obtained by increasing the pressure or lowering the temperature. Thus, according to Merz,

$$\Delta P_s/P_s = -g(\Delta V/V) \tag{91}$$

where ΔV is the volume change and g is a constant which can be determined from either volume compressibility or thermal expansion data.

From the measured P_s *versus* p data and taking for the compressibility $(1/V)(\mathrm{d}V/\mathrm{d}p)$ a value $\simeq -5\times 10^{-3}$ kb^{-1}, Merz obtains $g \simeq 20$. On the other hand, the P_s *versus* T data, combined with a volume expansion coefficient $(1/V)(\mathrm{d}V/\mathrm{d}T) \simeq 5\times 10^{-3}$°K^{-1}, yield $g \simeq 25$. Thus, it is seen that the change in P_s is indeed proportional to the change in volume. However, what is more important is the fact that the observed change in P_s is $\sim$20–25 times larger than would be expected from an increase in the density of dipoles, for in the latter case g would $=1$.

One is thus led to conclude that the dipole moments in GASH must become a great deal larger with decreasing unit cell volume. This is an unusual result, unlike that observed for most ferroelectrics. Merz suggested that the components of the H-bonds along the c-axis may be responsible for this behaviour, but, because of the complexity of the crystal structure, it is difficult to assess the validity of this suggestion and to provide a satisfactory explanation for the observed effect.

VIII. Miscellaneous Ferroelectrics

This section is devoted to a brief description of the properties of a few interesting, but unrelated, ferroelectrics which have received recent attention. For all these crystals, pressure turns out to be an important variable in that their properties are either strongly pressure dependent or that pressure is necessary to stabilize the FE phase. The crystals discussed are KNO_3, $NaNO_2$ and SbSI.

A. Potassium Nitrate

Potassium nitrate, KNO_3, is particularly interesting for our consideration since its ferroelectric phase is stable only at high pressure. At normal conditions it has the orthorhombic aragonite structure (space group D_{2h}^{16}–Pnma), which is referred to as phase II (see Fig. 32).

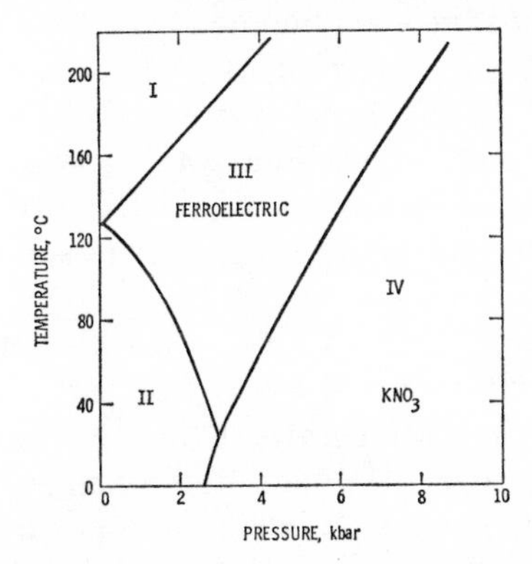

Fig. 32. Temperature–pressure phase diagram for potassium nitrate based on the data given by Bridgman (1916).

On heating at 1 b, it transforms at ~130°C to a rhombohedral phase (I) related to the calcite structure (space group $D_{3d}^{6}-R\bar{3}c$). On cooling, phase I does not revert directly into phase II, but, rather, it changes first into another phase (III) at ~125°C, and then back to phase II at ~115°C. Phase III, which is thus metastable at 1 b, was found to be ferroelectric by Sawada *et al.* (1958a, 1961). The structure of this phase is closely related to that of the PE phase I, the space group being $C_{3v}^{5}-R3m$. The nonpolar low-temperature phase II may be antiferroelectric, but this has not been definitely established.

Undoubtedly the ferroelectricity of phase III arises from the fact that the NO_3^- ion is not located exactly at the centre of the rhombohedral unit cell. Rather, it is displaced by ~0·5 Å along the *c*-axis (Barth,

1939), which turns out to be the polar direction. The NO_3^- ions can thus make a jumping motion between two equilibrium positions along the *c*-axis, each position corresponding to one orientation of the polarization vector.

Some of the atmospheric pressure dielectric properties of KNO_3 were recently summarized by Chen and Chernow (1967). The spontaneous polarization is ~ 6–$8\ \mu C/cm^2$ at 120°C. The dielectric constant ϵ_c is quite small and reaches a peak value of only ~ 70 at the I–III transition. In phase I, ϵ_c obeys a Curie–Weiss law, but different values for C and T_0 have been reported by various authors. Chen and Chernow's values are $C = 6{\cdot}1 \times 10^3$°K and $T_0 = 314$°K.

The pressure–temperature phase diagram of KNO_3 was studied in some detail by Bridgman (1916) long before the discovery of ferroelectricity in this material. (See also Rapoport (1965) for more recent results.) Bridgman's results are summarized in Fig. 32. It is seen that phase III becomes thermodynamically stable at very low pressures, and that the temperature range of its stability increases rapidly with increasing pressure. In this latter respect KNO_3 is similar to Rochelle salt, and, in addition, both crystals have an upper and a lower transition temperature. However, in the case of KNO_3 both transitions are first order, and the pressure dependences of the transition temperatures are different (compare Figs 30 and 32).

The effects of pressure, up to only $\sim 1{\cdot}3$ kb, on the FE properties of KNO_3 were recently reported by Taylor and Lechner (1968). The shifts of the transition temperatures are in good agreement with Bridgman's data; however, the most interesting result is the pressure dependence of the polarization which is shown in Fig. 33. Here isobars of the remanent polarization, P_r, *versus* temperature are presented. The spontaneous polarization, P_s, behaves similarly, but P_s values are $\sim 10\%$ larger than P_r's. The significant feature of the

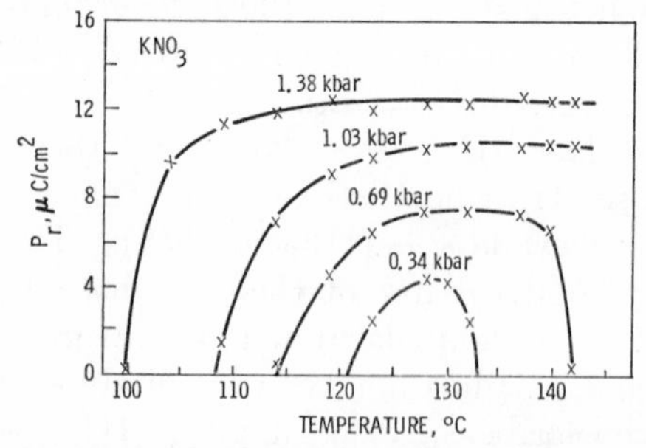

FIG. 33. Variation of the remanent polarization of potassium nitrate with temperature at various pressures. Reproduced from Taylor and Lechner (1968) by permission of the American Institute of Physics.

data in Fig. 33 is the strong dependence of the polarization on pressure. At a temperature near the centre of phase III this dependence is nearly linear with slope $dP_r/dp \simeq 7$ $\mu C/cm^2 kb$. This may reflect a large increase in the dipole moment with pressure—a behaviour similar to that for GASH. However, we suspect that this unusually large pressure effect is probably not intrinsic but rather is associated with the fact that, over the limited pressure range of the measurements, the FE phase is stable over a narrow temperature range between two transition temperatures. The increase in P with pressure may then be a manifestation of the increased effect of long-range ordering forces in aligning the dipole moments as the temperature range of stability of the FE phase increases, and P is measured at temperatures farther away from the transitions.

It is clear from the above that measurements at higher pressures would be of considerable interest especially for finding out how large the polarization becomes when saturation (with pressure) is finally achieved. In addition, the dielectric properties of phase IV, on which no data are available, may be quite revealing. Such studies by the present author are currently in process.

B. SODIUM NITRITE

Ferroelectricity in sodium nitrite, $NaNO_2$, was discovered in 1958 by Sawada and co-workers, years after the discovery of its phase transition. Although the crystal is relatively simple from both the chemical and structural point of view, the nature of its ferroelectric behaviour is still not well understood. The FE transition is believed to be an order–disorder transition of the molecular dipoles arising from the atomic configuration of the NO_2^-, $\left(\begin{smallmatrix} & N & \\ O & & O \end{smallmatrix}\right)^-$, ions. The measured total transition entropy, namely $1{\cdot}26 \pm 0{\cdot}08$ cal/mole–°K (Sakiyama *et al.*, 1965) which is nearly equal to $R \ln 2 = 1{\cdot}37$ cal/mole–°K, suggests that the orientational order of the NO_2^- ions, which is prevalent in the FE phase, is completely destroyed in the PE phase.

Careful studies of the properties of $NaNO_2$ near the Curie point have revealed that the crystal exhibits two transitions which are only 1–2°K apart at 1 b (see for example, Yamada *et al.*, 1963; Gesi *et al.*, 1965). The low temperature FE phase is orthorhombic, space group $C_{2v}^{20} - \mathrm{Im2m}$. The polar direction is along the *b*-axis, and the spontaneous polarization is relatively large, ~ 7 $\mu C/cm^2$. On heating, the crystal transforms at ~ 163°C, via a first-order transition, into a sinusoidal antiferroelectric (AFE) phase in which the order parameter of the dipoles changes sinusoidally along the a-axis with a period of ~ 8 layers (Yamada *et al.*, 1963). This phase transforms at $\sim 165°C$ into a PE phase with orthorhombic symmetry (space group D_{2h}^{25}–Immm), and the transition is of second order. In the PE phase the dielectric constant ϵ_b obeys a Curie–Weiss law with $C = 5 \times 10^3 °K$ and $T_0 = 433°K$.

Gesi and co-workers (1965) investigated the effects of hydrostatic pressure to 10 kb on the dielectric constant of $NaNO_2$ and observed that both the FE→AFE and AFE→PE transition temperatures, T_c and T_N, respectively, increase with pressure. The transition temperature *versus* pressure curves are slightly convex, but at pressures up to ~ 3 kb they are very nearly linear with slopes $dT_c/dp = 4{\cdot}9°C/kb$ and $dT_N/dp = 5{\cdot}6°C/kb$. The temperature range of the sinusoidal AFE phase increases from 1–2°C at 1 b to $\sim 8°C$ at 10 kb.

Since the FE→AFE transition in $NaNO_2$ is first order, the shift of T_c with pressure should obey the Clausius–Clapeyron equation. Sakiyama *et al.*'s (1963) volume expansion and specific heat measurements yield a volume change $\Delta V = 0{\cdot}027$ cm³/mole and a latent heat $Q = 56$ cal/mole. Substituting these values in eqn (39), one gets

$$dT_c/dp = 5°C/kb,$$

in good agreement with the measured value (Gesi *et al.*, 1965). The AFE→PE transition is second order, and dT_N/dp should obey the Ehrenfest relationship, eqn (38). An accurate test of this relationship is not possible in view of the fact that T_c and T_N are so close together, and values of $\Delta\alpha$ and ΔC_p at T_N are difficult to evaluate with confidence.

C. ANTIMONY SULPHO-IODIDE

Antimony sulpho-iodide, SbSI, exhibits an unusual combination of photoconductive, semiconductive, optical, and ferroelectric properties (Fatuzzo *et al.*, 1962). The crystal is orthorhombic space group D_{2h}^{16}–Pnam above and C_{2v}^{9}–Pna2$_1$ below the Curie point ($T_c \simeq 20°C$). Crystals grow in the form of bundles of thin single-crystal needles with the FE c-axis lying along the needle axis. The strong coupling between SbSI's various properties is emphasized by the observations that the piezoelectric strain and the dielectric constant along the c-axis vary with illumination by visible light (Tatsuzaki *et al.*, 1967), and that the fundamental absorption edge, E_g, shifts to shorter wavelength under the influence of a d.c. field (Kern, 1962). Furthermore, in the FE phase SbSI is the most strongly piezoelectric crystal known and exhibits large electromechanical coupling (Hamano *et al.*, 1965), and its FE properties should therefore be quite pressure sensitive.

Fridkin *et al.* (1966) measured the effect of hydrostatic pressure on the absorption edge of SbSI at 2°C and observed a large shift to longer wavelength ($\Delta E_g \approx -0{\cdot}04$ eV) between $\sim 0{\cdot}3$ and $0{\cdot}5$ kb. They attributed this to a pressure-induced FE→PE transition and estimated that the decrease of the Curie point with pressure is

$$(\mathrm{d}T_c/\mathrm{d}p) \simeq -50°\mathrm{C/kb}.$$

Similar results were obtained by the same authors from earlier measurements at 18°C.

The pressure dependence of the dielectric constant and Curie temperature of SbSI were recently reported by Samara (1968c). Two isotherms of ϵ_c *versus* pressure are shown in Fig. 34. These results

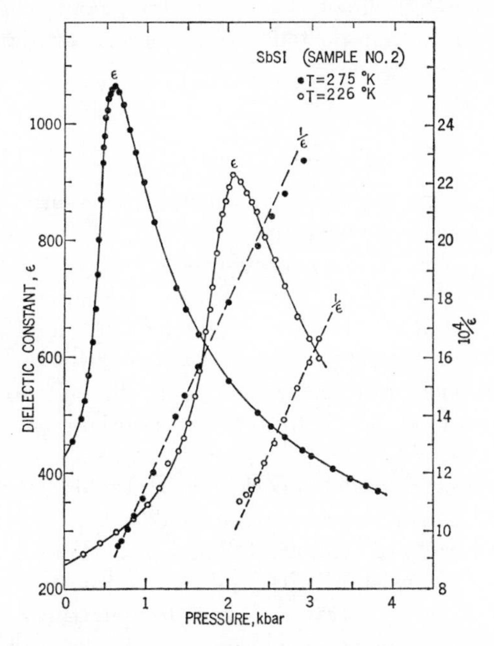

Fig. 34. The pressure dependence of the dielectric constant ϵ_c of SbSI at constant temperature. (After Samara, 1968c.)

show that the FE→PE transition can be induced by pressure, and the behaviour is qualitatively similar to other ferroelectrics. In the PE phase $1/\epsilon_c$ is a linear function of pressure over a limited pressure range as indicated. The Curie point, determined from both ϵ *versus* T isobars and ϵ *versus* p isotherms, decreases linearly with slope

$dT_c/dp = -37 \cdot 0 \pm 2 \cdot 0$°C/kb. This is the largest value yet reported for any ferroelectric (see Table I).

The FE–PE transition in SbSI is first order and of the displacive-type (Kikuchi *et al.*, 1967) analogous to that in $BaTiO_3$. The transition heat (taken as $\int c_p dT$ in the immediate vicinity of the transition) and entropy change (Mori *et al.*, 1965) are 58 cal/mole and 0·18 cal/mole–°K, respectively, as compared with 50 cal/mole and 0·12 cal/mole–°K for $BaTiO_3$.

Since the transition is first order, the shift of the Curie point of SbSI with pressure can be expected to obey the Clausius–Clapeyron eqn (39). Using the measured value of dT_c/dp, eqn (39) yields

$$\Delta V = V_{PE} - V_{FE} = -0 \cdot 31 \text{ cm}^3/\text{mole},$$

or a $\Delta V/V = -0 \cdot 57\%$, which is in good agreement with the value of $-0 \cdot 6\%$ estimated from reported lattice parameter data (Arndt and Niggli, 1964).

IX. Summary

In this chapter the effects of hydrostatic pressure on ferroelectric properties have been reviewed and discussed. It was seen that these properties are generally very pressure-dependent, and that, in the case of some ferroelectrics, for example $LiH_3(SeO_3)_2$ and KNO_3, pressure is an essential variable in elucidating the ferroelectric behaviour. Because different ferroelectrics exhibit widely different mechanisms for their ferroelectricity, it is difficult to draw general conclusions. However, it is helpful to give a brief recapitulation of the important features.

(1) Fairly successful model theories are available only for two types of ferroelectrics: (a) perovskites of the $BaTiO_3$-type and (b) hydrogen-bonded ferroelectrics of the KH_2PO_4-type. For both types the pressure results are consistent with and lend support to these theories. For the pervoskites, analysis of the temperature- and pressure-dependence of the dielectric constants (ϵ and n^2) in the PE phase in terms of both a macroscopic and a microscopic theory has shown that the intrinsic T-dependence of the infrared polarizability is the predominant factor in determining the dielectric behaviour. This is important in relation to the lattice dynamical theory which associates the PE→FE transition in the perovskites with the vanishing of the frequency, ω_T, of a long wavelength transverse optic phonon mode of the lattice. The results suggest that ω_T^2 should increase linearly with pressure. Some supporting evidence for this prediction comes

from results on the pressure dependence of the electrical conductivity of these materials.

For the KH_2PO_4-type ferroelectrics, the ordering of the protons in double-minimum potential wells along the H-bonds is the triggering mechanism for the FE transition. The length of the H-bond is crucial in determining whether or not ferroelectric ordering is possible. Increasing pressure decreases both the separation of the two potential minima and the height of the potential barrier between them. Both effects lead to a lower Curie point as is observed experimentally. At sufficiently high pressure, or for short bonds ($\leqslant 2{\cdot}4$ Å), only one potential minimum exists, and, therefore, no FE state is possible.

(2) The phenomenological (or thermodynamic) theory is very useful in correlating and interpreting ferroelectric properties, and the observed pressure effects are generally in good agreement with the predictions of this theory. In the case of $LiH_3(SeO_3)_2$ the theory is particularly useful because it makes it possible to calculate certain quantities which cannot be measured directly due to melting, for example the change in entropy and specific heat at the transition. Both quantities provide important information concerning the nature of the transition. For $BaTiO_3$, the first-order FE–PE transition acquires, with increasing pressure, the characteristics of a second-order transition and should become second-order above 25–30 kb.

(3) An examination of the effects of pressure on the Curie point, T_c, of various ferroelectrics (see Table I) reveals an interesting feature. For crystals in which the FE–PE transition is purely displacive in nature (for example perovskites and SbSI), T_c decreases with increasing pressure; whereas, for crystals in which the FE–PE transition is of the pure order–disorder type (for example TGS, RS, and $NaNO_2$), T_c increases with pressure. For KH_2PO_4-type ferroelectrics in which the transition involves both an order–disorder rearrangement of the protons and an accompanying displacement of the other ions, T_c decreases with pressure.

References

Anderson, P. W. (1960). *In* "Fizika Dielektrikov," ed. by G. I. Skanavi. Akad. Nauk SSSR, Moscow.

Arndt, R. and Niggli, A. (1964). *Naturwissenschaften* **51,** 158.

Bacon, G. E. and Pease, R. S. (1955). *Proc. R. Soc.* (*Lond.*) **A230,** 359.

Bancroft, D. (1938). *Phys. Rev.* **53,** 587.

Barker, A. S., Jr. and Tinkham, M. (1962). *Phys. Rev.* **125,** 1527.

Barrett, J. H. (1952). *Phys. Rev.* **86,** 118.

Barth, T. F. W. (1939). *Z. Phys. Chem.* **B43,** 448.

Beevers, L. A. and Hughes, W. (1941). *Proc. R. Soc.* (*Lond.*) A**177,** 251.
Blinc, R. (1960). *Physics Chem. Solids* **13,** 204.
Blinc, R. and Pintar, M. (1961). *J. Chem. Phys.* **35,** 1140.
Blinc, R. and Svetina, S. (1966). *Phys. Rev.* **147,** 423 and 430.
Blinc, R. and Žekš, B. (1968). *Phys. Lett.* **26**A, 468; also work to be published.
Bosman, A. J. and Havinga, E. E. (1963). *Phys. Rev.* **129,** 1593.
Bridgman, P. W. (1916). *Proc. Am. Acad. Arts Sci.* **51,** 581.
Busch, G. (1938). *Helv. phys. Acta* **11,** 269.
Busch, G. and Scherrer, P. (1935). *Naturwissenschaften* **23,** 737.
Cady, W. G. (1946). "Piezoelectricity." McGraw-Hill, New York.
Chen, A. and Chernow, F. (1967). *Phys. Rev.* **154,** 493.
Cochran, W. (1960). *Adv. Phys.* **9,** 387.
Cochran, W. (1961). *Adv. Phys.* **10,** 401.
Cook, W. R., Jr. (1967). *J. appl. Phys.* **38,** 1637.
Coulson, C. A. (1957). *Research, Lond.* **10,** 149.
Cowley, R. A. (1962). *Phys. Rev. Lett.* **9,** 159.
Devonshire, A. F. (1949). *Phil. Mag.* **40,** 1040.
Devonshire, A. F. (1954). *Adv. Phys.* **3,** 85.
Devonshire, A. F. (1957). *Phil. Mag.* **2,** 1027.
Di Domenico, M., Jr., Wemple, S. H. and Jayaraman, A. (1968). Ninth International Conference on the Physics of Semiconductors, Moscow. To be published.
Fatuzzo, E. and Merz, W. J. (1967). "Ferroelectricity." North-Holland Publishing, Amsterdam.
Fatuzzo, E., Harbeke, G., Merz, W., Nitsche, R., Roetschi, H. and Ruppel, W. (1962). *Phys. Rev.* **127,** 2036.
Forsbergh, P. W., Jr. (1954). *Phys. Rev.* **93,** 686.
Frazer, B. C., McKeown, M. and Pepinsky, R. (1954). *Phys. Rev.* **94,** 1435.
Fridkin, V. M., Gulyamov, K., Lyakhovitskaya, V. A., Nosov, V. N. and Tikhomirova, N. A. (1966). *Soviet Phys. solid St.* **8,** 1510.
Fröhlich, H. (1949). "Theory of Dielectrics", Appendix 3. Clarendon Press, Oxford, England.
Gavrilova-Podol'skaya, G. V., Gabuda, S. P. and Lundin, A. G. (1967). *Soviet Phys. solid St.* **9,** 911.
Gesi, K., Ozawa, K. and Takagi, Y. (1965). *J. phys. Soc. Japan* **20,** 1773.
Ginzburg, V. L. (1963). *Soviet Phys. Usp.* **5,** 649.
Goswami, A. K. (1966). *J. phys. Soc. Japan* **21,** 1037.
Goswami, A. K. and Cross, L. E. (1968). *Phys. Rev.* **171,** 549.
Hablützel, J. (1939). *Helv. Phys. Acta* **12,** 489.
Hamano, K., Nakamura, T., Ishibashi, Y. and Ooyane, T. (1965). *J. phys. Soc. Japan* **20,** 1886.
Hegenbarth, E. (1964). *Solid St. Phys.* **6,** 333.
Hegenbarth, E. and Frenzel, C. (1967). *Cryogenics* **7,** 331.
Hegenbarth, E. and Ullwer, S. (1967). *Cryogenics* **7,** 306.
Holden, A. N., Merz, W. J., Remeika, J. P., and Matthias, B. T. (1956). *Phys. Rev.* **101,** 962.
Hulm, J. K., Matthias, B. T., and Long, E. A. (1949). *Phys. Rev.*, **79,** 885.
Jona, F. and Shirane, G. (1960). *Phys. Rev.* **117,** 139.

Jona, F. and Shirane, G. (1962). " Ferroelectric Crystals." Macmillan, New York.

Kabalkina, S. S. and Vereschchagin, L. F. (1962). *Soviet Phys. Dokl.* **7,** 310.

Kabalkina, S. S., Vereschchagin, L. F. and Shulenin, B. M. (1962). *Soviet Phys. Dokl.* **7,** 527.

Känzig, W. (1957). *In* " Solid State Physics ", ed. by F. Seitz and D. Turnbull, Vol. 4, pp. 1–197. Academic Press, New York.

Kern, R. (1962). *Physics Chem. Solids* **23,** 249.

Khanna, R. K., Decius, J. C. and Lippincott, E. R. (1965). *J. chem. Phys.* **43,** 2974.

Kikuchi, A., Oka, Y. and Sawaguchi, E. (1967). *J. phys. Soc. Japan* **23,** 337.

Klimowski, J. (1962). *Soviet Phys. solid St.* **2,** 456.

Klimowski, J. and Pietrzak, J. (1960). *Proc. phys. Soc.* **75,** 456.

Kobayashi, K. K. (1968). *J. phys. Soc. Japan* **24,** 497.

Landau, L. D. and Litshitz, E. M. (1962). " Statistical Physics." Pergamon Press, London.

Lawless, W. N. and Gränicher, H. (1967). *Phys. Rev.* **157,** 440.

Leonidova, G. G. and Polandov, I. N. (1963). *Soviet Phys. oslid St.* **4,** 1916.

Leonidova, G. G. and Volk, T. R. (1966). *Soviet Phys. solid St.* **7,** 2694.

Leonidova, G. G., Netesova, N. P. and Volk, T. R. (1967). *Soviet Phys. solid St.* **9,** 454.

Lippincott, E. R. and Schroeder, R. (1955). *J. chem. Phys.* **23,** 1099.

Lyddane, R. H., Sachs, R. G. and Teller, E. (1941). *Phys. Rev.* **59,** 673.

Lytle, F. W. (1964). *J. appl. Phys.* **35,** 2212.

Mason, W. P. (1950). " Piezoelectric Crystals and their Application to Ultrasonics", Chapter 11. Van Nostrand, New York.

Mazzi, F., Jona, F. and Pepinsky, R. (1957). *Z. Kristallogr.* **108,** 359.

Megaw, H. D. (1957). " Ferroelectricity in Crystals." Methuen, London.

Merz, W. J. (1950). *Phys. Rev.* **77,** 52.

Merz, W. J. (1956). *Phys. Rev.* **103,** 565.

Merzbacher, E. (1961). " Quantum Mechanics." John Wiley, New York.

Minomura, S., Kawakubo, T., Nakagawa, T. and Sawada, S. (1964). *J. appl. Phys. Japan* **3,** 562.

Mitsui, T. (1958). *Phys. Rev.* **111,** 1259.

Moreno, M. and Gränicher, H. (1964). *Helv. phys. Acta* **37,** 625.

Mori, T., Tamura, H. and Sawaguchi, E. (1965). *J. phys. Soc. Japan* **20,** 281.

Mueller, H. (1935). *Phys. Rev.* **47,** 175.

Mueller, H. (1940). *Phys. Rev.* **58,** 565 and 805.

Mylov, V. P., Polandov, I. N. and Strukov, B. A. (1966). *Soviet Phys. JETP Lett.* **4,** 172.

Mylov, V. P., Polandov, I. N. and Strukov, B. A. (1968). *Soviet Phys. solid St.* **9,** 2375.

Pepinsky, R. and Vedam, K. (1959). *Phys. Rev.* **114,** 1217.

Pirenne, J. (1949). *Helv. phys. Acta* **22,** 479; also *Physica* **15,** 1019.

Pockels, F. (1906). " Lehrbuch der Kristalloptik." Teubner, Leipzig.

Polandov, I. N. and Mylov, V. A. (1964). *Soviet Phys. solid St.* **6,** 393.

Polandov, I. N., Strukov, B. A. and Mylov, V. P. (1967). *Soviet Phys. solid St.* **9**, 1153.

Rapoport, E. (1966). *Phys. Rev. Lett.* **17**, 1097.

Reese, W. (1968). To be published.

Reid, C. (1959). *J. chem. Phys.* **30**, 182.

Richard, M. (1961). *Ann. Phys.* **8**, 333.

Sakiyama, M., Kimoto, A. and Seki, S. (1965). *J. phys. Soc. Japan* **20**, 2180.

Samara, G. A. (1965a). *Physics Chem. Solids* **26**, 121.

Samara, G. A. (1965b). *Bull. Am. Ceram. Soc.* **44**, 638.

Samara, G. A. (1966). *Phys. Rev.* **151**, 378.

Samara, G. A. (1967). *Phys. Lett.* **25**A, 664.

Samara, G. A. (1968a). *Phys. Rev.*, **173**, 605.

Samara, G. A. (1968b). *Physics Chem. Solids* **29**, 870.

Samara, G. A. (1968c). *Phys. Lett.* **27**A, 232.

Samara, G. A. and Anderson, D. H. (1966). *Solid State Comm.* **4**, 653.

Samara, G. A. and Giardini, A. A. (1965). *Phys. Rev.* **140**, A954.

Sawada, S., Nomura, S. and Fujii, S. (1958a). *J. phys. Soc. Japan* **13**, 1549.

Sawada, S., Nomura, S., Fujii, S. and Yoshida, I. (1958b). *Phys. Rev. Lett.* **1**, 320.

Sawada, S., Nomura, S. and Asao, Y. (1961). *J. phys. Soc. Japan* **16**, 2486.

Shirane, G. and Takeda, A. (1952). *J. phys. Soc. Japan* **7**, 1.

Shirane, G., Frazer, B. C., Minkiewicz, V. J., Leake, J. A. and Linz, A. (1967a). *Phys. Rev. Lett.* **19**, 234.

Shirane, G., Nathans, R. and Minkiewicz, V. J. (1967b). *Phys. Rev.* **157**, 396.

Silsbee, H. B., Uehling, E. A. and Schmidt, V. H. (1964). *Phys. Rev.* **133**, A165.

Slater, J. C. (1941). *J. chem. Phys.* **9**, 16.

Slater, J. C. (1950). *Phys. Rev.* **78**, 748.

Spitzer, W. G., Miller, R. C., Kleinman, D. A. and Howarth, L. E. (1962). *Phys. Rev.* **126**, 1710.

Takagi, Y. (1948). *J. phys. Soc. Japan* **3**, 271.

Tatsuzaki, I., Itoh, K., Ueda, S. and Shindo, Y. (1966). *Phys. Rev. Lett.* **17**, 198; (1967) **18**, 453.

Taylor, G. W. and Lechner, B. J. (1968). *J. appl. Phys.* **39**, 2372.

Tokunaga, M. and Matsubara, T. (1966). *Prog. theor. Phys., Osaka* **35**, 581.

Triebwasser, S. (1957). *Physics Chem. Solids* **3**, 53.

Ubbelohde, A. R. and Gallagher, K. J. (1955). *Acta Cryst.* **8**, 71.

Umebayashi, H., Frazer, B. C., Shirane, G. and Daniels, W. B. (1967). *Solid State Comm.* **5**, 591.

van den Hende, J. H. and Boutin, H. P. (1963). *Acta Cryst.* **16**, A184.

von Arx, A. and Bantle, W. (1944). *Helv. phys. Acta* **17**, 298.

von Hippel, A. (1963). Technical report 178, Laboratory for Insulation Research, Massachusetts Institute of Technology, Cambridge, Massachusetts (unpublished).

Valasek, J. (1921). *Phys. Rev.* **17**, 475.

Vedam, K., Okaya, Y. and Pepinsky, R. (1960). *Phys. Rev.* **119**, 1252.

Wemple, S. H., Jayaraman, A. and Di Domenico, M. (1966). *Phys. Rev. Lett.* **17**, 142.

West, J. (1930). *Z. Kristallogr.* **74**, 306.

Wieder, H. H. and Parkerson, C. R. (1966). *Physics Chem. Solids* **27**, 247.
Yamada, Y., Shibuya, I. and Hoshino, S. (1963). *J. Phys. Soc. Japan* **18**, 1594.
Zheludev, I. S., Tikhomirova, N. A. and Fridkin, V. M. (1962). *Kristallografiya* **7**, 795.

CHAPTER 4

Electrical Conductivity and Electronic Transitions at High Pressures

N. H. MARCH

Department of Physics, The University, Sheffield, England

I. Outline

In discussing the modifications to be expected when we squeeze solids, two approaches appear to be open to us, familiar from text-book accounts of band theory. The first is to consider the broadening of the atomic energy levels as we assemble the crystal by bringing up the atoms from infinity. Here, caution is required while the atoms are far apart, for band theory is not good and the Coulomb correlation keeps electrons on their own atoms, so to speak. However, by the time the stable lattice spacing of many solids at atmospheric pressure is reached, one can usefully describe the system in terms of its energy bands, and one might attempt to calculate these, using atomic wave functions, by the so-called tight-binding method. Even for Na, where these wave functions for the outer ($3s$) electrons overlap greatly, this method has been tried and can yield sensible results, although with undue labour in this case.

The second approach, which at present is largely academic, is to think of such enormous pressures applied to the crystal that even the core electrons can no longer be thought of as associated with particular atoms and all the electrons belong to the crystal as a whole. Here, the atomic correspondence discussed above no longer affords a useful description but the kinetic energy of the electrons is so high compared with potential energy terms that the free electron correspondence is more appropriate. In the extreme limit of pressure ionization all the electrons will be in wide bands, which above some pressure will become completely overlapping, all forbidden energy regions having disappeared.

In this high pressure regime, the statistical theory of Thomas and Fermi is able to give us a good deal of information. It leads us to expect properties to vary smoothly with atomic number for instance, and although, will as be seen, truly enormous pressures are required to cause such pressure ionization, it is useful to enquire as to the consequences of such behaviour in the extreme high pressure limit. Here all solids are expected to become metallic, and first the electrical resistivity will be discussed, both for liquid as well as solid metals, under such extreme conditions.

These predictions are then confronted with the observed behaviour of simple metals, and are found to be qualitatively different. However, one can then enquire when the atomic core structure will begin to re-establish itself as the atoms are separated from the high pressure limit. This is done by using a model of pseudoatoms (ion plus electronic screening cloud, resulting in a neutral entity) and enquiring when a bound state is formed round one of these. With a finite energy gap for the bound state around a single pseudoatom, the wave functions overlap from one atom to the next, giving the " bound " state now a finite width. When the energy gap is greater than roughly half the band width, a forbidden energy region will develop, and one can speak, crudely, of the 1*s* shell re-establishing itself.

In a simple solid like He, with just two electrons per atom, the 1*s* band would then be full, and separated by a forbidden energy region from the conduction band. Insulating behaviour should be apparent at sufficiently low temperature. Looking at the problem the other way round, bringing the atoms together from a large separation, the filled valence band will eventually overlap the empty conduction band, and we shall have a metallic transition. We show below that the pressure at which such a transition occurs, which may loosely be called " breaking open " the 1*s* shell, is of the order of 10^5 kb in face centred-cubic He.

This discussion of He prompts us to consider next the pressures at which crystals formed from heavier noble gases will have overlapping valence and conduction bands; that is the pressure at which an initially forbidden energy region shrinks to zero. Gandel'man (1965) and Ross (1968) have independently considered this problem. Plotting the critical pressures estimated by Ross for Ar and Xe, and adding the above result for He, we can interpolate to obtain results for Ne and Kr.

Using the arguments from the high pressure regime, we then try to assess when valence and conduction bands will overlap for the alkalis, in order to understand the really high pressure regime of resistivity, studied experimentally by Drickamer and his co-workers. Some estimates of the way the electrical resistivity might be affected when bands overlap have been made by Stocks and Young (1969) for the alkalis. However, these estimates seem to be valid only at very much higher pressures than those explored so far by Drickamer. Even then, the behaviour they find is vastly different from that to be expected in the pressure-ionized regime.

Amassing the information known about the core electrons in solids, the available band structures and some pseudoatom results for bound states, a very primitive estimate has been made of when forbidden energy regions will shrink to zero for various solids. One would expect the correlation that has been attempted with atomic number to become seriously blurred out as soon as the low pressure regime is reached.

This leads one then to discuss the effect of pressure on the details of band structure and on Fermi surface shape at low pressures. Since the review article by Dugdale (1968), interesting developments have occurred in the theory for the noble metals and some experimental work on the alkaline earths has also appeared.

So far (Sections I–V), the article concerns itself entirely with the band theory interpretation of electrical conductivity and electronic transitions. However, as Wigner (1938) and Mott, in a somewhat different context and in a whole series of papers, have pointed out, in low electron density systems Coulomb repulsion between electrons can lead to insulating behaviour in cases for which band theory would predict metallic conduction. For example Li metal, with one conduction electron per atom at the normal spacing, would be expected to become insulating at sufficiently large lattice spacing, and it would undergo a Mott transition (for a recent survey of work on the Mott transition and related aspects, the reader should consult the October 1968 issue of *Reviews of Modern Physics*). Because of the present fundamental interest in the role of electron correlations, a survey is made of recent progress in three areas in which the one-electron description is inadequate. In particular, consider:

(1) The search for so-called excitonic phases. Here Yb and Sr are considered as possible candidates, as discussed in some detail in Section VI.
(2) The somewhat related problem of spin and charge density waves. Here Cr and Cr based alloys are discussed, with a few remarks also on K.
(3) The role of pressure in affecting superconductivity.

Finally, and very briefly, it is speculated that under the extreme high pressures referred to in the first part of this article, a plethora of low symmetry crystal structures might be expected, due to the rather curious form of the interionic force law in this regime.

II. Pressure Ionization and Electronic Transitions

A. Band Theory Study of Helium

To study the behaviour of a simple solid under such extreme conditions of pressure that an inner shell is broken open, Simcox and March (1962) considered the overlap of the first two energy bands in face-centred cubic helium. The Brillouin zone for this lattice is shown in Fig. 1.

The philosophy adopted in this work was to start from the metallic regime, establish an approximately self-consistent field as though we were dealing with a divalent metal, and then study in this field the variation of the valence and conduction band edges with lattice spacing. At the same time, using this metallic model, the equation of state could be obtained.

Referring to Fig. 2, the important energy levels are an s-like state at W, W_s, and a p-like state at L, L_p', which lie in the lowest and next highest band, respectively. These were found to become equal for $r_w = 0{\cdot}91'\, a_0$,

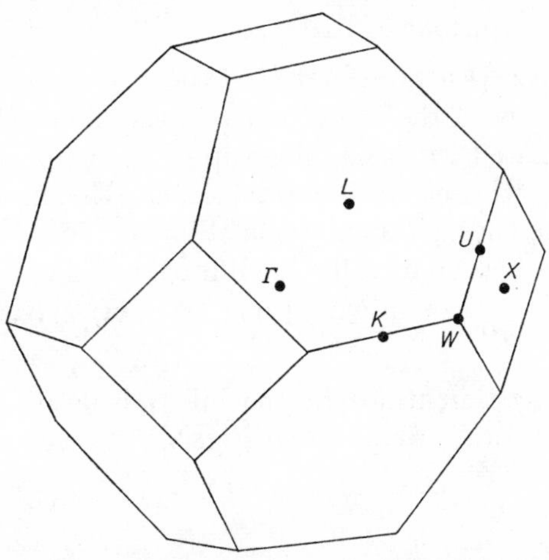

FIG. 1. Brillouin zone for f.c.c. direct lattice, showing important symmetry points Γ, W, U, K, X and L.

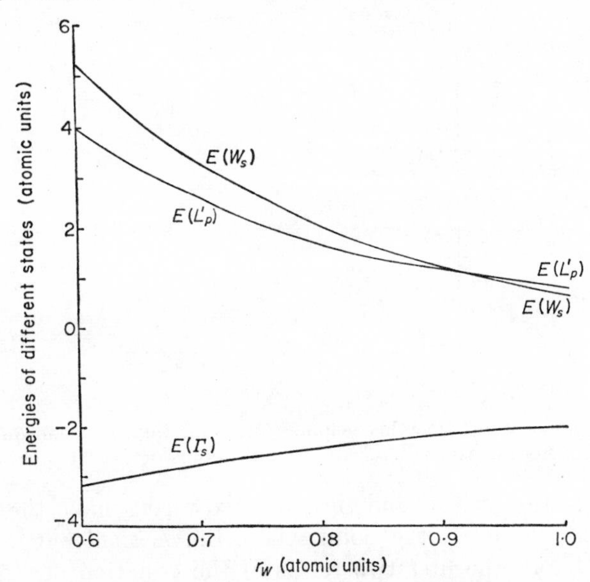

FIG. 2. Pressure dependence of electronic levels Γ_s, W_s and L'_p for f.c.c. helium near metallic transition.

r_w being the radius of the Wigner–Seitz equal-volume sphere, and a_0 the Bohr radius. The variation of these two levels with r_w is displayed in Fig. 2 together with Γ_s, corresponding to $\mathbf{k}=0$. A calculation of the level X_s verified that W_s was the valence band edge, and not X_s.

The total electronic energy of the metal, at absolute zero, can be written as the sum of four contributions,

$$E_{\text{total}} = E_0 + E_{\text{Fermi}} + E_{\text{exchange}} + E_{\text{correlation}}, \tag{1}$$

E_0 being the energy of the lowest state in the band, E_{Fermi} describing the energy associated with the filling up of the band, and the last two terms arise from the electron interactions. Estimating these terms from their band theory calculations, Simcox and March found the equation of state for the metallic modification of f.c.c. He as

$$10^{-8}p = \frac{1{\cdot}620}{r_w^5} - \frac{1{\cdot}098}{r_w^4} - \frac{0{\cdot}0144}{r_w^3} + \frac{0{\cdot}0316}{r_w^2} + \cdots \tag{2}$$

Here the pressure p is in atmospheres while r_w is measured in units of a_0, and we show the form of eqn (2) in Fig. 3. As we saw above, at the

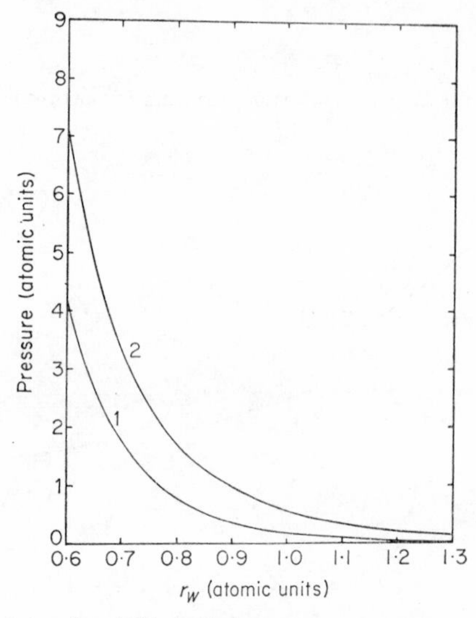

FIG. 3. Equation of state of metallic helium. Curve 1 is obtained from eqn (2); Curve 2 is the free Fermi gas result. 1 atomic unit of pressure $= 2{\cdot}90 \times 10^8$ atm.

metallic transition $r_w \sim 1$, and the first two terms make the dominant contribution. Thus, at the point where the K shell merges into the conduction band, the first two terms of the equation of state (2) are adequate. In fact, these agree with the equation of state given earlier by the author (March, 1955) from Thomas–Fermi theory, for a general atomic number Z, namely

$$p = \frac{h^2}{5m}\left(\frac{3}{8\pi}\right)^{2/3}\frac{Z^{5/3}}{v^{5/3}}$$

$$\times\left\{1 - \frac{2\pi me^2}{h^2}(4Zv)^{1/3} - \frac{10\pi me^2}{3^{2/3}h^2}(4Zv)^{1/3}\left(\frac{6^{1/3}}{4(\pi Z)^{2/3}}\right) + \cdots\right\} \tag{3}$$

which is probably adequate for calculating the pressures at which the K shell is broken open, at least for the light elements, once the critical volume is known from a band theory calculation such as that reported above for He. Ten Seldam (1957) had earlier made a study of band overlap in He, starting however from an atomic-like picture. The transition volume was given reasonably well by this alternative approach, and gives confidence in the estimate we have for the transition pressure of $1{\cdot}0 \times 10^8$ atm.

B. CRUDER APPROACH TO METALLIC SCREENING

The self-consistent field of Simcox and March was constructed by the cellular method, whereas, as mentioned above, Ten Seldam started essentially from a superposition of neutral atoms. The type of band calculations reported above are quite lengthy and a cruder method is therefore useful; such a method has been developed recently by the author (March, 1968a). The idea is suggested by the agreement between the approach of Ten Seldam, based on superposition of localized potentials, and the self-consistent cellular calculations reported above, and additionally by the model of pseudoatoms in vogue in the theory of metals (see for example, Ziman, 1964). If the periodic self-consistent field is calculated by, say, the Thomas–Fermi theory which has been widely used in high pressure studies (see for example, March, 1957, and for more recent work Gilvarry, 1968), for a chosen atomic number $Z > 2$, one would again be involved in a full band-structure calculation. These Thomas–Fermi fields have however been used to calculate the bands for a few metals: see Sections III and V. Therefore, we shall adopt the point of view that each ion is screened by the Fermi gas. In calculating the screening in the pressure ionized regime, all Z electrons in each atom are to be included. Then, dealing with this screening by the linearized Thomas–Fermi theory, the potential around a single ion is given by

$$V = -\frac{Z}{r}\exp(-qr): \quad q^2 = \frac{4k_f}{\pi a_0} \tag{4}$$

where k_f is the Fermi wave number, related to the electron density $n(\equiv Z/v$, where v is the atomic volume) by

$$n = \frac{k_f^3}{3\pi^2}. \tag{5}$$

Then if the sites in the crystal lattice are denoted by $\mathbf{R}_n$, a first approximation to the crystal potential in the pressure ionized regime is taken as

$$V_{\text{periodic}} = \sum_{\mathbf{R}_n} \frac{-Z}{|\mathbf{r}-\mathbf{R}_n|} \exp(-q|\mathbf{r}-\mathbf{R}_n|). \tag{6}$$

It must be stressed, however, that a given periodic potential can be split up into a sum of localized potentials in essentially an infinite number of ways. This is because the only non-zero Fourier components of V_{periodic} are $V_{\mathbf{K}_n}$, when $\mathbf{K}_n$ denotes the reciprocal lattice vectors, and therefore the Fourier transforms of the localized potentials are only defined at these values of $\mathbf{k}$, leaving a great deal of ambiguity. However, eqn (6) is the simplest representation of the periodic potential we can construct in the pressure ionized regime which has approximate self-consistency built in, through the k_f dependence of the screening radius q^{-1} defined by eqn (4). In a more proper, non-linear, Thomas–Fermi theory, q^{-1} would become a function of atomic number Z.

When core electrons are present, which are known to be little perturbed from atomic wave function form, eqn (6) would not be adequate and the Hartree–Fock fields of the core should obviously be used. Even when pressure ionization has occurred, Ten Seldam's work suggests that a superposition of atomic potentials might still be a useful approximation to V_{periodic}. This is, as has been stressed, not saying anything about the similarity of the screened Coulomb potential in eqn (6) and an atomic potential, except for their Fourier components at the reciprocal lattice vectors $\mathbf{K}_n$. This point is worth further numerical study.

The above picture of superposition of localized potentials centred on every lattice site is useful, as will be seen, not only for estimating the pressures at which electronic transitions occur but also in calculating the electronic conductivity of metals (cf Sections II.D and IV) and for discussing interatomic forces (cf Section IX).

C. FORMATION OF BOUND STATES ROUND SCREENED IONS

We now wish to consider the formation of bound states around screened point ions described by eqn (4) as the density is reduced from the high-pressure regime in which the atoms are completely ionized.

It is known that, for $Z=1$, the critical value of the screening radius q^{-1} for bound state formation is $0{\cdot}80\,a_0$ (see for example, Isenberg, 1950). However, we shall try to obtain the Z-dependence of the first electronic transition as we come from the pressure ionized regime, and in order to do this the author has used a (less accurate) variational method based on the hydrogen $1s$ wave function

$$\phi=\left(\frac{\alpha}{\pi}\right)^{3/2}\exp\left(-\alpha r\right) \tag{7}$$

where α is used as a variational parameter. Then the critical value of the screening radius is given by

$$q_{\text{bound}}^{-1} = \frac{a_0}{Z}. \tag{8}$$

This overestimates the screening radius by about 20% for $Z=1$. Using eqn (8) in conjunction with eqn (4) we find

$$k_{f\,(\text{bound})} = \frac{\pi Z^2}{4a_0}. \tag{9}$$

The above estimate eqn (9) cannot represent physically the electronic transition, which is of course dependent on the periodic potential assumed, and clearly cannot be affected by the lack of uniqueness in the choice of the localized potential, whereas eqn (9) depends specifically on the choice of eqn (4). The reason why eqn (9) is not the answer sought is because overlap of the wave functions eqn (7) on the various lattice sites $\mathbf{R}_n$ now occurs, and this overlap broadens the bound state. If this overlap were accounted for exactly, all choices of the localized potential would lead in the end to the same answer for the critical Fermi wave number. With the approximate method used here this will not be exactly true.

Thus, the argument proceeds as follows. With k_f given by eqn (9), a bound state just breaks off from the conduction band. However, the wave function eqn (7) is then very diffuse, and great overlap occurs. The bound state is thereby merged again with the band. Thus, a $k_f < k_{f\,(\text{bound})}$ must be chosen to yield a finite energy gap E_g between the bound state and the bottom of the conduction band. For this energy, the variational parameter α in eqn (7) is known, and this determines the extent of the bound state wave function. Including the overlap by the tight-binding method (see for example Mott and Jones, 1936; and for the details of the present application see March, 1968a) we can say that the condition for the bound state to form is roughly that the band width E_b is such that

$$E_g \sim \tfrac{1}{2} E_b. \tag{10}$$

Carrying through the calculations, and representing the critical value of k_f by a power law, as suggested by eqn (9), we find

$$k_{f\,(\text{critical})} \approx \frac{0{\cdot}6 Z^{5/3}}{a_0}. \tag{11}$$

This result was obtained for small Z only and may need refining for large atomic numbers. To see the main term in the Z-dependence of the pressure p at which the $1s$ shell breaks open, we restrict ourselves to the Fermi gas term in eqn (3), and then we find

$$p_{(\text{critical})} \propto k_f^5{}_{(\text{critical})} \propto Z^{25/3}, \tag{12}$$

showing a rapid rise with atomic number.,

Two points are worth noting about the result in eqn (11). First, the coefficient in eqn (11) is relatively insensitive to structure, differing by only 5% between face-centred and body-centred cubic structures. Secondly, the critical k_f for $Z=2$ according to eqn (11) is $1{\cdot}90/a_0$, whereas the more accurate band overlap calculation of Simcox and March gave a value of $2{\cdot}66/a_0$. This suggests that a more accurate value of the coefficient of $Z^{5/3}/a_0$ in eqn (11) will be 0·7 or 0·8. Using the value 0·8, the transition pressure at which the K shell in Li will break open is around 5×10^6 kb from eqn (3). This value is to be compared with the He value of 10^5 kb.

D. Electrical Conductivity of Pressure-Ionized Metals†

While the pressures at which even the light elements undergo pressure-ionization are enormously high, it is of some interest to enquire as to what will be the qualitative trends of the electrical conductivity in this regime compared with the low pressure experiments for simple metals (see Section IV). The theory outlined below is valid for temperatures T greater than the Debye temperature Θ.

1. *Theory for Liquid Metals*

It will be convenient to present the theory first in a form suitable for dealing with the liquid state. If ρ_0 is the number of ions per unit volume and Z is the atomic number, then ρ_0 is related to the number of electrons per unit volume, n, by

$$n = Z\rho_0. \tag{13}$$

The ions scatter electrons and the scattering potential energy for a single ion is denoted by $U(K)$, where we have chosen to work with the Fourier transform of the potential energy $U(r)$. $U(K)$ is chosen so that

$$U(0) = -\tfrac{2}{3}E_f \tag{14}$$

where E_f is the Fermi energy; this follows only in the linear theory we are using. Later $U(r)$ will be identified with the screened coulomb field in eqn (4) round the nucleus of charge Ze in the pressure-ionized regime, but this identification must *not* otherwise be made.

† Note added in proof. The behaviour of liquids in the neighbourhood of the critical point has been reviewed recently by D. Greenwood and R. G. Ross (to appear in *Progr. Mater. Sci.*).

The structure of the liquid is also needed. This gives the instantaneous picture of the distribution of the other ions around an ion which is chosen as origin. This is generally described by the radial distribution function $g(r)$, such that $g(r) \to 1$ as $r \to \infty$. The structure factor $S(K)$ is the Fourier transform of $g(r)-1$. For liquids under normal conditions, $S(K)$ can be measured by X-ray or neutron experiments. Such measurements are, of course, not feasible at the high pressures of interest here. Fortunately, an approximate theory can be given in terms of $U(K)$, as will be seen below.

The basic formula given by Ziman (1961) for the resistivity ρ_l of the liquid, in the Born approximation, is then

$$\rho_l = \frac{3\pi}{\hbar e^2 v_f^2 \rho_0} \int_0^1 S(K)|U(K)|^2 4 \left(\frac{K}{2k_f}\right)^3 \mathrm{d}\left(\frac{K}{2k_f}\right) \tag{15}$$

where $v_f = \hbar k_f/m$ is the Fermi velocity, and m the electron mass. For a given temperature the problem is characterized then by the Fermi wave number k_f and the number of conduction electrons per atom Z, ρ_0 being known from eqns (13) and (5).

It will be convenient to write $S(K)$ and $U(K)$ in terms of the dielectric function $\epsilon(K)$ of the Fermi gas in the extreme high-pressure regime. $\epsilon(K)$ gives us the linear response of the Fermi gas to a charged ion, and in terms of it we have (March, 1968a)

$$U(K) = -\frac{2}{3}\frac{E_f q^2}{K^2 \epsilon(K)} \tag{16}$$

and

$$S(K) = \frac{1}{1-(ZU(K)/k_B T)}. \tag{17}$$

It must be stressed that eqn (17) needs drastic modification when core electrons are present and must only be used when all shells are broken down.

Two expressions for $\epsilon(K)$ are available (see for example, Kittel, 1963, or March, 1968b).

(1) Using semiclassical theory

$$K^2 \epsilon(K) = K^2 + q^2. \tag{18}$$

This follows immediately by taking the Fourier transform of eqn (4) and comparing the result with eqn (16).

(2) Using wave theory

$$K^2 \epsilon(K) = K^2 + \frac{k_f}{\pi q_0} g\left(\frac{K}{2k_f}\right) \tag{19}$$

where

$$g(x) = 2 + \frac{(x^2-1)}{x} \ln \left| \frac{1-x}{1+x} \right|. \tag{20}$$

The wave theory includes the de Broglie wavelength $2\pi/k_f$ for electrons at the Fermi surface in an essential way (π/k_f is the precise wavelength which dominates the problem).

Results for resistivities of liquid metals calculated in this way are recorded by March (1968a), where a primitive theory of the melting temperature is also considered. Since the main interest here is in solids, and since the principal conclusion from the calculations (eqn (28) below) is the same for solids and liquids, we turn immediately to the problem of the resistivity of solid metals at high temperatures.

2. *Solid Metals at High Temperatures*

If we were dealing with scattering from independent ions, the structure factor $S(K)$ in eqn (17) would simply be unity. In the solid, for temperatures high compared with the Debye temperature Θ, a semi-quantitative approximation is then obtained if we replace $S(K)$ by its long wavelength limit $S(0)$. This in turn is given by the thermodynamic result

$$S(0) = \rho_0 k_B T K_T \tag{21}$$

where K_T is the isothermal compressibility. Introducing the velocity of sound v_s, we shall write eqn (21) to sufficient accuracy for present purposes as (putting the ratio of specific heats γ equal to unity)

$$S(0) \sim \frac{k_B T}{M v_s^2} \tag{22}$$

where M is the ionic mass. Recalling that the elementary Debye theory gives (cf Mott and Jones, 1936)

$$\Theta = \frac{h}{k_B} \left(\frac{3}{4\pi} \right)^{1/3} \frac{v_s}{v^{1/3}} \tag{23}$$

where v is the atomic volume, eqn (15) can be written in the approximate form

$$\rho_s = \frac{K_i k_B T}{M \Theta^2} \tag{24}$$

where K_i clearly depends on the scattering from a single ion, and from eqn (15) is proportional to

$$\int_0^1 |U(K)|^2 \left(\frac{K}{2k_f} \right) \mathrm{d} \left(\frac{K}{2k_f} \right). \tag{25}$$

Consistent with the model of dielectric screening of ions, we can also construct a first-order theory for the Debye temperature Θ. This was done in essence by Bohm and Staver (1952) who pointed out that the velocity of sound† is given by

$$v_s = \left(\frac{Zm}{3M}\right)^{1/2} v_f \tag{26}$$

where m is the electronic mass. Combining eqns (23) and (26) gives

$$\Theta = \frac{3^{1/6}}{2^{5/3}\pi^{2/3}} \frac{\hbar^2}{mk_B} \left(\frac{m}{M}\right)^{1/2} \frac{Z^{5/6}}{v^{2/3}}. \tag{27}$$

ρ_s has been evaluated (March, 1968a) using both eqns (18) and (19) for the dielectric function $\epsilon(K)$. The semiclassical results based on eqn (18) can then be expressed in universal form, by plotting $C\rho_s k_f/T$ against $k_f a_0$, where $C = 4\hbar e^2/27\pi^3 mk_B$.

Results are shown in Fig. 4 and we have verified that for large k_f the results are affected only in a minor way by using the wave theory eqn

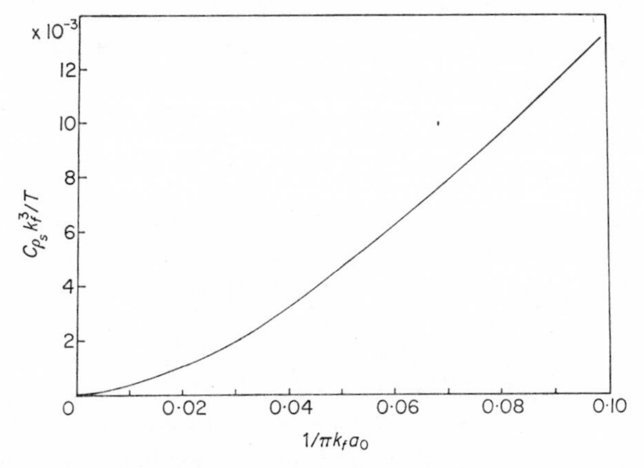

FIG. 4. High-temperature resistivity ρ_s of pressure-ionized metals. Constant C is given by $4\hbar e^2/27\pi^3 mk_B$, k_f is the Fermi wave number and $a_0 = \hbar^2/me^2$.

(19). The essential point to stress is that the dependence of ρ_s on volume v, in the extreme high-density limit $v \to 0$, is given by

$$\rho_s \propto v^{5/3} \ln v. \tag{28}$$

This result, although with different constants, is also true for the liquid

† The Bohm–Staver theory is a linear theory. A non-linear theory for point-ions is sketched out in Appendix I, where a generalization of eqn (26) to metallic alloys is also effected. Some other properties of alloys are considered in Section VIII.

just above the melting temperature. In terms of the pressure p, if we use simply the free Fermi gas equation of state given by the first term in eqn (3), we find

$$\rho_s \propto p^{-1} \ln p. \tag{29}$$

The presence of the logarithmic term in eqn (28), while weak compared with $v^{5/3}$ term, is interesting theoretically, as this transport property cannot then be expanded as a Taylor series in the lattice parameter or the average interelectronic spacing. The result in eqn (28) indicates that the resistivity should decrease with increasing pressure and, while we expect this to become true at sufficiently high pressures, if we examine measurements on, for example, the alkalis under pressure, then up to about 100 kb the behaviour is quite different, as discussed in Section IV.

III. Metallic Transitions at Lower Pressures

Clearly, to attack the regime of pressures attainable in the laboratory, or even in shock wave experiments, we have to deal now with the core electrons. In all solids, other than hydrogen, as has been seen, the K shell will be closed for pressures $< 10^5$ kb, and at that pressure for solids other than He will probably constitute quite a narrow band. One of the matters of interest then is whether the nature of the conduction process, and/or the number of conduction electrons per atom, can be changed by pressure. Following the work on He, some interesting studies have been carried out by Gandel'man (1965) and Ross (1968) on the heavier noble gas crystals, and these calculations give us further information about the disappearance of forbidden energy regions.

In practice, as a result of explosive shock experiments, it is now possible to compress liquid inert gases to densities of two or three times the normal densities (van Thiel and Alder, 1966). It is this improvement in really high pressure techniques which has motivated further work on metallic transitions.

A. Metallic Transitions in Heavier Noble Gas Crystals

Ross (1968) used the Wigner–Seitz method in his calculations for argon and xenon. Here the Schrödinger equation is solved in a spherical cell having a volume equal to the atomic volume. The method basically was designed to calculate the wave function for $\mathbf{k}=0$ but, as discussed below, it is also useful for estimating band widths. For the spherical potential assumed within the cell, the boundary conditions on the radial wave function $R(r)$ at the surface of the sphere, $r = r_W$, are given by

$$R'(r_w) = 0, \qquad l \text{ even}$$

and

$$R(r_w) = 0, \qquad l \text{ odd},$$

where l is the orbital angular momentum quantum number, as discussed by Brooks (1948). The solutions with $R'(r_w) = 0$ for l odd, and $R(r_w) = 0$ for l even are often useful for estimating the energy at the maximum **k** values in the band.

The crystal potential used by Ross was obtained by solving the Thomas–Fermi–Dirac equation for the appropriate cell volume. The crystal potential $V(\mathbf{r})$ was then constructed from this potential, minus the exchange part. If we denote the resulting potential by V_-, then exchange was added separately by writing

$$V(\mathbf{r}) = V_-(\mathbf{r}) - 6\left[\frac{3\rho(\mathbf{r})}{8\pi}\right]^{1/3}, \tag{30}$$

where $\rho(\mathbf{r})$ is taken from the Thomas–Fermi–Dirac theory. Ross took $V_-(\mathbf{r})$ to be zero at the cell boundary in his work.

1. *Argon*

Gandel'man (1965) used a non-linear Thomas–Fermi potential and a slightly more sophisticated form of the Wigner–Seitz method in his calculations on argon. By studying band extrema as a function of lattice spacing, he predicted that Ar should become metallic at about 1300 kb at 0°K, and he estimated the corresponding critical volume as 7 cc/mole.

There are, however, some objections to the use of the Thomas–Fermi potential in this case. Firstly, Gandel'man predicts a band gap of only 6 eV under normal conditions; but Baldini (1962) has measured the ultraviolet spectra of Ar, Kr and Xe and from his experiments he deduces that the conduction band lies about 14 eV above the valence band for Ar. Also, Gandel'man finds from his calculations that the gap appears between $3p$ and $3d$ like bands. From other evidence (for example, the more accurate band theory studies of Matthiess, 1964) it seems that the bottom of the conduction band is $4s$-like.

Although Gandel'man's choice of potential should be a better approximation at high pressures, the facts discussed above suggest that his initial estimate of the metallic transition pressure may be too low and the recent work of Ross has shown this to be so. In addition to his

Wigner–Seitz calculations, Ross has also carried out augmented plane wave (APW) calculations, although these are not yet published. The general form of the band structure is shown in Fig. 5.

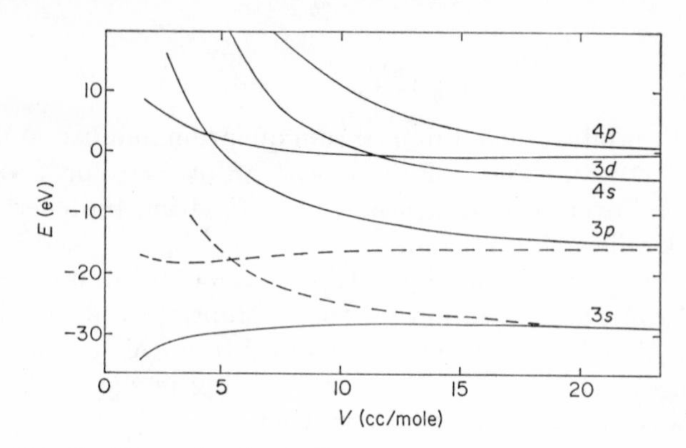

FIG. 5. Energy bands of solid argon according to Ross (1968).

It will be seen that while the 4s-like band lies below the 3d band at large lattice spacing, a crossover occurs at about 1|cc/mole. The forbidden energy gap between the 3p and 3d bands can be seen to disappear at about 4·5 cc/mole, which is about 60% of the volume predicted by Gandel'man. Using the Thomas–Fermi–Dirac equation of state, the critical pressure for the metallic transition turns out to be 4400 kb at 0°K.

The factor of three between these estimates of the critical pressures seems to stem from Gandel'man's use of a Thomas–Fermi field, rather than from detailed differences in the calculational procedures. That the band gaps are sensitive to the choice of field was shown by Ross's APW calculations. The work of Matthiess already referred to yields an even larger band gap at the normal lattice spacing than that found by Ross. Thus, pressure experiments are almost certainly going to throw light on the central problem of energy band theory—the choice of the crystal potential. We can test different theoretical models by investigating whether the variation of crystal potential with lattice spacing which they imply leads to physical predictions in agreement with experiment. This has already been commented on in the study of He in Section II.A, and will appear again in the discussion on the energy bands in copper in Section V.C.

For the face-centred cubic structure, the APW calculation of Ross (private communication), using the crystal potential of eqn (30), predicts that the bottom of the conduction band is at Γ (see Fig. 1) and is an s-like state. A d-like state at X lies above it, and a further, almost pure, d state lies slightly higher still. However, with increasing pressure, the energy of the s-like state at Γ is raised, while the first d-like state at X becomes the bottom of the conduction band. The bands touch at a lattice spacing which is in quite good agreement with that given by the Wigner–Seitz method discussed above.

2. *Xenon*

The energy levels for xenon as calculated by the Wigner–Seitz method are shown in Fig. 6. The bands are again labelled according to the atomic correspondence, and the calculations take no account of crystal field splitting. They assume a single s state, the p state to be triply degenerate, and so on.

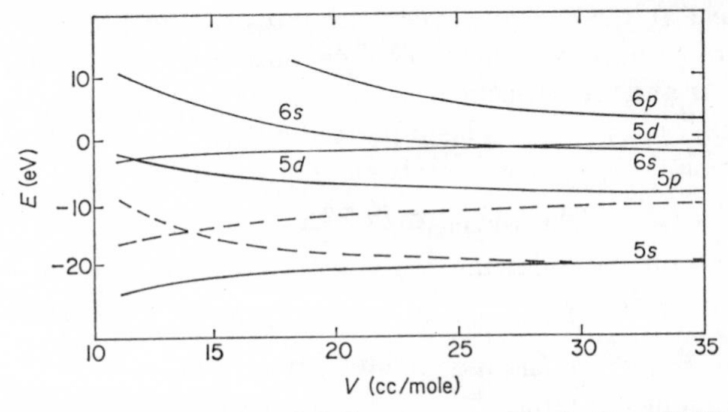

FIG. 6. Energy bands of solid xenon according to Ross (1968).

At normal density, the 6s band is the lowest conduction band, lying about 7 eV above the valence band. The bottom of the 6s and 5d bands cross at about 27 cc/mole. This is like the transition which occurs in caesium when the highest occupied 6s levels at the Fermi surface intercept the 5d band (cf Sternheimer, 1950; the transition occurs at a somewhat larger volume (35 cc/mole) than that predicted by the simple crossing of the two bands). The conclusions of Ross agree with Sternheimer's calculation on Cs.

The calculations predict that the valence band will intersect the conduction band at about 11·7 cc/mole. Again using the Thomas–Fermi–Dirac equation of state, the critical pressure at 0°K is about 700 kb. Ross (unpublished work) has also made augmented plane wave calculations for xenon at very high pressures, and the results are generally consistent with the bands shown in Fig. 6.

It is relevant to point out here that Flower and March (1962) studied the variation of the band structure with pressure for the heavy ionic crystal Cs I, with atoms adjacent to Xe in the periodic table. However, these calculations were restricted to the *s*-like state at the bottom of the conduction band. Ross (private communication) has suggested to the writer that, as for Xe, eventually the *d* levels will lie lowest. This may turn out to resolve the discrepancy between theory and experiment, for whereas the shock wave measurements indicated a metallic transition in Cs I, no strong variation of the band gap was found from the theory over a range of 250,000 atmospheres.† It would be of interest to extend the calculation of Flower and March to deal with the variation with lattice parameter of the lowest *d*-like state in the conduction band, to see whether Ross's suggestion is quantitatively successful.

3. *Estimated Metallic Transitions in Neon and Krypton*

Similar calculations are not known to exist for the other noble gas elements Ne and Kr. However, estimates of the pressures at which metallic transitions will occur can be obtained by combining Ross's results for Ar and Xe with the work of Simcox and March for He. Thus, in Fig. 7, $\ln_{10} p_{\text{transition}}$ has been plotted against $\ln_{10} Z$. It is seen that the three points lie approximately on a straight line yielding

$$p_{\text{transition}} \propto Z^{-3/2}.$$

From the graph the following rough estimates can be made

$$p_{\text{transition}} \doteqdot 10^4 \text{ kb for Ne}$$

$$p_{\text{transition}} \doteqdot 1400 \text{ kb for Kr,}$$

which will be useful later.

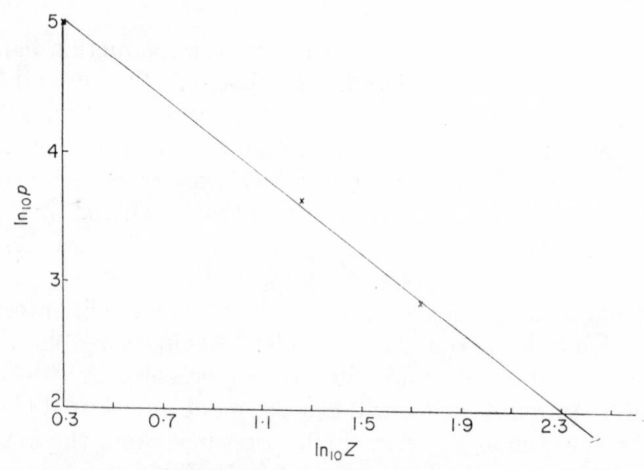

FIG. 7. Metallic transition pressure for noble gas crystals *versus* atomic number Z. Crosses show calculated results for He, Ar and Xe.

†Note added in proof. Dr. M. Ross has noted that more recent results exist on the conductivity of CsI under pressure. B. J. Alder presents conductivity data ("Solids Under Pressure", ed. by W. Paul and D. M. Warschauer (McGraw Hill, New York, 1963, p. 385) for CsI of 0·2 ohm cm, at $V/V_0 = 0{\cdot}61$, $T = 3900°$K and a forbidden energy gap of 3·4 eV. The forbidden gap is still finite and no metallic state has been reached. Further correspondence with Dr. M. Ross prompts the author to remark that the

(*continued*)

B. CRUDE CLASSIFICATION OF FORBIDDEN ENERGY REGIONS AS FUNCTION OF ATOMIC NUMBER

While in the long run it is unlikely that there will be any quantitative substitute for detailed band structure calculations for individual solids until we reach the regime of Section II, some additional comments are desirable on the vanishing of forbidden energy regions, as a function of atomic number.

We can speak in He of the metallic transition as the " breaking open " of the K shell. Similarly, in Ne we are dealing, at the metallic transition, with the breaking open of the p subshell (at least) of the L shell. The $2p$ band must be overlapping the $3s$ band here.

As we trace the disappearance of forbidden energy regions, there must, clearly, be some dependence on crystal structure, and it must be emphasized that only a very primitive classification is being made therefore.

1. L *Shell*

While, in principle, the calculation outlined in Section II.C for the K shell can be repeated in order to study the " excited " bound states, this has not yet been done. It will, of course, be important in doing so to use physically reasonable wave functions and potentials for the bound electrons in the core.

In work of Meyer, Nestor and Young, however (1967; referred to below as MNY) a discussion is given of bound states breaking off from the band for monovalent metals, although it is only used by these authors to set a limit of validity to their " pseudoatom " picture. From their results, which screen a proper Hartree–Fock field for the ion-core with a shell of electronic charge placed at a radius chosen to satisfy the Friedel sum rule (cf eqn (31) below), we can estimate that if the lattice of Li metal were expanded, such that the Fermi wave number k_f became as low as $0{\cdot}27\,a_0$, a_0 being the Bohr radius for hydrogen ($k_f \sim 0{\cdot}55$ atomic units under normal conditions), then their one-centre potential would have a bound $2s$ state. This gives us a lower bound to the critical atomic volume (corresponding to their k_f, this is ~ 1550 atomic units) for $Z=3$, the overlapping of the $2s$ wave functions on neighbouring lattice sites causing the bound state to merge again with the band.

There is not enough data here to make any quantitative predictions. However, combining the above result with the estimated transition pressure for Ne, it seems fairly clear that we would have to go to pressures $> 10^4$ kb before worrying whether in simple metals like

emphasis on band theory here is appropriate while temperature effects are unimportant. In the next decade many very high pressure experiments will be carried out by the shock wave method, with resulting high temperatures. It might be anticipated that large electrical conductivity could be observed at significantly larger volumes than the actual closure volume. Also, the problem of what happens to the forbidden energy regions when the lattice temperatures are of the order of electron volts has received little attention so far and is worthy of study.

Na ($Z=11$) and Mg ($Z=12$) the number of conduction electrons per atom is different from one and two respectively. At pressures substantially greater than 10^4 kb, presumably we would eventually reach valencies of 7 and 8 respectively.

It should be mentioned here that a valuable quantitative study of the variation of important electronic eigenvalues with lattice spacing was made by Ham (1962) for the alkali metals, including Li ($Z=3$), in the relatively low pressure regime. Referring to the Brillouin zone for the body-centred cubic structure shown in Fig. 21, Ham considers the variations of the eigenvalues with lattice parameter at N, H and P explicitly. It would obviously be of considerable interest to devise experiments to test these results.

Unfortunately, the low temperature modification of Li is not the body-centred cubic structure and the classical experiments on Fermi surfaces are not applicable. But soft X-ray emission, photo-emission and optical properties under pressure would be experiments which it would be interesting to have. Particularly in the former case, the role of the hole in the K shell is of current interest. As suggested by our discussion of bound states, the change in the soft X-ray emission spectrum with pressure might be a useful tool for gaining more understanding on the influence of the hole on the emission intensity. (It is also possible that the role of electron–electron effects might become clearer because of the variation of the plasma frequency with pressure.) The work of Stott and March (1966, 1968) points to the fact that the hole is playing only a minor role and that the band energies and wave functions dominate the problem, whereas other theories (Goodings, 1965; Allotey, 1967) lay much emphasis on the hole. A pressure experiment might help us to decide between these different approaches.

Although it is not strictly a matter relevant to an article on pressure behaviour, it would be interesting if one could get metal–ammonia solutions or compounds sufficiently dilute to check the above prediction concerning the critical k_f for Li.† This does not seem impossible with present techniques, but it should be borne in mind that electron correlations may become so important in modifying the band theory that new states (for example, the spin or charge density wave states proposed by Overhauser, which are discussed in Section VII) may be induced before that density.

† A recent paper on Na in liquid ammonia should be mentioned here (Schwindewolf *et al.*, 1966).

2. M *Shell*

Here we start again from the estimate by Ross that the $3p$ subshell for Ar ($Z=18$) will " break open " at a critical pressure of ~ 4400 kb at 0°K.

The work of MNY is again useful. It seems that, for Na ($Z=11$), the $3s$ state becomes bound in their one-centre potential for $k_f=0{\cdot}350$, with the corresponding atomic volume 690 a.u. If the lattice of Na could be expanded in such a manner, the Coulomb correlations might again induce a charge density wave or spin density wave state (see Section VII.B). However, the above value is useful in giving us a crude addition to our band theory knowledge. Taking this, together with Ross's information, we are able to say, with some confidence, that the $3p$ subshell in K will not break open until a critical pressure >5000 kb is reached. Whereas the explanation of the rapid rise in resistivity observed by Stager and Drickamer (1963) above 200 kb motivated work by Stocks and Young (1969) on potassium metal with seven conduction electrons per atom, it seems to the author that the number of conduction electrons per atom must remain at unity over the pressure range explored to date. However, there is certainly a dramatic decrease in mean free path, from around 750 Å at 0°C and normal pressures to 25 Å at 600 kb.

Ross has pointed out that the crossing of the $3d$ and $4s$ levels for Ar shown in Fig. 5 at a volume of about 11·5 cc/mole suggests that, at this sort of volume, potassium would be expected to undergo an electronic transition similar to the one observed in caesium. However, this has not been observed since a compression to about four times the initial density would be required. This effect, that the d levels are coming down relative to the s levels, must surely be a major cause of the dramatic shortening of mean free path found from the results of Stager and Drickamer.

3. N *Shell*

For the N shell, we have the estimate that the metallic transition in Kr ($Z=36$) occurs around 1400 kb, the $4p$ subshell presumably being broken open here.

The bound state calculations of MNY give 0·280 as an estimate for the value of k_f at which a $4s$ bound state breaks off from the continuum for K which corresponds to an atomic volume of 1350 a.u. In this case, we also have a value for Cu ($Z=29$), where $k_f=0{\cdot}504$ is evaluated, with atomic volume 232 a.u. A crude plot of $\ln v_{\text{critical}}$ *versus* $\ln_{10} Z$ is shown in Fig. 8. This suggests that for Rb there will not be band

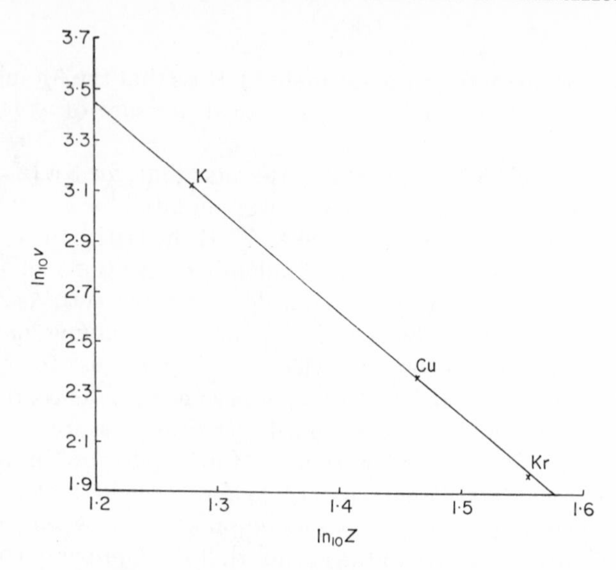

FIG. 8. Atomic volume v at which N shell breaks open *versus* atomic number Z.

overlap between the $4p$ and $5s$ shells until pressures in excess of 1500 or 1600 kb are applied.

4. 5S *and* 5P *Subshells*

The bound states calculated by MNY for Rb and Ag occur at k_f values of 0·286 and 0·492, with corresponding atomic volumes of 1270 a.u. and 250 a.u. respectively. The relevant metallic transition is now Xe for which, as we have seen, the $5p$ and $5d$ bands separate at a critical volume of around 11·7 cc/mole, or an atomic volume of 130 a.u. The work of Ross indicates that the $5s$ and $5p$ bands in Xe probably separate at about 14 cc/mole.

It does not seem profitable to pursue the argument now in much greater detail. Clearly, as more data accumulates, transition plots like that shown in Fig. 8 might be made and refined and attention perhaps can be given to classifying the s–p separations within a shell. However, we must not be too surprised if the smooth dependences on Z used here turn out to be a serious oversimplification.

From this we may deduce that many, and probably almost all, of the properties of solids are going to be dominated by detailed band structure considerations up to the highest pressures attainable. Nevertheless, particularly within groups, it seems likely that pattern exists in the disappearance of forbidden energy regions, and it will be surprising if the pattern of metallic transitions in the inert gases does not conform, at least approximately, to the results shown in Fig. 7.

Needless to say, at low pressures, all sorts of subtle effects come in, and in some cases forbidden energy gaps, over a limited pressure range, are known to increase with pressure rather than to decrease (for example in Ge and Si).

This discussion has put hardly any emphasis on structure, although it is known, of course, that an important quantity like the electronic density of states $N(E)$ will depend critically on the structure. We also believe this will be true of electrical resistivity in most cases. The discussion of conduction in the pressure-ionized regime, given in Section II.D, may therefore remain largely academic for a long time to come.

C. COLLAPSE OF COVALENTLY BONDED STRUCTURES INTO METALLIC STATES

The phase changes which occur in the group 4 elements C, Si, Ge and Sn, and in some 3–5 and 2–6 compounds, such as InSb and CdTe should be briefly mentioned here. (See for example Austin, 1966.)

The normal form of these materials is the diamond structure or the zinc blende lattice. But under pressure, a sharp transition occurs to a metallic state, with a substantial decrease in volume. Just what significance the metallic state has here is not yet clear. The most reasonable explanation offered to date is that a short wavelength vibrational mode becomes unstable (cf Austin, 1966). Further work in this interesting area is called for.

IV. Electrical Conductivity of Alkali Metals

The resistivity of the alkali metals has been considered in detail by Dugdale (1968) in the previous volume in this series, and therefore we shall stress only those points which are relevant to the present theme, where we are seeking an overall picture of the electrical properties of metals over the entire range of pressures.

Immediately, the experiments are seen not to agree at all with the qualitative trends required by eqn (28) for the pressure-ionized regime, and the presence of bound electrons in the core must be crucial. Dickey and co-workers (1967) have considered this problem in some detail; especially the scattering of the Fermi electrons off pseudoatoms. These are now described, however, not by the simple screened Coulomb field, eqn (4), but by a single-centre potential constructed as follows:

(1) Take the Hartree–Fock–Slater fields for Li^+ . . . to Cs^+.

(2) Screen these ions, not by means of the dielectric functions (18) or (19), but with a shell of electrons at a distance s from the ion, with s chosen to satisfy the Friedel sum rule

$$\frac{2}{\pi}\sum_{l}(2l+1)\eta_l(k_f) = 1 \tag{31}$$

for the alkali metals (generally, the valency appears on the right-hand side of eqn (31). Here the η_l's are the phase shifts of the partial waves, evaluated at the Fermi momentum. The radius s is determined for each value of the lattice parameter, and hence again the change in the self-consistent field with compression is at least crudely estimated.

It will be clear from the above discussion that the scattering off a single ion is calculated beyond the Born approximation, and hence, in eqn (24), K_i is obtained from the phase shifts directly. In addition, Dickey and co-workers use the semiempirical estimates made by Dugdale for the variation of the Debye temperature with volume, rather than the formula (27), and thus a comparison of the quantity K_i in eqn (24) with experiment can be made. Scaling the results in terms of K_0, the value of K_i at atmospheric pressure, the " experimental " results at room temperature, based on Bridgman's work and taken from a paper by Dugdale (1961; for a more recent discussion see Dugdale and Phillips, 1965), are shown in Figs 9 and 10, and the theory for K_i/K_0 in Fig. 11. While the agreement with experiment is not fully quantitative

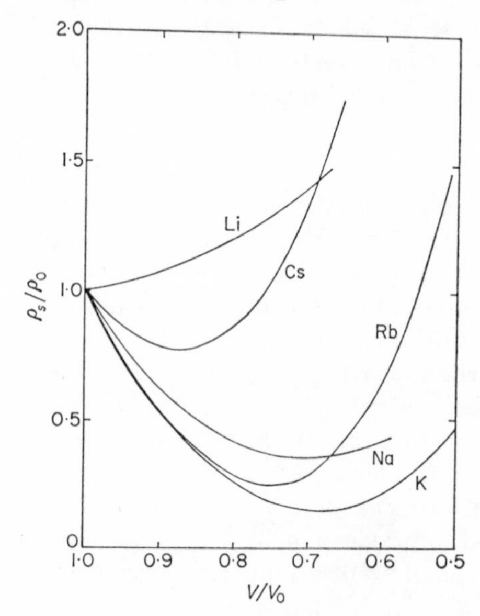

FIG. 9. Measured resistivities of alkali metals *versus* pressure.

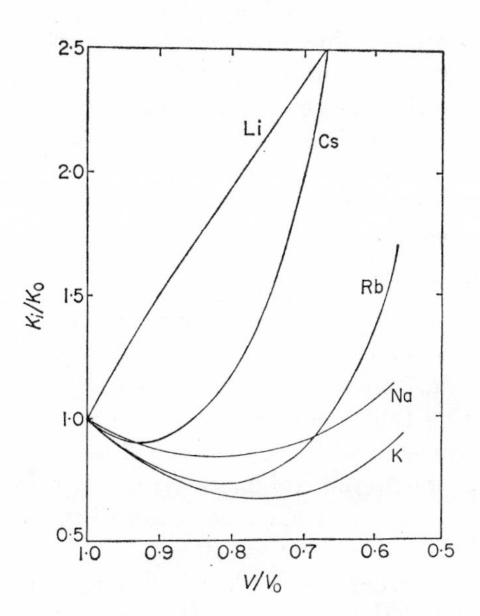

FIG. 10. "Measured" K_i/K_0 for alkali metals. K_i is defined in terms of resistivity and Debye temperature in eqn (24).

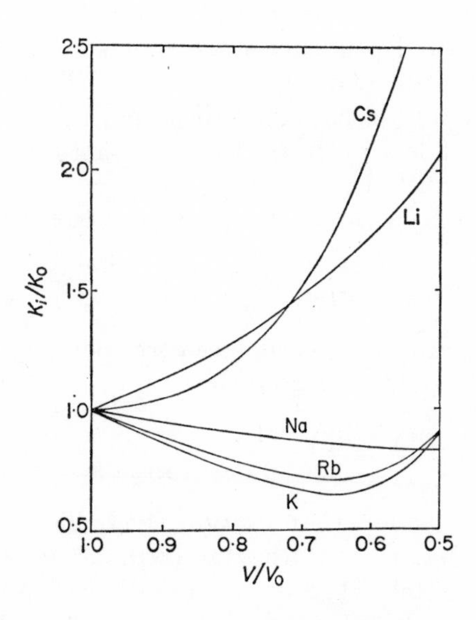

FIG. 11. Theory of K_i/K_0 for alkali metals.

(neither are the absolute values of ρ_s very good, partly because of the use of $S(0)$ as discussed just before eqn (21)), the main features of the experiments are clearly revealed. Obviously, then, the influence on the conduction electrons of electrons in bound states is a dominant feature in this pressure regime. Some qualitative difficulties remain, apart from the quantitative discrepancies referred to above. Two of these difficulties are listed below.

(1) The resistivity of Li drops suddenly, by a factor of about four, at about 100,000 atm according to Drickamer (1965). If this is a crystal structure change, it is hard to understand it within the above framework unless the Fermi surface is dramatically altered, perhaps even to the extent of contacting the zone boundary. On the other hand, it could mean that the treatment of the phonons will have to be refined, along lines discussed by Greene and Kohn (1964) and Darby and March (1964).
(2) The resistivity of K increases by a factor of the order of 100 in the pressure range $10^5 - 5 \times 10^5$ atm as mentioned in Section III.B. We shall not elaborate further on this here, although we say a little more about the variation of the energy bands in K with pressure in Section V.A.

A. PRESSURE-DEPENDENCE OF HALL EFFECT

In connection with the resistivity results of Dickey and co-workers (1967), Paul has recently drawn the attention of the author to measurements of the pressure-dependence of the Hall constant at room temperature by Deutsch, Paul and Brooks (1961). The Hall constant R is written as $1/nex$, where, as usual, n is the number of conduction electrons per unit volume while x expresses the deviation from the free-electron value of the Hall constant.

For spherical energy surfaces, but with an anisotropic relaxation time τ defined by

$$\tau = \tau_0(1 + CK_4 + C_1K_6 + \ldots) \tag{32}$$

where K_4 and K_6 are cubic harmonics corresponding to $l=4$ and $l=6$, it has been shown by Davis (1939) that

$$x = 1 - \frac{4}{21} C^2 - \frac{8}{13} C_1^2 \tag{33}$$

which must clearly be less than unity. Over this range of pressure, x decreases for Li, Na, K and Rb, but increases for Cs beyond about 11,000 or 12,000 kg/cm^2. It follows for Cs that we must have, in addition to anisotropy in the relaxation time, non-spherical energy surfaces, since x begins to exceed unity when the pressure is above about 1,000 kg/cm^2. This is in general agreement with the band structure of caesium under pressure, as calculated by Ham (1962).

There is similarity between the pressure variation of the conductivity σ and of x, from the measurements of Deutsch and co-workers. However, to order of K_4, there is no C^2 term in the conductivity. Thus, it may perhaps be valid to calculate the conductivity with a spherical approximation to the relaxation time, even though the free-electron Hall constant is not found. This argument is less well founded for Cs, where the departure of the energy surfaces from spherical symmetry should be considered more carefully. In this case, spherical harmonic expansions like eqn (32) may not afford the best approach, and an alternative way to calculate the Hall constant may be through the work of Allgaier (1967).

In summary, we can say that very complex phenomena are to be expected in the variation of ρ_s with pressure before the asymptotic form of equation (29) will be remotely relevant. Some understanding can be achieved by a treatment of the electron–ion interaction along the lines of Dickey and co-workers. We should not, however, be surprised if a detailed account has to be taken eventually of both the detailed anisotropy of the Fermi surface, and of the marked phonon anisotropy in the alkali metals.

V. Pressure-dependence of Band Structures of Simpler Metals

A. Potassium

From the discussion of potassium we have already given, it seems clear that it is a particularly interesting metal from the point of view of pressure experiments in the future.

Two studies have been made by now on the pressure dependence of the band structure. We have referred already to Ham's work on the alkalis; within the low pressure regime of his calculations, they have a validity which cannot be matched by the Wigner–Seitz type of calculation we discussed for Ar and Xe. However, such calculations give us a useful overall picture of the way the bands vary over a vast pressure range and we shall therefore briefly report later work of Gandel'man (1967). While starting again from a Thomas–Fermi potential, Gandel'man has carried out a Hartree self-consistent field calculation within the Wigner–Seitz approximation. However, he has not included exchange in the crystal potential and some of the objections to his potential for Ar still apply. We shall therefore only briefly summarize his energy band results, which are shown in Figs 12–15, for ratios of the normal atomic volume to the compressed atomic volume, v_0/v, of 1, 3, 5, and 10. We

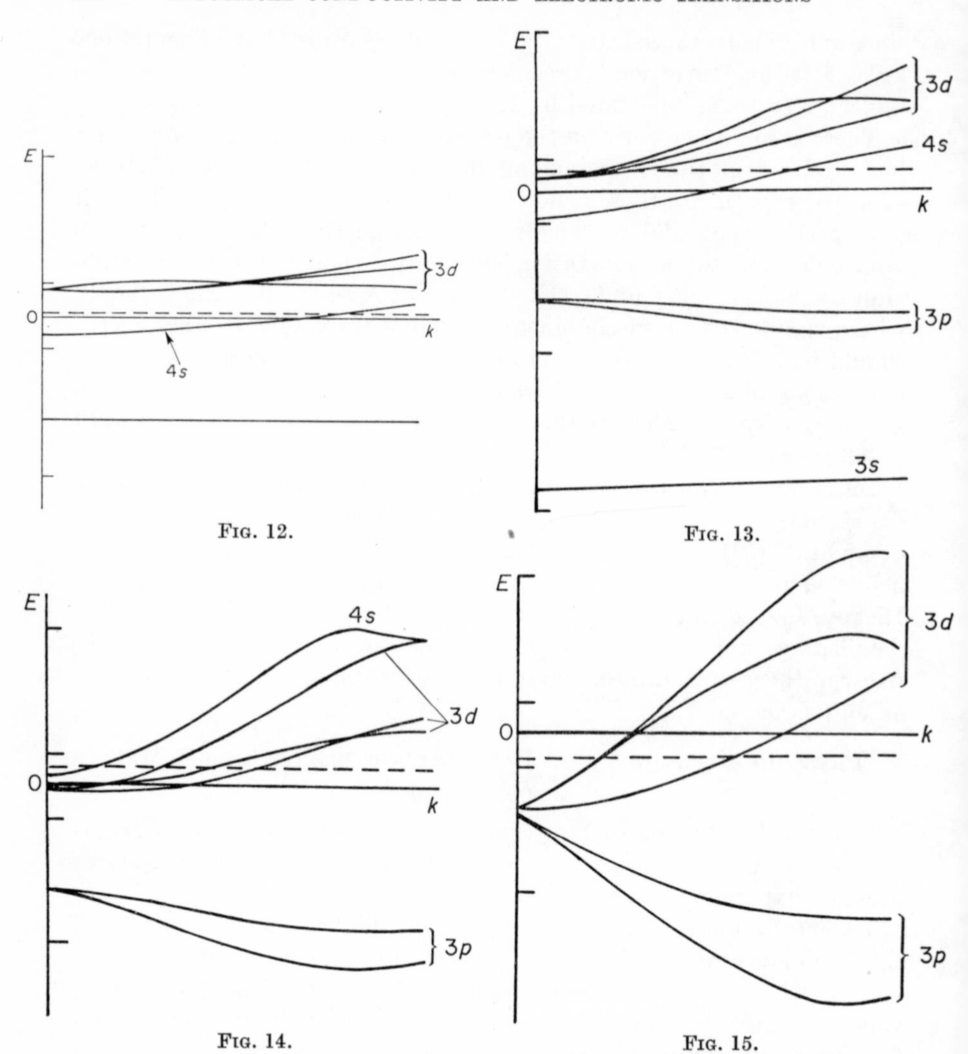

Fig. 12.

Fig. 13.

Fig. 14.

Fig. 15.

Figs 12–15. Schematic form of energy bands in potassium for different relative compressions Fig. 12 $(\nu_0/\nu)=1$, Fig. 13 $(\nu_0/\nu)=3$, Fig. 14 $\nu_0/\nu)=5$, Fig. 15 $(\nu_0/\nu)=10$. Broken line shows Fermi level in each case.

wish to stress two points from these band calculations. Firstly, the lowering of the $3d$ bands relative to the $4s$ bands with increasing compression, which we discussed for Ar, is again clear. At $v_0/v=5$, the $4s$ band is hardly occupied, while at $v_0/v=10$, only the $3d$ band lies below the Fermi level. This lowering of the d bands relative to the s bands under pressure appears to occur in many materials.

Secondly, Gandel'man has used the energy bands shown in Figs 12–15 to calculate the electronic Gruneisen constant γ_e. This can be done as a function of temperature, at relatively low temperatures, since a knowledge of the band structure $E(\mathbf{k})$ can be turned into a calculation of the chemical potential μ as a function of temperature. The thermal contributions to the energy and pressure can then be obtained and γ_e is found to have the form shown in Fig. 16.

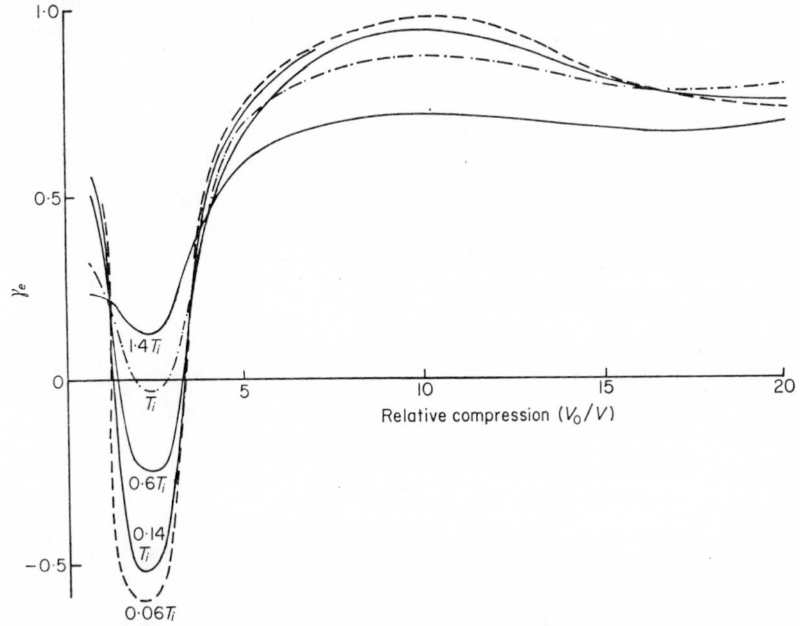

Fig. 16. Gruneisen's constant γ_e *versus* pressure for potassium. T_i is $\sim(20–30)\times10^3$ °K. Temperatures of other curves are shown in terms of T_i.

Usually, with increasing density, the energy bands broaden and lead to a decrease in the density of states at the Fermi level. This leads to a positive γ_e. However, in some cases, the density of electronic energy levels increases with compression, which leads to a negative γ_e. Figure 16 shows that such a region exists in K. The reason for its appearance is that for $v_0/v \sim 2–3$, the unfilled $3d$ band begins to overlap with the $4s$ band, leading to an increase in the density of states at the Fermi surface. When the temperature increases to $(20–30)\times10^3$ °K, the curve labelled T_i is reached, and the calculated value of γ_e changes sign. The Hugoniot obtained from Gandel'man's work appears to be in reasonable agreement with experiment, but we must refer to the original paper for further details.

B. ALUMINIUM

Gandel'man finds also in the case of Al a region of negative Grüneisen constant γ_e for $1 < v_0/v < 2{\cdot}4$. There appears to be some experimental evidence from shock wave experiments that there is a corresponding anomaly near $v_0/v \sim 2$ in the low temperature pressure volume relation of Al (Skidmore and Morris, 1962; Altshuler *et al.*, 1960).

The bands from which these results were obtained are shown in Fig.

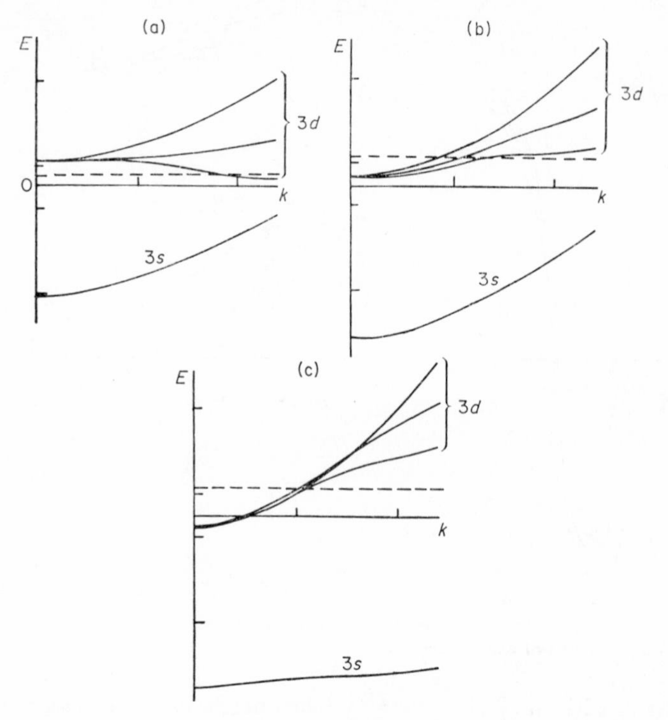

Fig. 17. Schematic form of energy bands for Al in spherical approximation. (a) $v_0/v = 1{\cdot}5$, (b) $v_0/v = 3$, (c) $v_0/v = 4{\cdot}2$. Broken line shows Fermi level in each case.

17, and again it may be remarked that the influence of exchange on the wave functions and the potential was not included, which may mean that the results are, at best, semi-quantitative. The lowest $3d$ band in Fig. 17 changes its character markedly under compression. This will influence the density of states at the Fermi surface greatly and hence conductivity and electronic specific heat, among other properties.

Whereas the pressure range covered by Gandel'man's calculations is vast, we now turn to discuss some other aspects of the behaviour of Al under relatively low pressure. Here the discussion can be made much more quantitative. The work of Melz (1966) on the Fermi surface has been well surveyed by Dugdale (1969). We shall therefore focus attention mainly on some recent related measurements by Burton and Jura (1968) on electron momenta, as they again throw light on the variation of band structure with pressure.

1. *Positron Annihilation*

Whereas most experiments designed to study the electronic structure of metals need to be carried out at helium temperatures, it is well known that positron annihilation measurements can be made at room temperature. This is particularly important when one has the additional complication of high pressures. The experiments are still difficult to carry out and furthermore the positron perturbs the metal, the electrons piling up round the positron to screen out its field (cf Section II.B), which sometimes leads to problems of interpretation. Nevertheless, valuable results have been obtained using this technique, and we shall discuss the measurements of Burton and Jura (1968) on Al here.

In the pressure range 0–100 kb studied by these workers, Al does not exhibit any phase transition as far as is known. The measurements of the annihilation spectra were carried out at five different pressures and were analysed to yield the variation of the Fermi momentum as a function of pressure. The results are shown in Fig. 18. The elementary free electron relation, eqn (5), for the Fermi momentum may be used in conjunction with measured p–V data for Al to convert to a Fermi momentum–pressure relation, and this is shown as the solid line in Fig. 18. It can be seen that the Fermi momentum of Al is well described by the free electron model over this pressure range.

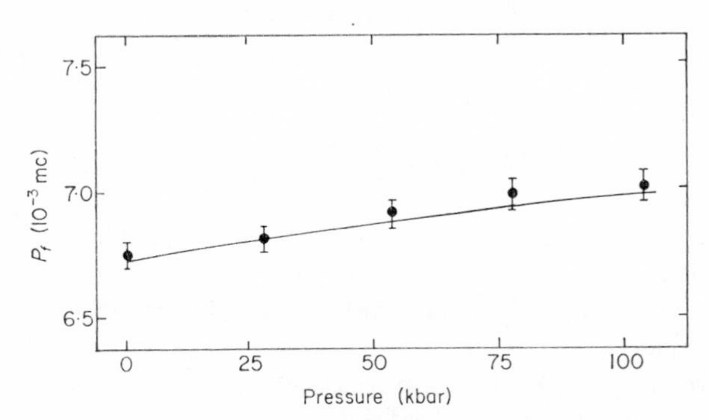

FIG. 18. Measurements of Fermi momentum *versus* pressure for Al. Points denote measurements of Burton and Jura. Solid curve is obtained from free electron relation (5), together with measured equation of state for Al.

It might seem, at first sight, that this result is in conflict with the work of Melz (1966) who found that in experiments to 7 kb the Fermi surface of Al does not expand in the manner predicted by the free-electron model. However, using the pressure derivatives of the pseudopotential coefficients given by Melz (see Dugdale, 1969, for a simple discussion of the pseudopotential) Burton and Jura have calculated the average momenta at the Fermi surface. Over the entire range of pressure they find agreement with the free electron model to within an accuracy of 0·1%. This difference is well within their experimental error, which is 1%. Although we know from the work of Melz that the band gaps of Al vary with pressure, rather little of the Fermi surface lies close to the zone boundaries. Thus, the average momentum at the Fermi surface does not depend strongly on the band gaps as long as these gaps are small.

It would be of considerable interest if positron annihilation could be performed on the alkali metals, particularly Li and Cs. We might expect to see the electronic transition in Cs reflected in the positron annihilation experiments, and it would be nice to know whether the peak in the electrical resistivity observed by Drickamer in Li is due to a phase change which leads to a very different Fermi surface. This might be picked up in the positron experiments.

2. *Thermoelectric Power*

Some interesting experiments on the thermoelectric power and electrical resistance of Al (and also Au, Ni and Pt) have recently been reported by Bourassa and co-workers (1968) up to 4 kb. Their main interest was in the thermoelectric power and the merit of their technique was the extension of the temperature range over which the measurements could be carried out (room temperature to 900°K for Al).

In a metal at temperature $T > \Theta$, the thermoelectric power S is given by (see for example, Ziman, 1960)

$$S = (\pi^2 k_B^2 T / 3eE_f) \left[\frac{\partial}{\partial \ln \epsilon} \rho(\epsilon) \right]_{E_f} \tag{34}$$

where $\rho(\epsilon)$ is the resistivity as a function of energy. The thermoelectric power is therefore telling us about the energy dependence of the resistivity ρ at the Fermi surface and is more sensitive to the scattering mechanism than ρ.

The interesting point which Bourassa *et al.* stress is that, since in the face-centred cubic metals the predominant thermally activated defect is the single vacancy, the thermoelectric power of a metal at high temperatures might be used as a tool to study vacancies. Furthermore, it seemed feasible that a small change in concentration of vacancies induced by pressure might produce a larger effect on the thermopower, due to its sensitivity to the scattering mechanism. The experiment can be made at high temperatures and avoids problems due to non-equilibrium and due to phonon drag.

However, the method is not without its difficulties, as the authors point out, for pressure causes changes in the Fermi level, and changes the electron–phonon interaction and the phonon distribution. These effects were at least partly avoided by using low-temperature results as a reference.

The experimental results for the change in thermoelectric power ΔS with pressure as a function of temperature are shown in Fig. 19. A

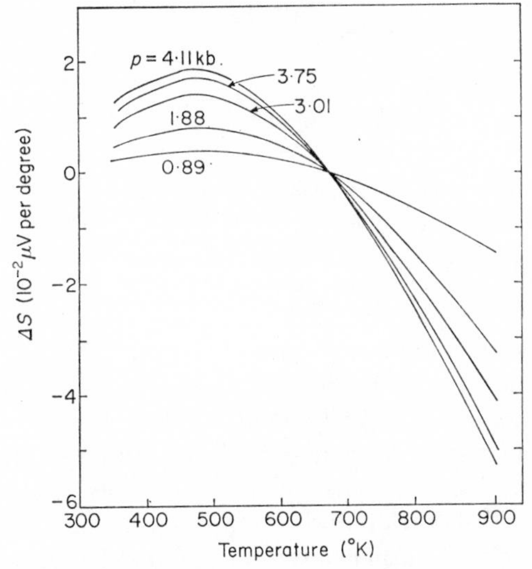

FIG. 19. Change in thermoelectric power with pressure *versus* temperature for Al, from work of Bourassa and co-workers (1968).

change of sign occurs at about 670°K and, to a good approximation, this is independent of pressure. The next step is to extract the vacancy contribution to ΔS and the results for this are shown in Fig. 20. The simple model given by Bourassa *et al.* predicts the straight line shown, the slope being the vacancy formation energy expressed in °K. Excellent agreement with the vacancy energy of 0·77 eV given by Simmons and Balluffi (1960) is obtained at low temperatures. However, it is clear from Fig. 20 that the whole of the effect is unlikely to be due to vacancies.

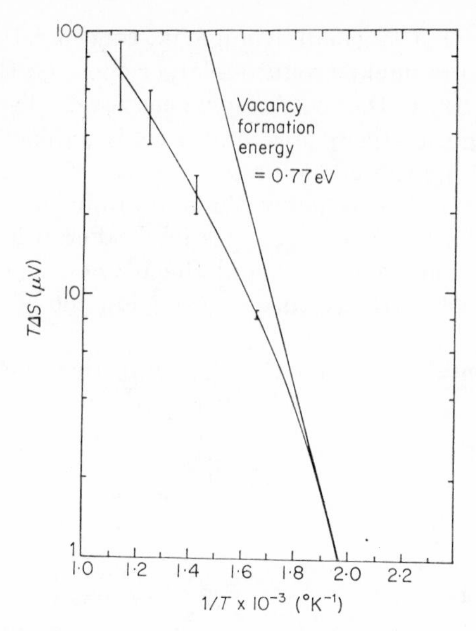

Fig. 20. Change in thermoelectric power due to defects *versus* temperature for Al. Straight line is result of simple model invoking only vacancies.

Assuming other defects are present (impurity-vacancy pairs and divacancies) further progress can be made, and in particular some evidence is obtained for the existence of a stable divacancy in Al, with a binding of around $\frac{1}{2}$eV. On the other hand, in Au, it appears that the divacancy may have a relatively small binding energy.

From available results on pair potentials, the large binding energy of $\frac{1}{2}$eV for Al, assuming it is subsequently confirmed, points to very long range forces, large relaxation and/or major electronic distribution round the divacancy. On the other hand, for Au, a value of say 0·1 eV would not be difficult to understand from a simple pair potential model, with a moderate range for the pair force. Even for Au, the position is far from final though, the recent work of Wang and co-workers (1968) giving a value of 0·4 eV.

Pressure experiments, along the general lines laid down by this work, might prove very useful as a tool for further study of intrinsic defects in metals.

C. COPPER

As we have stressed earlier, one of the interests in studying the variation of band structure with lattice constant is to throw light on the way to choose the crystal potential in band structure calculations.

1. *Construction of Potential*

The method developed by Korringa (1947) and Kohn and Rostoker (1954) was used in the work of Davis and co-workers (1968) on copper, which we shall now discuss. Although not completely essential to the theory, the use of a muffin-tin potential greatly simplifies the calculations. This means that the potential is taken as spherical within the unit cell, but having a finite range. Such an assumption is probably not very serious with regard to its effect on electronic band structure.

The construction of the crystal potential $V(\mathbf{r})$ then proceeded as follows:

$$V(\mathbf{r}) = V_{\text{coulomb}}(\mathbf{r}) + V_{\text{exchange}}(\mathbf{r}). \tag{35}$$

Both the coulomb and exchange terms were then calculated using wave functions of the free atom. At a chosen site in the crystal, the Coulomb part is taken from the atomic potential, plus contributions from the same potential situated on near neighbour sites. These contributions were spherically averaged about the chosen site. The exchange contribution to the potential was obtained using Slater's prescription, based on the ideas of Dirac for introducing exchange into the Thomas–Fermi theory. Thus, just as in eqn (30) we write

$$V_{\text{exchange}}(r) = -6\left[\frac{3\rho(r)}{8\pi}\right]^{1/3} \tag{36}$$

but, unlike eqn (30), ρ is constructed as a superposition of Hartree–Fock atomic charge densities, and then its s term is extracted in the spherical harmonic expansion about the site under consideration.

Clearly, as the lattice is compressed, the potential changes according to the above prescription solely because the superposition of atomic densities and potentials is carried out on the sites appropriate to each pressure. The method has the same philosophy as that of pseudoatoms, but the effect of the environment on the atom is treated less carefully. Refinements will clearly be called for later but, for the moment, the method affords a useful starting point.

In the choice of atomic charge densities, Davis and co-workers were guided by their earlier work on the Fermi surface of Cu at normal lattice spacing (Faulkner *et al.*, 1968). Here they investigated the effect of different atomic charge densities. Of the three investigated, the two which started from a $3d^{10}4s^1$ configuration gave severe disagreement with the Fermi surface of Cu as established by experiment. The third potential, based on the atomic Hartree–Fock wave functions calculated by Watson (1960) for a $3d^94s^2$ configuration, was found to give quite satisfactory agreement with experiment.

Thus, in the study of the pressure variation of the band structure, the potentials were generated in each case by the use of Watson's wave functions inserted in a $3d^{10}4s^1$ configuration. This procedure is clearly arbitrary in that there is no reason to believe that calculating wave functions in this manner has any special relevance to metallic copper. The following results indicate, however, that once an " atomic " charge density has been found which yields a reasonable potential function for the normal lattice spacing when inserted in eqns (35) and (36) then the effects of small changes in the lattice spacings can be predicted using the same charge density in the same equations. In copper, similar results could not be obtained by a simple scaling of the potential or a free-electron treatment.

2. *Eigenvalues as Function of Lattice Spacing*

Using these potentials, calculations were carried out for lattice spacings $0{\cdot}995a$ and $0{\cdot}99a$, in addition to the calculations available for the normal lattice parameter a. Linearity with lattice spacing was found to within a few ten-thousandths of a Rydberg. In addition to the determination of the Fermi energy, the density of states at the Fermi level $\rho(E_f)$ and the low temperature electronic specific heat γ ($\rho(E_f)$ and γ were found to be linear in the lattice spacing to three figure accuracy), the linear dimensions and cross-sectional areas of a given constant energy surface were obtained. These are, of course, accessible to experiment through magneto resistance, magnetoacoustic, de Haas–van Alphen and cyclotron resonance experiments.

From these dimensions of the constant energy surfaces, some frequencies occurring in the de Haas–van Alphen effect and cyclotron resonance masses were calculated. These are shown in Table I. The masses vary linearly again with lattice spacing and, for the potentials chosen, should be accurate to better than two decimal places.

The results of Templeton (1966) on the change with pressure of the de Haas–van Alphen frequencies for the neck and the [111] belly orbits of the copper Fermi surface have been extended recently by O'Sullivan and Schirber (1968). These workers have made measurements of the pressure derivatives of five cross-sectional areas of the Fermi surface of copper. In their experiments the de Haas–van Alphen oscillations were detected using the low-frequency field-modulation technique in fields up to 55 kOe. There is quite remarkable agreement between their measurements and the predictions of Davis and co-workers, as can be seen from Table II, where the earlier results of Templeton are also included.

TABLE I. Results of Davis, Faulkner and Joy for frequencies occurring in the de Haas–van Alphen effect and cyclotron resonance masses

Lattice spacing	a	$0\cdot995a$	$0\cdot99a$
Frequency of [100] belly orbit in units of 10^8 G	5·920	5·979	6·036
Frequency of [111] belly orbit	5·731	5·785	5·837
Dog's bone orbit	2·462	2·483	2·505
Neck orbit	0·280	0·289	0·298
m^*_{100} in units of m	1·429	1·433	1·440
m^*_{111}	1·515	1·521	1·529
m^*_{db}	1·228	1·243	1·250
m^*_{neck}	0·417	0·418	0·420

TABLE II. Comparison of experimental and theoretical results for logarithmic pressure derivatives of cross-sectional areas of Fermi surface of copper, in units of 10^{-4} kb^{-1}

	O'Sullivan and Schirber	Templeton	Theory of Davis, Faulkner and Joy
$[111]_{belly}$	$4\cdot25\pm0\cdot2$	4·29	4·35
$[111]_{neck}$	18 ± 2	19·70	15·30
$[110]_{dogbone}$	$4\cdot0\pm0\cdot2$		4·04
$[100]_{belly}$	$4\cdot6\pm0\cdot2$		4·62
$[100]_{rosette}$	$4\cdot3\pm0\cdot3$		

Other detailed comparisons made with experiment confirm the usefulness of the potentials on which these calculations were based. However, the changes in lattice parameter considered here are quite small, and care will have to be exercised, no doubt, before using this prescription for the crystal potential over very wide ranges of pressure.

D. ALKALINE EARTH METALS

We now turn to discuss the change in the electrical properties of Ca, Sr and Ba under pressure.

The experimental findings may be summarized as follows:

(1) In Ca, there is a pronounced increase in resistivity and a negative temperature coefficient of resistivity at 300–400 kb, which corresponds to half the zero pressure atomic volume.
(2) Sr shows a negative temperature coefficient of resistivity at 35 kb in the f.c.c. phase, before the transition occurs to a b.c.c. structure.

(3) Ba, which is b.c.c. at zero pressure, has a positive temperature coefficient of resistivity at all pressures.

The recent work of Vasvari *et al.* (1967) indicates that the f.c.c. phase becomes a semimetal for small atomic volumes, not a semiconductor, because of a line of degeneracy between the lowest two bands which is split only by spin-orbit coupling. This is discussed further in Section VI.B. Also it appears that there are not just simple *sp* bands, but the band structure involves a large admixture of *d* components near the Fermi level.

In contrast, for the b.c.c. structure, the two bands are degenerate along all zone edges PH (see Fig. 21). The b.c.c. structure is expected to be a metal at all pressures.

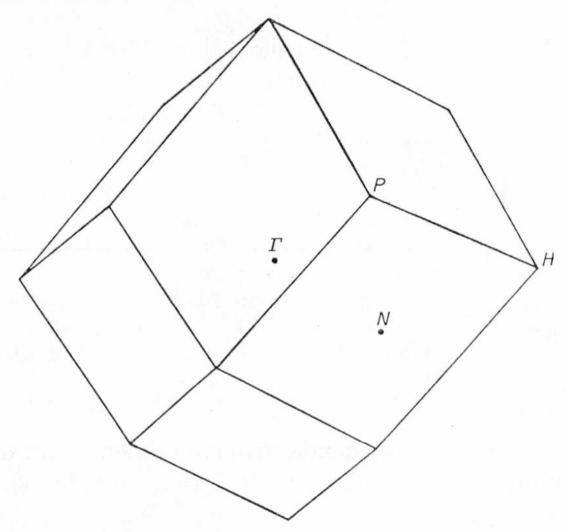

FIG. 21. Brillouin zone for b.c.c. direct lattice with high symmetry points marked.

In the actual band structures (see Fig. 30 below for Ca), the bands are basically nearly-free electron-like. However, it is not possible for the band structure of the alkaline-earth metals to be described, near the Fermi level, by constant (111) and (200) pseudopotential matrix elements, which work well for *sp* bands. The states near E_f already have considerable *d* components, as was indeed apparent from early work of Manning and Krutter (1937).

1. *Electrical Resistivity*

Vasvari and Heine (1967) have made some estimates of the resistivity of f.c.c. Ca based on the above band structures.

In the metallic region ($v/v_0 > 0{\cdot}55$), the conductivity for a cubic metal can be calculated from (Ziman, 1960)

$$\sigma = (e^2/12\pi^3\hbar) \int v_f \tau \, dS \tag{37}$$

where the integration is over the Fermi surface, v_f is the modulus of the Fermi velocity and τ is the relaxation time.

Based on the bands shown in Fig. 30, Vasvari and Heine approximate the electron surface by spherical calottes on each hexagonal face of the zone shown in Fig. 1. The complicated hole surface is approximated by an ellipsoid, with effective masses estimated from the $E(\mathbf{k})$ relation around the point W.

Next the mean free path must be estimated. This can be done by dealing with the electron–phonon interaction by the deformation model of Bardeen and Shockley (1950). Standard theory then allows τ to be expressed in terms of two deformation parameters for electrons and holes, effective mass components and elastic constants. Putting these together into eqn (37), Vasvari and Heine estimate the room temperature resistivity at atmospheric pressure as $2{\cdot}5 \times 10^6 \Omega$ cm, which is over 60% of the measured value for Ca in the f.c.c. structure.

The resistivity calculated as a function of v/v_0 by this method is shown in Fig. 22. It seems in fair agreement with the observed resistance for Ca as a function of pressure, plotted in Fig. 23. It seems then that the structure of Ca is likely to be f.c.c. up to about 400 kb. This is followed by a semi-metallic region with a small negative

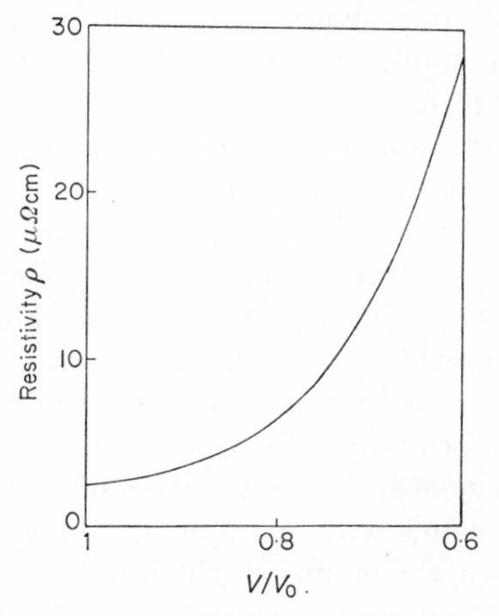

FIG. 22. Calculated resistivity ρ for divalent metal calcium in f.c.c. phase at room temperature.

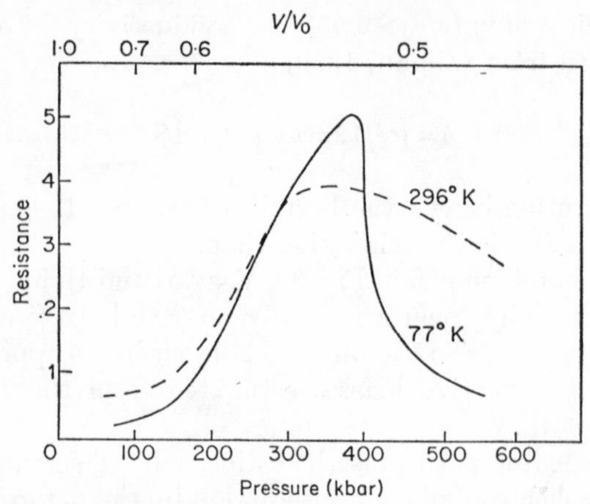

FIG. 23. Observed resistance of Ca as function of pressure (Drickamer, 1965). As drawn, the resistance is in arbitrary units. Volume scale plotted is very rough (see Vasvari and Heine, 1967).

temperature coefficient of resistance. Qualitative understanding appears to be emerging. But particularly the behaviour of Ba is very complex.

We turn now, from the interpretation of electrical resistance and electronic transitions in terms of simple band theory, to discuss recent progress in some areas where electron correlation seems to play an important role.

VI. Excitonic Phases: Transcending Band Theory

A new type of insulating phase has recently been predicted, although it has not yet been found experimentally (see Section VI.C). This phase can be expected to occur in semiconductors with a small band gap, or semimetals with very small band overlap, at very low temperatures.

The observation was made by Mott (1961) that in a semimetal it might be possible, under certain conditions, to form bound electron-hole pairs, leading then to a nonconducting state. Two years later, Knox (1963) pointed out that, in an insulator, the binding energy of an exciton $|E_x|$ might possibly exceed the energy gap E_g. In this case, the normal insulating state with a fully occupied valence band would be unstable against the formation of excitons.

Following these suggestions, a number of authors have recently pointed out that for solids with small energy gaps, there may exist, at very low temperatures, a new phase which has been termed by Jérome and co-workers (1967) an excitonic insulator. As suggested by Kozlov and Maksimov (1965, 1966), this may have a phase diagram of the kind

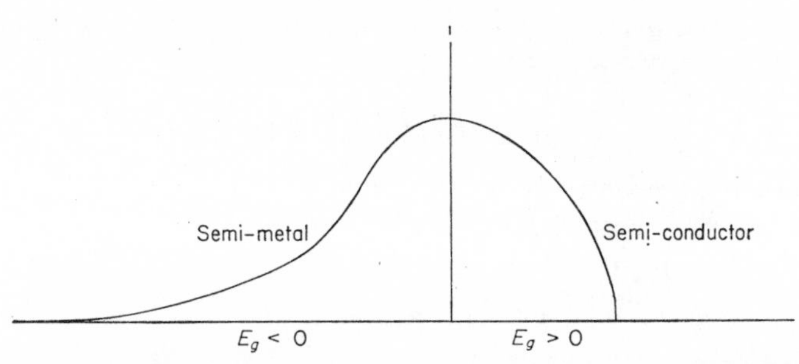

FIG. 24. Phase diagram of excitonic insulator according to Kozlov and Maksimov (1966).

shown schematically in Fig. 24. The properties of this phase have been studied extensively, especially by Jérome *et al.* (1967). The ground state can be described by a theory like that of Bardeen–Cooper–Schrieffer for superconductivity (see for example, Rickayzen, 1965).

The mathematical description of the ground state for the case of a single valence band maximum at $\mathbf{k}=0$ and a single conduction band minimum at $\mathbf{k}=\mathbf{w}$ will be briefly summarized here. Although, as will be seen, there is a close formal similarity between the theory of the excitonic insulating state and the superconducting state (see Appendix II for some further details) the physical properties of the states are, of course, vastly different and in particular there is no Meissner effect in an excitonic insulator.

For the case referred to above the single-particle energies may be written

$$\epsilon_a(\mathbf{k}_a) = -\tfrac{1}{2}E_g - (2m_a)^{-1}k_a^2$$
$$\epsilon_a(\mathbf{k}_b) = \tfrac{1}{2}E_g + (2m_b)^{-1}k_b^2$$

where $\mathbf{k}_a$ and $\mathbf{k}_b$ refer to the respective band extrema. Energies are measured from the centre of the gap and E_g may be positive or negative, m_a and m_b being the appropriate effective masses.

Then for the insulating ground state in simple band theory we can write

$$\Phi = \prod_{\mathbf{k}} a_{\mathbf{k}}^{\dagger}|\rangle \tag{38}$$

where $|\rangle$ is the state with no electrons, $a_{\mathbf{k}}^{\dagger}$, $a_{\mathbf{k}}$ create and destroy electrons in band a with wave vector $\mathbf{k}$ and the values of $\mathbf{k}$ in eqn (38) run over the Brillouin zone.

The possible formation of bound pairs suggests that a new Hartree–Fock ground state wave function of the form

$$\Psi = \prod_{\mathbf{k}} \alpha_{\mathbf{k}}^{\dagger}|\rangle \tag{39}$$

may have a lower energy, where $\alpha_{\mathbf{k}}^{\dagger}$ creates an electron in a linear combination of band a and band b states, say

$$\alpha_{\mathbf{k}} = u_{\mathbf{k}} a_{\mathbf{k}} - v_{\mathbf{k}} b_{\mathbf{k}}; \quad |u_{\mathbf{k}}|^2 + |v_{\mathbf{k}}|^2 = 1, \tag{40}$$

$b_{\mathbf{k}}^{\dagger}$ and $b_{\mathbf{k}}$ being respectively creation and destruction operators in band b, with wave vector $\mathbf{w}+\mathbf{k}$.

From eqns (38)–(40), Ψ may be written

$$\Psi = \prod_{\mathbf{k}} (u_{\mathbf{k}}^* - v_{\mathbf{k}}^* b_{\mathbf{k}}^{\dagger} a_{\mathbf{k}})\Phi \tag{41}$$

which shows, in analogy with the Bardeen–Cooper–Schrieffer theory, that in this state a hole in $(a, \mathbf{k})$ and an electron in $(b, \mathbf{k}+\mathbf{w})$ are either both present or both absent.

The relation of this theory to the usual excitonic equations is developed a little further in Appendix II where it is verified that, provided the energy gap is less than the excitonic binding energy, eqn (41) leads to a lower energy than eqn (38). Jérome and co-workers have calculated the finite-temperature d.c. conductivity of this excitonic phase and for the semi-metallic situation they find the situation depicted in Fig 25, and this is also discussed in more detail in Appendix II.

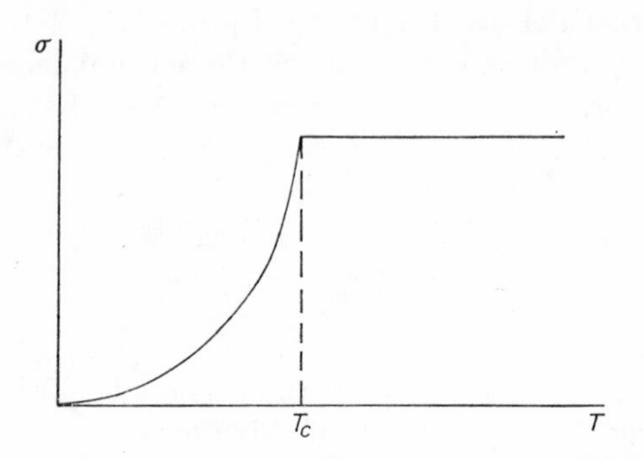

FIG. 25. Schematic form of d.c. conductivity σ as function of temperature in semi-metallic limit.

As has already been indicated, the scale on the phase diagram shown in Fig. 24 is determined by the exciton-binding energy in the normal insulating phase. This is given by the hydrogenic formula

$$|E_x| = \tfrac{1}{2}(\mu^*/m)\left(\frac{1}{\epsilon_0^2}\right) Ry \tag{42}$$

where μ^*/m is the ratio of reduced effective mass to free-electron mass and ϵ_0 is the static dielectric constant. The cut-off shown for positive E_g

in Fig. 24 is given by

$$E_{g\ \text{Cut-off}} = |E_x|. \tag{43}$$

For negative E_g there is an exponential decrease of transition temperature whose decay is determined by $|E_x|$. The maximum transition temperature according to Kozlov and Maksimov (1966) is given, in order of magnitude, by

$$k_B T_{\text{transition}} \sim |E_x|. \tag{44}$$

The question of whether this excitonic insulator can be realized in practice is obviously one of considerable interest. Since, as has been seen, we must have small band gaps, and we need to vary the gap continuously through zero, the application of hydrostatic pressure to suitable materials immediately suggests itself. Since calcium, strontium and ytterbium appear to change from metallic to insulating behaviour under pressure (Drickamer, 1965; see also Section V.D) these are obvious materials to study. Jérome *et al.* (1967) estimate that, for Sr, the exciton binding energy may be $\sim 4 \times 10^{-3}$ eV, and hence from eqn (42), with $(\mu^*/m) \sim 0{\cdot}25$ and $\epsilon_0 \sim 30$,

$$T_{\text{transition}} \sim 20\,^\circ\text{K}.$$

They stress that this is at best an order of magnitude estimate.

The semimetals As, Sb and Bi also suggest themselves as candidates, although the exciton-binding energy is now $\sim 10^{-5}$ eV and

$$T_{\text{transition}} \sim 0{\cdot}05\,^\circ\text{K}.$$

Finally, the pressure range over which the new phase can exist is only the order of 2×10^{-3} kb which makes the experiment impracticable at present.

A. METAL-SEMICONDUCTOR TRANSITION IN YTTERBIUM AND STRONTIUM

These considerations suggest that the most favourable materials in which to search for excitonic phases are ytterbium and strontium. If the excitonic phase exists, then, according to the theoretical predictions, it will produce a lattice distortion, or a spin density wave will be formed (cf Section VII.B). The additional periodicity introduced will lead to energy gaps in the band structure, which is another way of arguing that an anomaly in the resistivity as a function of pressure should occur if an excitonic phase is formed.

The predictions of Jérome *et al.* (1967) have led to further investigation by McWhan and co-workers (1969), which we shall now discuss in some detail. No positive evidence for the existence of the excitonic phase is obtained, but some interesting contact with the band theory studies of Vasvari *et al.*, discussed in Section V.D, can be made.

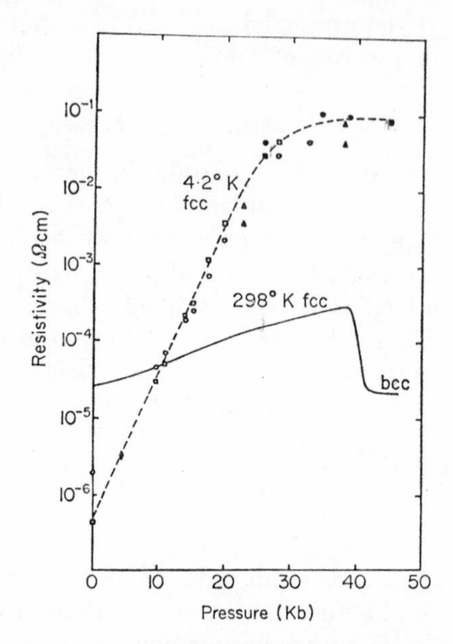

FIG. 26. Measured resistivity of ytterbium as function of pressure at 4·2°K and 298°K. Samples *A*, *B* and *C* used by McWhan and co-workers correspond to purities of 97·8, 99·99 and 99·9% respectively. □■ Sample *A*, △▲ Sample *B*, ○● sample *C*.

The measured resistivity of Yb is shown in Fig. 26, over a range of pressures from 0–50 kb and for temperatures of 4·2°K and 298°K. At low temperatures the structure is f.c.c. over the entire pressure range. The resistivity ρ at 4·2°K is seen from Fig. 26 to increase by a factor of 6×10^4 as a pressure of 25 kb is applied, and then to saturate at higher pressures. In contrast, at room temperature the transition to a b.c.c. structure at 40 kb which had been observed by earlier workers, is clearly reflected in the measurements of McWhan and co-workers shown in Fig. 26. The resistance ratio $\rho_{4.2°K}/\rho_{298°K}$ at 25 kb is about 220. In the high pressure b.c.c. phase, Yb is again a good metal, with a resistivity ratio similar to that at atmospheric pressure. For samples *B* and *C* of Fig. 26, Figs 27(a) and (b) show logarithmic plots of resistivity *versus* $1/T$ for different pressures. For sample *B* an exponential temperature dependence is observed over a range from 100–300°K at pressures of 19·7 and 27·8 kb. McWhan and co-workers also observe

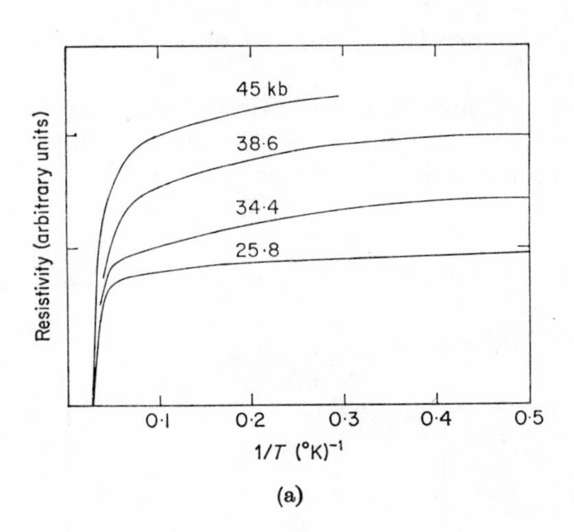

(a)

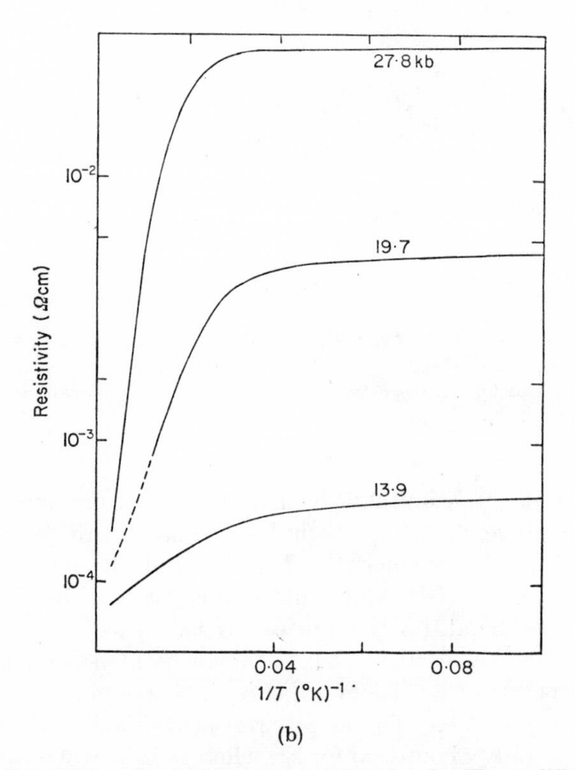

(b)

Fig. 27. (a) Logarithm of resistivity *versus* $(1/T)$ plot for Yb at different pressures (sample C of Fig. 26). (b) Logarithm of resistivity *versus* $(1/T)$ plot for Yb at different pressures (sample B of Fig. 26).

that for Yb the temperature coefficient of resistivity becomes negative at about 11 kb.

These results can now be compared with those for Sr, although the measurements here were made on less pure samples than for Yb. Figure 28 shows the resistivity measurements made on three different

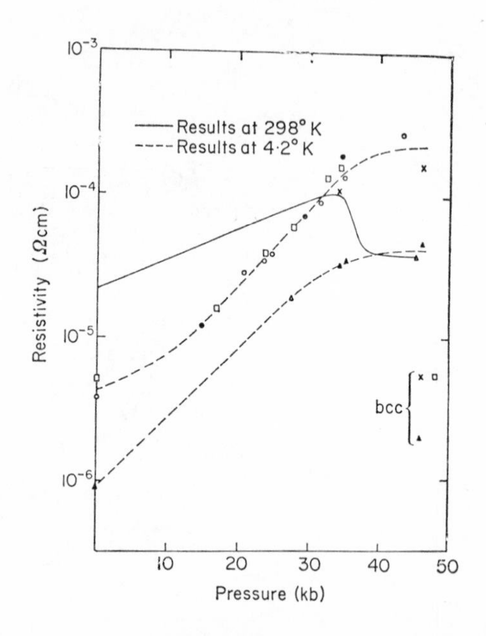

Fig. 28. Measured resistivity of strontium as function of pressure at 4·2°K and 298°K. Samples *D*, *E* and *F* used by McWhan and co-workers correspond to purities of 98·1, 98·8 and 98·6% respectively. × ○● Sample *D*, □ sample *E*, △ sample *F*. All at 4·2°K.

samples of Sr at 4·2°K and 298°K as a function of pressure, again over the pressure range 0–50 kb. Samples *D* and *E* are seen to behave similarly to the Yb specimens (cf Fig. 26), but the magnitude of the resistivity changes is very different. Thus, with increasing pressure, the resistivity of Sr at 4·2°K increases by two orders of magnitude, in contrast to a factor $\sim 10^5$ for Yb. As can be seen from Fig. 29, the negative temperature coefficient appears for Sr around 30 kb, instead of around 10 kb as in Yb. The largest rise in resistivity with decreasing temperature is only about two for Sr, whereas it is ~ 300 for Yb. The way in which these results can be understood in terms of the band structures of divalent f.c.c. metals will now be considered.

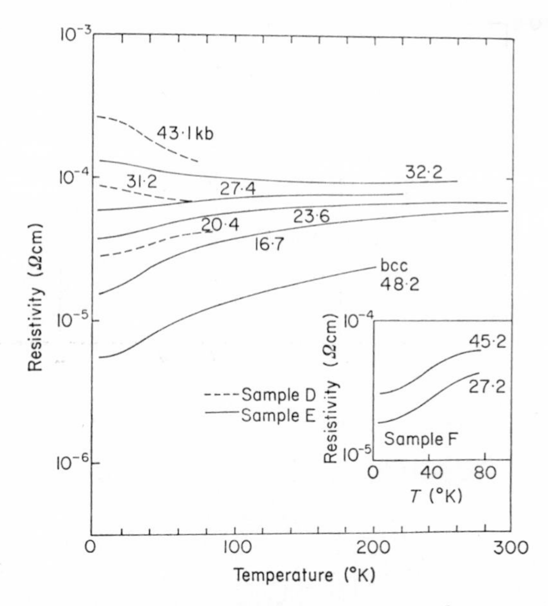

FIG. 29. Measured resistivity of strontium as function of temperature at various pressures. Samples D, E and F as in Fig. 28. Results for sample F are shown in inset.

B. RELATION TO BAND STRUCTURES OF DIVALENT FACE-CENTRED-CUBIC METALS

In the original model put forward by Mott and Jones (1936), a crystal made up of divalent atoms is insulating at large interatomic spacing, with a filled s valence band and an empty p conduction band. When such a crystal is squeezed, the forbidden energy region separating these allowed energy bands shrinks and eventually becomes zero (cf the discussion of He in Section II.A). The crystal then becomes metallic and can become semiconducting under further compression, depending on the variation of the band gaps.

However, the recent work of Vasvari and co-workers which was referred to earlier, has indicated that d states can play a major role in some cases. To demonstrate this, their band structure for Ca is shown in Fig. 30. The important new feature is that a line of degeneracy exists between the first and second bands between W and L. As the pressure is increased, it can be seen from Fig. 30 that the energy difference between the states W_2' and L_2' becomes smaller and the Fermi surface shrinks. Because of the degeneracy, the Fermi surface cannot shrink to zero, but will take the form of a ring of connected " sausages " of alternating

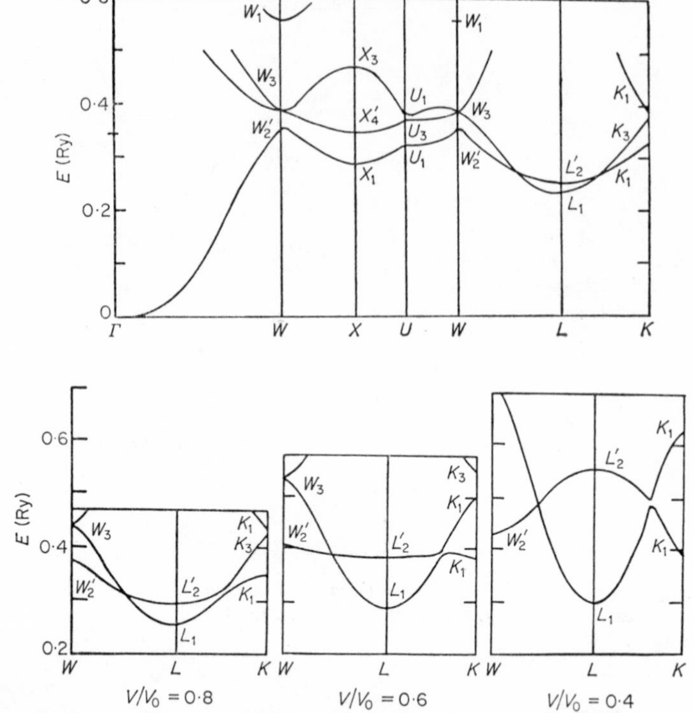

FIG. 30. Electronic band structure for Ca in f.c.c. phase according to Vasvari and co-workers (1967). For symmetry points, refer to Fig. 1.

pockets of electrons and holes on the hexagonal face of the Brillouin zone of Fig. 1.

This model predicts that the divalent f.c.c. metals become semimetals under high pressure, but not semiconductors. However, the discussion of Vasvari and co-workers did not include the effects of spin-orbit coupling, which removes the degeneracy and disconnects the "sausages" (Halperin and Rice, 1968). If the energy gaps introduced by the spin-orbit coupling are larger than the width of each "sausage", the Fermi surface will shrink to zero and the material will become a semiconductor.

Although the band structure of Fig. 30 is for Ca, it seems probable that the electronic states of Sr should resemble those of Ca at a reduced volume. A band structure calculation for f.c.c. Yb is in progress in the author's Department but, in the absence of results for the energy bands at the present time, there is evidence from the similarity of numerous physical properties of Yb and Sr that their band structures are not too dissimilar. However, the role that the 4*f* bands play is not yet clear in Yb.

The experimental resistivities of Yb and Sr, shown in Figs 26–29, are compatible with the band structure discussed above, if spin-orbit coupling is included. The residual resistivity increases with decreasing volume, suggesting a decrease in carrier concentration as the Fermi surface shrinks. Whereas in Sr, the small negative temperature coefficient of resistivity suggests semimetallic behaviour, the exponential behaviour shown in Fig. 27 suggests that Yb becomes a semiconductor at high pressures, which is consistent with its much larger spin-orbit interaction.

It should be noted here that at least two other explanations have been put forward to explain the properties of Yb.

(1) Rocher (1962) suggested that a virtual $4f$ state passed through the Fermi level as the pressure was increased, and that this resulted in an increase in the resistivity; McWhan and co-workers emphasize, however, that this model will not account for the temperature dependence of the resistivity.
(2) A two-band model with a very narrow $4f$ band lying near the Fermi energy in a free electron band (Coqblin and Blandin, 1968) has been considered, and this model could lead to the development of a semi-conducting energy gap with increasing pressure. Coqblin and Blandin then offer as an explanation of the subsequent f.c.c.–b.c.c. transition the promotion of a $4f$ electron to the conduction band, in a manner similar to that occurring in cerium. Hall and co-workers (1963) interpreted their X-ray data on the basis of an electronic transition, but as McWhan and Jayaraman (1963) have emphasized the percentage change in the atomic radii of each phase (calculated on a hard sphere model) is the same for both Yb and Sr and results from a change in coordination number.

If an unfilled $4f$ band in the b.c.c. phase exists, it might perhaps be expected to exhibit magnetic order. But the resistivity of b.c.c. Yb down to 2°K shows no anomalies which could be associated with spin disorder scattering or other magnetic phenomena. Thus, there seems no experimental evidence for an unfilled $4f$ band in b.c.c. Yb. The similarity between Sr and Yb and the agreement with the band structure work on Ca, when account is taken of spin-orbit coupling, suggests that the $4f$ electrons do not play an important role in f.c.c. Yb.

C. EVIDENCE FOR EXCITONIC PHASE

The temperature dependence of the resistivity of Yb down to 2°K in the pressure range from 10–20 kb shows no abrupt change in slope which might reflect the formation of the excitonic phase. However,

McWhan and co-workers point out that it would be entirely consistent with their results if the critical temperature for the formation of the excitonic phase is well below their experimental limit of 2°K, the binding energy of an exciton being of the order of 0·5°K.

As in chromium, to be discussed in Section VII.B, where a loss of carriers occurs due to the formation of a spin density wave, one might expect to see a transition to an excitonic state reflected in a sharp change in the resistivity at the transition. No definitive evidence is found for this. Pure materials will be required before a final conclusion can be drawn as to the existence or otherwise of an excitonic phase above 2°K in Yb and Sr.

VII. Effect of Pressure on Transition Temperatures of Cooperative Phenomena

A. Superconducting Transition Temperature†

The formal analogy between the theory of the excitonic insulator and the Bardeen–Cooper–Schrieffer theory of superconductivity was spoken of above. An interesting question is whether or not superconductivity can be destroyed by the application of a sufficiently high pressure. Quite a number of experimental studies have been made on the effect of pressure on the superconducting transition temperature T_c, but before reviewing these results it will be useful to consider a phenomenological approach to the pressure dependence of T_c.

1. *Phenomenology of Pressure Dependence of Transition Temperature*

The Bardeen–Cooper–Schrieffer theory of superconductivity yields a transition temperature which may be written approximately (for a derivation see for example, Rickayzen, 1965; March *et al.*, 1967)

$$T_c = \Theta \exp\left(-\frac{1}{g}\right) \tag{45}$$

where Θ is as usual the Debye temperature and g measures the strength of the electron–phonon interaction. Hence we have

$$\ln T_c = \ln \Theta - g^{-1} \tag{46}$$

and

$$\frac{\partial \ln T_c}{\partial \ln v} = \frac{\partial \ln \Theta}{\partial \ln v} + \ln\left(\frac{\Theta}{T_c}\right)\frac{\partial \ln g}{\partial \ln v}. \tag{47}$$

The Gruneisen parameter γ is simply $-\partial \ln \Theta / \partial \ln v$ and to understand the volume dependence of T_c we require an estimate of

† Note added in proof. It has been suggested recently that metallic hydrogen may be a super conductor with a high transition temperature (N. W. Ashcroft (1968). *Phys. Rev. Lett.* **21**, 1748).

$\partial \ln g / \partial \ln v$. The observed values of $\partial \ln g / \partial \ln v$ are shown in Table III, while in Fig. 31 explicit results for T_c as a function of volume are also plotted.

Ziman (1962) has proposed a form for g which may be written

$$g \approx 3Z \left(\frac{m}{M}\right) (\overline{U}/k_B\Theta)^2 \tag{48}$$

where Z is the valency, M is the ionic mass and $\overline{U}$ is a quantity related to the resistivity of the liquid metal, discussed in Section II.D. We may then write

$$\frac{\partial \ln g}{\partial \ln v} = 2\,\frac{\partial \ln U}{\partial \ln v} - 2\,\frac{\partial \ln \Theta}{\partial \ln v}. \tag{49}$$

While the theory of liquid metals is now sufficiently developed for a

TABLE III. Variation of electron–phonon interaction parameter g with volume v

	Pb	Hg	Sn	In	Tl	Al	Ga	Zn	Cd
$\left(\dfrac{\partial \ln g}{\partial \ln v}\right)_{\text{observed}}$	2·1	1·7	2·3	2·3	0	3·4	1·8	2·0	2·9
2γ	5·7	6·0	4·5	5·0	4·5	4·4	2·9	4·1	4·6
$\left(\dfrac{\partial \ln g}{\partial \ln v}\right)_{\text{calculated}}$	2·7	2·4	2·4	2·8	2·6	3·2	2·1	2·9	3·3

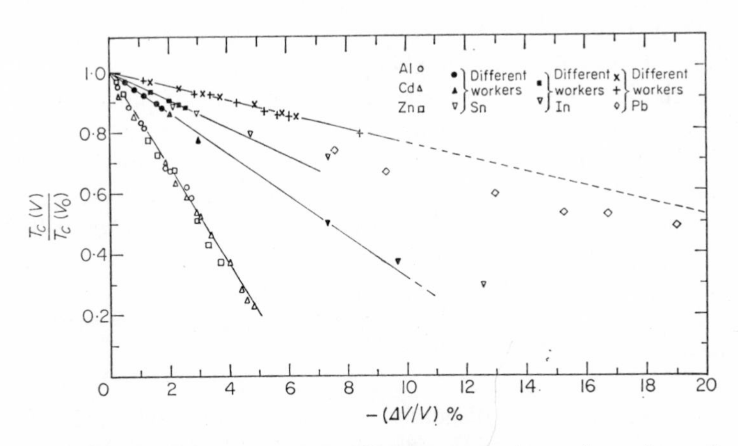

FIG. 31. R duced transition temperature T_c *versus* percentage volume change for superconductors, according to Smith and Chu (1967).

start to be made on the calculation of the first term on the right-hand side of eqn (49), detailed estimates are not yet available. If therefore, we assume tentatively that the second term dominates, then eqn (49) yields $(\partial \ln g/\partial \ln v) \sim 2\gamma$. Values of 2γ are much larger than the observed values of $\partial \ln g/\partial \ln v$ as Table III shows clearly. Thus, until the variation of $\bar{U}$ with v is calculated, eqn (49) is not quantitatively useful and an alternative approach is needed.

This is afforded by the recent work of McMillan (1968), who has used the strong coupling theory of superconductors to calculate the superconducting transition temperature as a function of the electron–phonon and electron–electron coupling constants. Reference will have to be made to the original paper for details, but the important result in the present context is that the quantity g in eqn (45) may be written in the form

$$g = \frac{\lambda}{1+\lambda} \tag{50}$$

where $\lambda = C/M\Theta^2$. As Olsen and co-workers (1968) have pointed out, the constant C seems from empirical arguments to depend only weakly on volume and if we neglect this dependence then eqn (50) leads to

$$\frac{\partial \ln g}{\partial \ln v} = \frac{2\gamma}{1+\lambda}. \tag{51}$$

The estimates of λ made by McMillan from his theory, including an approximate account of Coulomb correlations, then bring theory and experiment into reasonable agreement for non-transition metal superconductors, except for the highly anisotropic metals Hg and Tl, as can be seen from Table III. Accounting for the anisotropy, some measure of agreement is again found in these cases also.

Although the above argument allows the effect of pressure on non-transition metal superconductors to be understood in at least a semi-quantitative way, the situation is much less satisfactory for the transition metals. Here the role of Coulomb correlations between electrons as well as the effect of spin fluctuations must be taken fully into account (Andres and Jensen, 1969; see comments in Olsen *et al.*, 1968). Inclusion of these effects leads to the sort of correlation between the pressure effect and the isotope effect that has been observed (Olsen and co-workers, 1964).

2. *Experimental Results*

The two approaches discussed above involve uncertainties. In particular, the variation of U with v in eqn (49) is not yet known, while the volume dependence of the quantity C on which λ in eqn (50) depends has so far been neglected. Thus, these methods do not as yet afford a really reliable way of answering the question we raised as to the destruction of superconductivity by means of the application of high pressure.

It is therefore worthwhile to review briefly attempts which have been made to extrapolate directly from measurements made over a limited pressure range to obtain estimates of the critical pressures needed to destroy superconductivity. Numerous experimentalists certainly agree that the effect of pressure is in general to lower the superconducting transition temperature, but, so far, no example is known in which pressure has destroyed the superconducting state.

It must be said that the estimates of different workers differ considerably. One possible empirical approach follows the work of Chester and Jones (1953), who found that a plot of T_c against v gave a good straight line. This method has been pursued more recently by Smith and Chu (1967) who, in addition to making measurement of T_c as a function of pressure for Pb, have plotted the available data for a number of superconductors as a function of volume. The results of their investigation are displayed in Fig. 31, where it can be seen that, over a limited pressure range, T_c appears to be linear with volume. Of course, to extrapolate linearly to find the critical pressure is somewhat arbitrary, but by doing this the critical pressures for the destruction of superconductivity appear to be about 70 kb for Al and near to 40 kb for Cd and Zn.

In concluding this brief summary, it is worth recording that the high-pressure phases of numerous substances, including Te, Se, Si and Ge have been found to be superconducting, with transition temperatures usually from 4–10°K.

B. EFFECT OF PRESSURE ON SPIN DENSITY WAVE STATES

Due to electron interactions, it was seen in Section VI how a new phase might arise by formation of excitons. According to the theory, this phase can produce a spin density wave state and such a state is characterized by densities for upward and downward spin densities which are out of phase. That is, we can write

$$\rho\uparrow(\mathbf{r}) = \tfrac{1}{2}n[1 + A\cos(\mathbf{Q}.\mathbf{r} + \phi)] \tag{52}$$

and

$$\rho\downarrow(\mathbf{r}) = \tfrac{1}{2}n[1 + A\cos(\mathbf{Q}.\mathbf{r} - \phi)]. \tag{53}$$

Evidently, the mean density is n, A measures the amplitude of the spin density wave, while **Q** is its wave vector. A pure spin density wave state corresponds to $\phi = \pi/2$.

Although the original prediction of spin density wave states by Overhauser suggested that such states should occur rather widely, it is only in one case, metallic chromium, that concrete experimental evidence is available for their existence. However, Overhauser argues for their existence in postassium (or related charge density waves: this is the case when $\phi = 0$ in eqns (52) and (53) above) and we shall discuss this metal briefly below.

1. *Chromium*

It seems from a variety of experiments that the spin-density wave state proposed by Overhauser exists in Cr metal because of the rather exceptional nature of the Fermi surface. We feel it would take us too far from the main theme of this article to deal at all fully with the detailed arguments and we refer the reader to the papers by Lomer (1962) and Overhauser (1962). We shall instead deal specifically with the work of McWhan and Rice (1967) who have studied the pressure dependence of the itinerant antiferromagnetism. Their results for the electrical resistance as a function of pressure are shown in Fig. 32.

As Overhauser (1962) pointed out, when Coulomb interactions are included in an electron gas in the Hartree–Fock approximation, the antiferromagnetic or spin-density wave state has a lower energy than the paramagnetic state. However, the inclusion of electron correlation, in addition to the exchange terms treated in the Hartree–Fock approximation, almost certainly suppresses the instability of the paramagnetic state against spin-density wave formation in a uniform electron gas.

With the inclusion of a realistic band structure, however, it becomes clearer that the criterion for antiferromagnetism in a metal depends very strongly on the nature of its Fermi surface (Lomer, 1962; see also Fedders and Martin, 1966). In particular, as Lomer has proposed, Cr should be antiferromagnetic by a two-band mechanism, and we denote the two bands by a and b. Then, following Lomer, the Fermi surface of Cr can be characterized schematically in the (100) plane as in Fig. 33. The shaded areas show the occupied parts of each band. It can be seen that the occupied part of band a is similar in shape to the unoccupied region of band b. If now the band b is shifted by q shown in Fig. 33, then the new displaced Fermi surfaces are as shown in Fig. 34.

The condition that the band structure is unstable against a spin-density wave state with a wave-vector **Q** is just that there are two neighbouring bands whose energies are close to each other over an appreciable region of k space. Evidently, when $\mathbf{q} = \mathbf{Q}$ then a situation

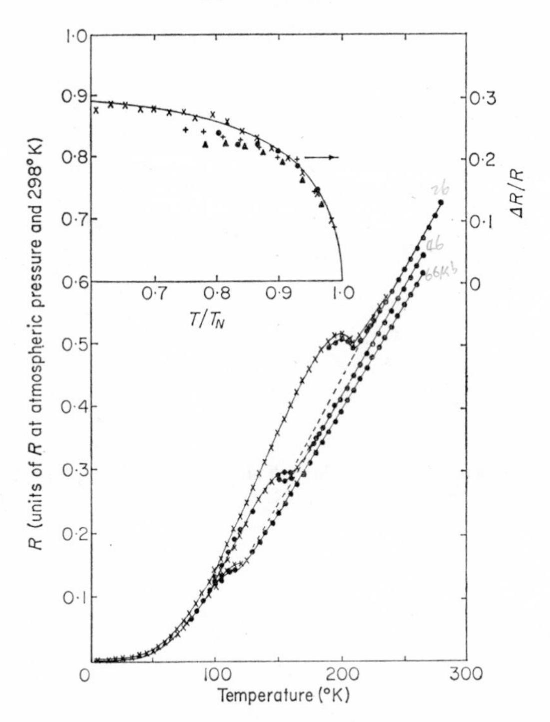

FIG. 32. Measured resistivity of chromium as function of temperature for various pressures. ×, ● = two different samples, A and B respectively. From top, the pressures are approximately 26, 46 and 66 kb in each case. – – – extrapolated paramagnetic resistance, denoted by R_p. If R_a is resistance in antiferromagnetic phase, insert shows $(R_a—R_p)/R_a = \Delta R/R$ *versus* temperature. —— = calculated by McWhan and Rice from two-band model. +, ▲ = sample A at 27 and 46 kb respectively. ● = sample B at 46 kb. + = a third sample C at 46 kb. McWhan and Rice only give information about samples that were single-crystal chromium (Battelle Iochrome) with samples A and B being from same ingot.

which is very favourable for antiferromagnetism in Cr arises, the area A of close contact of the two Fermi surfaces being the shaded area in Fig. 34.

With this qualitative discussion we turn immediately to discuss the consequences of the results of McWhan and Rice (1967) shown in Fig. 32. From the electrical resistivity, they extract the volume dependence of the Néel temperature and the results are plotted in Fig. 35.

There is a very marked contrast with the pressure dependence which has been observed experimentally for the magnetic ordering temperatures in Ni, Fe and some rare-earth metals. Here, a linear variation has been found for similar changes of volume (see below for some references).

As McWhan and Rice emphasize, these results, for the Néel temperature

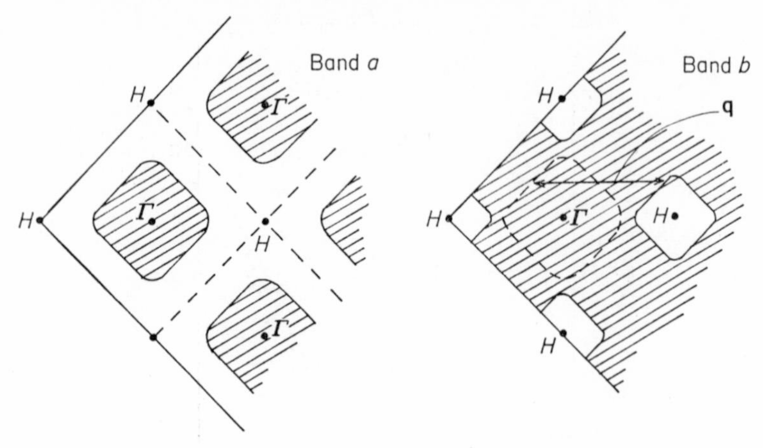

FIG. 33. Schematic form of Fermi surface of chromium in (100) plane, following Lomer. For symmetry points, refer to Fig. 21.

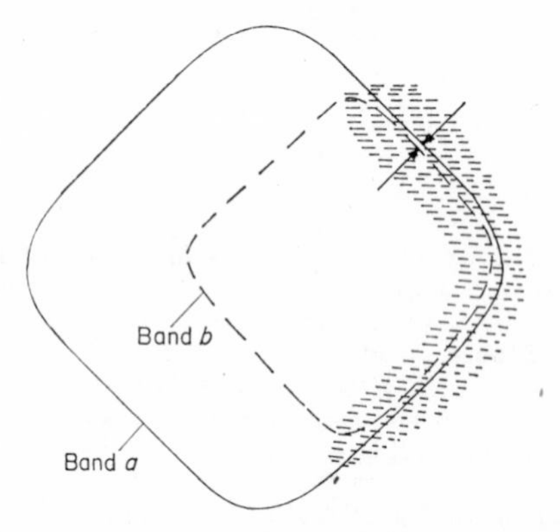

FIG. 34. Fermi surfaces displaced by wave vector **q**.

T_N as well as the magnetic contribution to the resistivity, can be interpreted using the itinerant two-band model of antiferromagnetism discussed above. Under the conditions shown in Fig. 34, where the electron Fermi surface and the hole Fermi surface displaced by a spin-density wave vector **Q** match well, the transition temperature will be determined by an equation like that of the Bardeen–Cooper–Schrieffer eqn (45) for the superconducting transition temperature, and the antiferromagnetic phase will have a temperature dependent energy gap.

Naturally, in applying eqn (45) to give the Néel temperature, we must re-interpret Θ and g. It is sufficient to say here that the Debye temperature Θ is replaced by the temperature corresponding to the

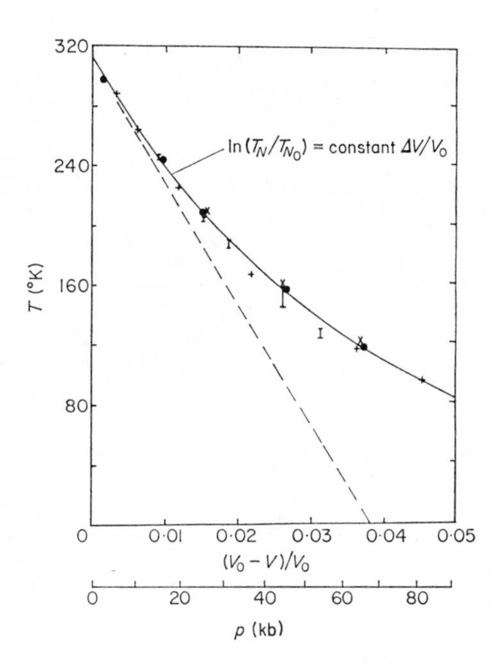

FIG. 35. Variation of Néel temperature of chromium with pressure. Samples A, B and C, referred to in Fig. 32, are crosses, solid circles and vertical bars respectively.

energy band width. Furthermore, g is proportional to the shaded area, A of Fig. 34, over which matching occurs, and inversely proportional to the average of the Fermi velocities of the electron and hole pockets.

McWhan and Rice then interpret the strong dependence of T_N on volume which is displayed in Fig. 35 as due to a rapid shrinkage of the area A as the pressure is increased. Clearly, reduction of A and hence g in eqn (45) will lower the Néel temperature T_N. A question similar to that raised in the previous section for superconductivity then arises: whether Cr will be antiferromagnetic at $T=0$ at all pressures! McWhan and Rice also give a theory of the fraction of their measured resistance which results from magnetic effects (see Fig. 32) but we must refer the reader to the original paper for the details.

It should be mentioned here that experimental work on the influence of pressure on magnetic ordering temperatures in the rare-earths Gd and Dy, and in some alloy systems, has been reported by Austin and Mishra (1967), while the theory of the pressure dependence of the Curie temperature of Ni and Ni–Cu alloys has been studied by Lang and Ehrenreich (1968).

2. *Possible Charge or Spin Density Wave State in* K

The situation with regard to likely instabilities of the band structure of potassium, which have been discussed earlier in this paper, with respect to spin density and/or charge density waves are now briefly summarized. Overhauser (1964) initially supposed that a spin density wave in K was required to explain the optical absorption as measured by Mayer and El Naby (1963). There has been some further discussion of this theory (Hopfield, 1965; Overhauser 1968) and the situation is not entirely clear at the present time. Some further results on the high-field magnetoresistance of K have, however, presented a puzzle (Penz and Bowers, 1968) and recently Reitz and Overhauser (1968) have proposed an explanation in terms of a charge-density wave state. This, as far as we know, is the only explanation so far offered for the anomalous magnetoresistance.

In the light of the work of McWhan and Rice on Cr, it might be interesting to explore the pressure dependence of the high-field magnetoresistance of K. It would seem intuitively that the effect of pressure will be to eventually suppress the charge density wave state, if it exists at all.

VIII. Alloys Systems

Increasing attention, we believe, will be focused on the properties of alloys under pressure during the next few years. With this in mind, we will now discuss very briefly two quite different investigations which have been carried out recently on alloys. The first, on Cr-based alloys has been chosen because of the relation to the discussion of the last section: the other because it touches on a rather controversial point in the theory of electron states in disordered materials.

A. Cr BASED ALLOYS

The first-order transition between a spin-density wave state, with periodicity not commensurate with the lattice, to a conventional anti-ferromagnetic state has been studied as a function of pressure (Syono and Isikawa, 1967). It had been observed previously that the magnetic properties of Cr were highly sensitive to alloying with quite small concentrations of solute. These workers investigated the alloy Fe-Cr at 2·3 atomic percentage of Fe, and Fig. 36 shows the results they obtained for resistance *versus* temperature for pressures in the range 0–24 kb. The transition temperature is easily identified, although the effect of pressure is to blur the sharpness of the transition. However, it

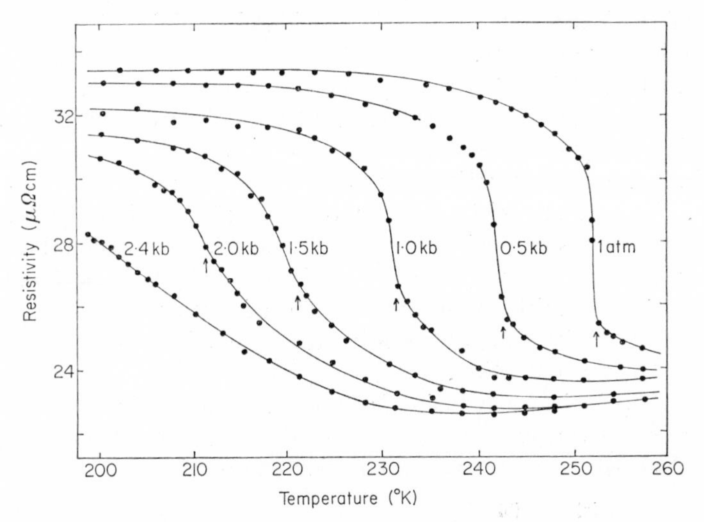

FIG. 36. Resistance *versus* temperature for Fe–Cr alloy with 2·3 atomic % of Fe. Transition temperatures for different pressures are indicated by arrows.

is quite clear that the transition temperature is lowered with increasing pressure for this alloy. Quite different results are obtained however, for Mn–Cr. Here, the transition is not identifiable from electrical resistance measurements, and therefore these workers studied the thermal expansion. An opposite pressure dependence of the transition temperature was found. A basic interpretation of these results is lacking at the present time.

In this general area, a theoretical study of the dependence of the periodicity of the spin-density waves in antiferromagnetic metals upon the number of conduction electrons has recently appeared (Falicov and Penn, 1967). In particular, they discuss Cr-rich Cr–Mn and Cr–Re alloys but it would take us too far from the main theme of this paper to go into further details here.

B. EFFECT OF PRESSURE ON RESISTIVITY AND THERMOELECTRIC POWER OF LIQUID MERCURY–INDIUM ALLOYS

The expected asymptotic high pressure behaviour of the resistivity of liquid metals was briefly mentioned in Section II.D. However, the detailed band structure was not considered there. One of the interesting questions in liquid metals at present concerns the form of the density of states for a divalent metal which, in the solid state, has a relatively small overlap between nearly full and nearly empty zones. Mott (1966) supposes that the density of states in the liquid has a minimum in the region of the Fermi energy and is considerably different from a free-electron density of states.

It is this point that Bradley (1966) has set out to test. To do so, measurements were made on the resistivity and thermopower of a series of liquid mercury–ndium alloys, up to 150°C and in a pressure range up to 6 kb. The concentrations covered were from pure Hg to 70% indium. The results obtained for the electrical resistivity and the thermopower are summarized in Figs 37 and 38. The anomalous peak at 3% indium at atmospheric pressure discovered by Cusack *et al.* (1964) is found to diminish with application of pressure. The pressure coefficients of resistivity fall rapidly from the value for pure Hg with increasing In concentration.

Bradley's principal conclusion is that both the effect of alloying, and the effect of compression, tend to lead to more free electron-like conditions. At least a qualitative explanation can be given on the basis of a

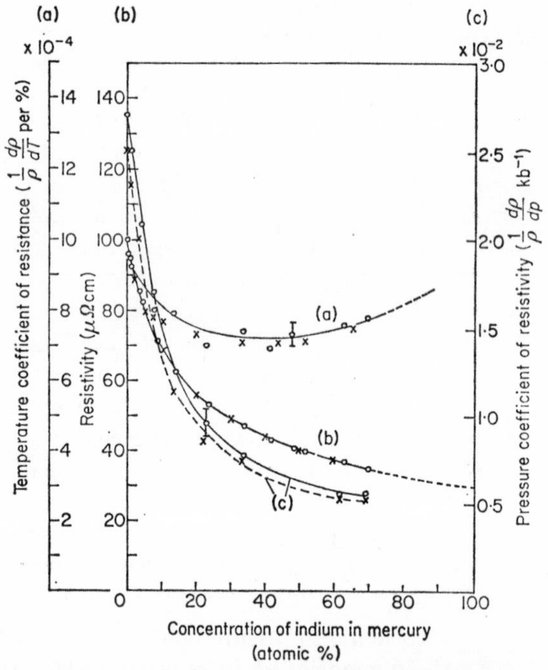

FIG. 37. Curve (a) Resistivity as function of concentration at atmospheric pressure. ○——○ Results of Bradley. ×——× Results of Cusack and co-workers (1964). Curve (b) Temperature coefficient of resistivity as function of concentration at atmospheric pressure. ○——○ Results of Bradley. ×——× Results of Schulz and Spiegler (1959). Curve (c) Pressure coefficient of resistivity as function of concentration. ○——○ Results of Bradley at 20°C. ×——× Results of Bradley at 100°C.

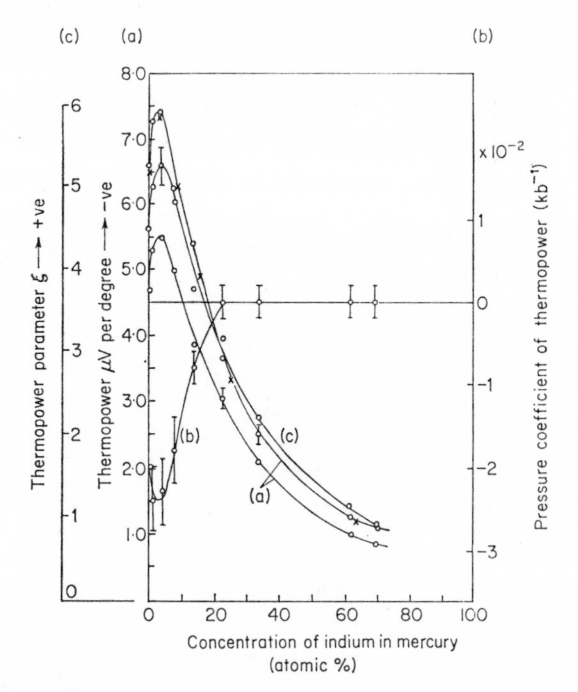

FIG. 38. Curve (a) Thermopower as a function of concentration at atmospheric pressure. O——O Results of Bradley at 20°C and 100°C. ×——× Results of Cusack and co-workers (1964). Curve (b) Pressure coefficient of thermopower *versus* concentration. O——O Results of Bradley at 20°C. Curve (c) Parameter ξ of eqn (54) as function of concentration at atmospheric pressure. O——O Results of Bradley at 20°C.

density of states curve of the form shown in Fig. 39. Following Mott, we suppose that the Fermi level is on the high energy side of the minimum. Then the large value of ξ, the thermopower parameter, defined by (cf eqn (34))

$$\xi = -E_f \left(\frac{\partial \ln \rho(\epsilon)}{\partial \epsilon} \right)_{\epsilon = E_f} \tag{54}$$

where $\rho(\epsilon)$ is the resistivity as a function of energy, is to some extent explained. In the case when there is a low density of states, the expression for the resistivity ρ is modified by a factor g^2, where g is the ratio of the real to the free electron density of states at the Fermi level. Thus we may write

$$\frac{1}{\rho} = \frac{e^2 S g^2 \lambda}{12\pi^3 \hbar} \tag{55}$$

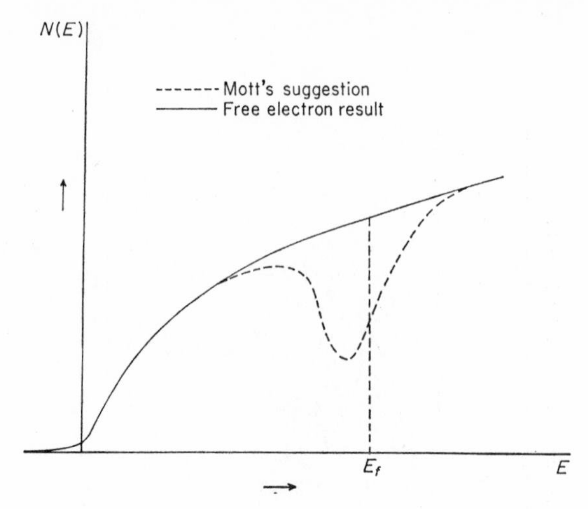

Fig. 39. Schematic form of density of states in liquid mercury according to Mott model. No attempt has been made to correct for normalization here.

where S is the Fermi surface area, λ the mean free path, and from eqns (54) and (55) it may be seen that ξ is a function of $(\partial N(\epsilon)/\partial\epsilon)|_{\epsilon=E_f}$. With regard to the Fermi energy, increasing In concentration and increasing pressure have similar effects. Thus, for small concentrations $(1/E_{\mathrm{F}})(\mathrm{d}E_f/\mathrm{d}c) \sim +0{\cdot}4$ (or $\sim 1{\cdot}2$ if one assumes the data for the Hall coefficient from Cusack *et al.*) and

$$\frac{1}{E_f}\frac{\mathrm{d}E_f}{\mathrm{d}\rho} \sim +2{\cdot}5\times 10^{-6}\ \mathrm{b}^{-1}.$$

It should be remarked at this point that the above result for the change in Fermi level with concentration should be used with caution, because, as Friedel has shown, the shift in E_f with c is very small at small c, and is certainly not linear in c. However, the picture of localized screening leading to the above results, and an argument more akin to a "rigid band" model in which the Fermi level shifts, are in agreement in certain respects and Bradley's conclusions probably remain intact.

Proceeding then with the argument, Mott suggests that the effect of alloying and increasing pressure on the density of states will tend to decrease the minimum and the factor g will become more nearly unity. A qualitative explanation is thereby afforded for the drop in ρ as g increases by alloying and increasing pressure. The two effects are complementary and it is to be expected, as indeed is observed, that $(v/\rho)(\partial\rho/\partial v)$ falls rapidly with increasing In concentration.

To understand the initial positive value of $\partial\xi/\partial c$, Mott proposed that alloying causes the Fermi energy E_f to move to higher energy and to a steeper part of the density of states $N(E)$ shown in Fig. 39. Thus, an increase in ξ could occur before g has increased sufficiently to make the minimum more shallow and decrease $(\mathrm{d}N/\mathrm{d}E)_{E=E_f}$; but with increasing concentration, the latter effect prevails and there is a fall in ξ. However increasing pressure does not cause ξ to increase and it must then be assumed that either $(1/E_f)(\partial E_f/\partial c)$ must be substantially larger than $(1/E_f)(\partial E_f/\partial \rho)$ or pressure must increase g much more rapidly than alloying.

The experimental results shown in Fig. 38 seem to indicate that after about 20% In concentration, application of pressure is having little or no effect on ξ and it seems probable that g has approached unity at this point.

Fundamental justification for Mott's ideas is still lacking. However, preliminary work by Rousseau, Stoddart and the author (to be published) suggests that the density of states at E_f in divalent metal Be is not greatly changed as we go from the crystal range form to the short order characteristic of the liquid structure. On the other hand, as we go to a " gaseous " like structure factor, very marked changes occur in the density of states (for a brief discussion of this latter theory, see March and Stoddart, (1968) although the low energy tail is similar to that for the liquid.

Photoemission studies have, to date, not revealed any dip in the density of states for liquid Hg (Enderby and Stevenson, private communication), and we do not know how to resolve this apparent discrepancy with the above conclusions from Bradley's work. A very recent paper by Adams (1968) suggests, however, that Mott's model is not appropriate, and that the simple free-electron model of Faber and Ziman (1965), which gives a generalization of Ziman's formula (15) for alloys, can explain the electrical resistivity of Hg and its alloys, provided the Hg pseudopotential (that is the quantity denoted by $U(K)$ in eqn (15)) is energy dependent. Adams points out that knowledge of the volume dependence of the Hall coefficient at constant temperature of pure Hg would clarify a situation which is rather confused at present.

IX. Speculations on Crystal Structures in Extreme High Pressure Regime

We shall conclude by considering the crystal structures which might be met in the extreme high pressure regime.

From the model of dielectric screening of point ions in the pressure-ionized regime, the interionic forces can be calculated, as described by

Corless and March (1961) and Johnson and March (1963). In spite of the presence of the electron gas, a pair potential $\phi(r)$ can be defined (quite precisely when the Born approximation applies) and its asymptotic form for large r is given by

$$\phi(r) \sim A \frac{\cos 2k_f r}{r^3}. \tag{56}$$

These oscillations arise from the " kink " at $2k_F$ in the wave theory form of eqn (19) for the dielectric function $\epsilon(K)$. The presence of the subsidiary minima in the potential curve, and the facts that (a) the first minimum is not simply related to the near-neighbour distance, and (b) the first minimum is not always the deepest, lead one to speculate that the conventional close-packed structures may not be the stable ones in the extreme high-pressure regime. Structures with a lower coordination number, and chosen such that the subsidiary minima are carefully exploited, may well turn out to be of interest, and further work might be worthwhile on this aspect of the problem.

X. Concluding Remarks

This paper has not only covered a vast pressure range, but also a vast variety of topics! The author has done this in the belief that some of the newer areas, particularly those on the possible new phases of Sections VI and VII ,are worthy of a great deal of development. It was already abundantly clear that pressure is a most valuable variable in studying energy band properties: in giving theorists a handle on the correct choice of crystal potential, and so on. What had not, perhaps, been emphasized quite so strongly until recently is the way in which we might hope to use pressure experiments to unscramble the role of band effects and electron–electron interaction terms. Indications from the work on excitonic phases and on charge and spin density wave states are that an astonishing variety of phenomena may, in principle, occur under conditions in which favourable band gaps, Fermi surface matching of electron-hole surfaces, and so on, are brought about by application of pressure. However, the time has clearly come when the theoretical models (perhaps a fair word to use, because almost none of the theories are genuine many-body theories, with the exception of the Bardeen–Cooper–Schrieffer theory, and the precise physical nature of many-body states is a question of fearsome complexity) need confronting with experiment. Undoubtedly, the improvement of high pressure techniques and the increasing variety of the physical phenomena which can be studied under pressure is going to have, in the next decade, a really major influence on our growing understanding of the many-body aspects of electrons in condensed matter.

Acknowledgements

It is a pleasure to thank my colleagues, Drs I. G. Austin, J. S. Faulkner and W. H. Young, for reading the ms. and for making numerous helpful suggestions.

I am indebted to Drs McWhan, Rice and Schmidt and to Dr Ross for making their work available to me before publication. Useful correspondence with Drs W. McMillan and T. F. Smith is also acknowledged. I wish to thank Professor G. Lehner for initially interesting me in the problem of the conductivity of metals at high pressures and temperatures and for sending me relevant reprints on electrical transport in a plasma.

References

Adams, P. D. (1968). *Phys. Rev. Lett.* **21,** 1324.
Allgaier, R. S. (1967). *Phys. Rev.* **158,** 699.
Allotey, F. K. (1967). *Phys. Rev.* **157,** 467.
Altshuler, L. V., Kormer, S. B., Bakanova, A. A. and Trunin, R. F. (1960). *Soviet Phys. JETP* **38,** 790.
Andres, K. and Jensen, N. A. (1969). To be published.
Austin, I. G. (1966). *Contemp. Phys.* **7,** 174.
Austin, I. G. and Mishra, P. K. (1967). *Phil. Mag.* **15,** 529.
Baldini, G. (1962). *Phys. Rev.* **128,** 1562.
Bardeen, J. and Shockley, W. (1950). *Phys. Rev.* **80,** 72.
Bohm, D. and Staver, T. (1952). *Phys. Rev.* **84,** 836.
Bourassa, R. R., Lazarus, D. and Blackburn, D. A. (1968). *Phys. Rev.* **165,** 853.
Bradley, C. C. (1966). *Phil. Mag.* **14,** 953.
Brooks, H. (1948). *Nuovo Cim. Suppl.* **7,** 186.
Burton, J. J. and Jura, G. (1968). *Phys. Rev.* **171,** 699.
Chester, P. F. and Jones, G. O. (1953). *Phil. Mag.* **44,** 1281.
Coqblin, B. and Blandin, A. (1968). *Adv. Phys.* **17,** 281.
Corless, G. K. and March, N. H. (1961). *Phil. Mag.* **6,** 1285.
Cusack, N. E., Kendall, P. and Fielder, M. (1964). *Phil. Mag.* **10,** 871.
Darby, J. K. and March, N. H. (1964). *Proc. phys. Soc.* **84,** 591.
Davis, L. (1939). *Phys. Rev.* **56,** 93.
Davis, H. L., Faulkner, J. S. and Joy, H. W. (1968). *Phys. Rev.* **167,** 601.
Deutsch, T., Paul, W. and Brooks, H. (1961). *Phys. Rev.* **124,** 753.
Dickey, J. M., Meyer, A. and Young, W. H. (1967). *Proc. phys. Soc.* **92,** 460.
Drickamer, H. G. (1965). " Solid State Physics ", ed. by Seitz and Turnbull, chapter 17, p. 1. Academic Press, New York.
Dugdale, J. S. (1961). *Science, N.Y.* **134,** 77.
Dugdale, J. S. (1969). " Advances in High Pressure Research ", ed. by R. S. Bradley, Vol. 2. Academic Press, London.

Dugdale, J. S. and Phillips, D. (1965). *Proc. R. Soc.* **A287,** 381.
Enderby, J. E. and Stevenson, A. Private communication.
Faber, T. E. and Ziman, J. M. (1965). *Phil. Mag.* **11,** 153.
Falicov, L. M. and Penn, D. R. (1967). *Phys. Rev.* **158,** 476.
Faulkner, J. S., Davis, H. L. and Joy, H. W. (1968). *Phys. Rev.* **161,** 656.
Fedders, P. A. and Martin, P. C. (1966). *Phys. Rev.* **143,** 245.
Flower, M. and March, N. H. (1962). *Phys. Rev.* **125,** 1144.
Gandel'man, G. M. (1965). *Soviet Phys. JETP* **21,** 501.
Gandel'man, G. M. (1967). *Soviet Phys. JETP* **24,** 99.
Gilvarry, J. J. (1968). NATO Conference on High Pressures, Newcastle.
Goodings, D. A. (1965). *Proc. phys. Soc.* **86,** 75.
Greene, M. P. and Kohn, W. (1964). *Phys. Rev.* **137,** 513.
Hall, H. T., Barnett, J. D. and Merrill, L. (1963). *Science, N.Y.* **139,** 111.
Halperin, B. I. and Rice, T. M. (1968). *Rev. Mod. Phys.*, **40,** 755.
Ham, F. S. (1962). *Phys. Rev.* **128,** 82.
Hopfield, J. J. (1965). *Phys. Rev.* **139,** A419.
Isenberg, I. (1950). *Phys. Rev.* **79,** 737.
Jérome, D., Rice, T. M. and Kohn, W. (1967). *Phys. Rev.* **158,** 462.
Johnson, M. D. and March, N. H. (1963). *Phys. Rev. Lett.* **3,** 313.
Kittel, C. (1963). " Quantum Theory of Solids." Wiley, New York.
Knox, R. (1963). " Solid State Physics ", ed. by Seitz and Turnbull, Suppl. 5, p. 100. Academic Press, New York.
Kohn, W. and Rostoker, N. (1954). *Phys. Rev.* **94,** 1111.
Korringa, J. (1947). *Physica,* **13,** 392.
Kozlov, A. N. and Maksimov, L. A. (1965). *Soviet Phys. JETP* **21,** 790.
Kozlov, A. N. and Maksimov, L. A. (1966). *Soviet Phys. JETP* **22,** 889.
Lang, N. D. and Ehrenreich, H. (1968). *Phys. Rev.* **168,** 605.
Lomer, W. M. (1962). *Proc. phys. Soc.* **80,** 489.
Manning, M. F. and Krutter, H. M. (1937). *Phys. Rev.* **51,** 761.
March, N. H. (1955). *Proc. phys. Soc.* **68A,** 726.
March, N. H. (1957). *Adv. Phys.* **6,** 1.
March, N. H. (1968a). NATO Conference on High Pressures, Newcastle.
March, N. H. (1968b). " Liquid Metals ". Pergamon Press, Oxford.
March, N. H. and Stoddart, J. C. (1968). " Reports on Progress in Physics." 31, 533.
March, N. H., Young, W. H. and Sampanthar, S. (1967). " The Many-Body Problem in Quantum Mechanics." Cambridge University Press.
Matthiess, L. F. (1964). *Phys. Rev.* **133,** A1399.
Mayer, H. and El Naby, M. H. (1963). *Z. Phys.* **174,** 289.
McMillan, W. L. (1968). *Phys. Rev.* **167,** 331.
McWhan, D. B. and Jayaraman, A. (1963). *Appl. Phys. Lett.* **3,** 129.
McWhan, D. B. and Rice, T. M. (1967). *Phys. Rev. Lett.* **19,** 846.
McWhan, D. B., Rice, T. M. and Schmidt, P. H. (1968). *Phys. Rev.* In Press.
Melz, P. J. (1966). *Phys. Rev.* **152,** 540.
Meyer, A., Nestor, C. W. and Young, W. H. (1967). *Proc. phys. Soc.* **92,** 446.
Mott, N. F. (1961). *Phil. Mag.* **6,** 287.
Mott, N. F. (1966). *Phil. Mag.* **13,** 989.
Mott, N. F. (1968). *Rev. mod. Phys.* **40,** 677.
Mott, N. F. and Jones, H. (1936). " The Theory of the Properties of Metals and Alloys ", p. 65. Clarendon Press, Oxford.

Olsen, J. L., Andres, K. and Geballe, T. H. (1968). *Phys. Lett.* **26A,** 239.
Olsen, J. L., Bucher, E, Corenzwit, E, Geballe, T. H., Levy, M, and Müller, J, (1964). *Rev. mod. Phys.* **36,** 168
O'Sullivan, W. J. and Schirber, J. E. (1968). *Phys. Rev.* **170,** 667.
Overhauser, A. W. (1962). *Phys. Rev.* **218,** 1437.
Overhauser, A. W. (1964). *Phys. Rev. Lett.* **13,** 190.
Overhauser, A. W. (1968). *Phys. Rev. Lett.* **167,** 691.
Penz, P. A. and Bowers, R. (1968). *Phys. Rev.* **172,** 991.
Reitz, J. R. and Overhauser, A. W. (1968). *Phys. Rev.* **171,** 749.
Rickayzen, G. (1965). " Theory of Superconductivity." Wiley, New York.
Rocher, Y. A. (1962). *Adv. Phys.* **11,** 233.
Ross, M. (1968). *Phys. Rev.* **171,** 777.
Schindewolf, U., Boddeker, K. W. and Vogelsgesang, R. (1966). *Ber. Bunsen. ges. Phys. Chem.* **70,** 1161.
Schulz, L. G. and Spiegler, P. (1959). *A.I.M.E. Ser.* **215,** 87.
Simcox, L. N. and March, N. H. (1962). *Proc. phys. Soc.* **80,** 830.
Simmons, R. O. and Balluffi, R. W. (1960). *Phys. Rev.* **117,** 52.
Skidmore, C. and Morris, E. (1962). Proceedings of the Symposium on High Pressures, Vienna, May 1962.
Smith, T. F. and Chu, C. W. (1967). *Phys. Rev.* **159,** 353.
Stager, R. A. and Drickamer, H. G. (1963). *Phys. Rev.* **132,** 124.
Sternheimer, R. (1950). *Phys. Rev.* **78,** 235.
Stocks, G. M. and Young, W. H. (1969). *J. Phys.* (*Proc. phys. Soc.*) *C,* **2,** 680.
Stott, M. J. and March, N. H. (1966). *Phys. Rev. Lett.* **23,** 408.
Stott, M. J. and March, N. H. (1968). " Soft X-ray Band Spectra and the Electronic Structure of Metals and Materials ", ed. by D. Fabian. Academic Press, London.
Syono, Y. and Ishikawa, Y. (1967). *Phys. Rev. Lett.* **19,** 747.
Templeton, I. M. (1966). *Proc. R. Soc.* **A292,** 413.
Ten Seldam, C. A. (1957). *Proc. phys. Soc.* **A70,** 97 and 529.
Van Thiel, M. and Adleı, B. J. (1966). *J. chem. Phys.* **44,** 1056.
Vasvari, B. and Heine, V. (1967). *Phil. Mag.* **15,** 731.
Vasvari, B., Animalu, A. O. E. and Heine, V. (1967). *Phys. Rev.* **154,** 535.
Watson, R. E. (1960). *Phys. Rev.* **119,** 1934.
Wang, C. G., Seidman, D. N. and Balluffi, R. W. (1968). *Phys. Rev.* **169,** 553,
Wigner, E. P. (1938). *Trans. Faraday Soc.* **34,** 678.
Ziman, J. M. (1960). " Electrons and Phonons." Oxford University Press.
Ziman, J. M. (1961). *Phil. Mag.* **6,** 1013.
Ziman, J. M. (1962). *Phys. Rev. Lett.* **8,** 272.
Ziman, J. M. (1964). *Adv. Phys.* **13,** 89.

Appendix I

Velocity of Sound in Metals and Metallic Alloys

The Bohm–Staver formula (26) can be obtained from the plasma frequency ω_p of the ions, carrying charge Ze, where Z is the valency. As is well known, ω_p is given by

$$\omega_p = \left\{\frac{4\pi\rho_0(Ze)^2}{M}\right\}^{1/2} \tag{I 1}$$

where M is the ionic mass. Now we use the relation $Z\rho_0 = n$, to obtain

$$\omega_p = \left\{\frac{4\pi nZe^2}{M}\right\}^{1/2}. \tag{I 2}$$

But eqn (I 2) is valid when the electron background is uniform. However, in practice, as discussed in Section IIB, the electrons pile up round the ions and shield them, so that the Coulomb potential Ze/r becomes, in semiclassical theory, $(Ze/r)\exp(-qr)$. Putting this in **k** (wave vector) space by taking Fourier transforms, the effect of screening may be expressed by

$$\frac{Z}{k^2} \to \frac{Z}{k^2+q^2}. \tag{I 3}$$

To obtain the velocity of sound, we take the long wavelength limit $k \to 0$ and eqn (I 3) becomes

$$Z \to \frac{Zk^2}{q^2}. \tag{I 4}$$

Substituting this in eqn (I 2) we find $\omega = v_s k$, and for the velocity of sound v_s, using eqn (4) for the screening radius q^{-1} we find

$$v_s = \sqrt{\frac{2ZE_f}{3M}}. \tag{I 5}$$

To illustrate the usefulness of this formula at atmospheric pressure, we have plotted in Fig. I 1 the square of the measured velocities of sound for liquid metals (chosen because of the anisotropy in the solid) in units of v_s^2 as given by eqn (I 5), as a function of valency Z.

While the scatter for $Z = 1$ is very large, for higher Z's the scatter is much less (the large core in Cu, $Z = 1$, is not accounted for in this theory) and we have drawn in the solid curve to pass through the point unity at $Z = 0$, since the Bohm–Staver formula is a small Z limit.

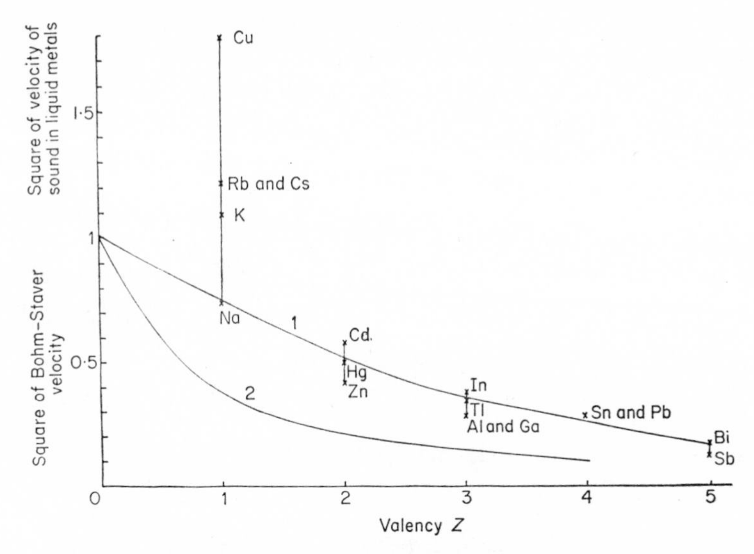

FIG. I 1. Square of measured velocity of sound in liquid metals divided by square of Bohm–Staver velocity. Curve 1, Line through experimental results which yields Bohm–Staver formula as $Z \to 0$. Curve 2, Non-linear plasma treatment for point-ions using Thomas–Fermi equation of state.

A. NON-LINEAR PLASMA

The formula (I 5) can be regained if we write

$$v_s^2 = \frac{\partial p}{\partial d} \tag{I 6}$$

where d is the mass density and take for p the equation of state of a free Fermi gas (cf Fig. 3). This suggests that some understanding of the curvature of the " experimental " results in Fig. I 1 might be gained by retaining the point-ion model but calculating p in eqn (I 6) from the Thomas–Fermi equation of state as given, for example by March (1955). The results are shown in the lowest curve of Fig. I 1. The non-linearity is now too strong, which is due to the point-ion assumption. We should, of course, use pseudopotentials, which take account of the core electrons: this will lead to a theory which is nearer to the linear result, which is simply a straight line parallel to the Z-axis in Fig. I 1.

B. EXTENSION OF BOHM–STAVER FORMULA FOR LIQUID METAL ALLOYS

If we work again in a linear theory, the Bohm–Staver formula can be readily generalized to mixtures. The argument given below is due to

Enderby and the author. It uses the formula (I 6), together with the equation of state of the free Fermi gas.

Thus, the relation between p and the number of electrons per unit volume n is given by

$$p = \frac{h^2}{5m}\left(\frac{3}{8\pi}\right)^{2/3} n^{5/3}. \tag{I 7}$$

If we have N_1 atoms of mass M_1, N_2 of mass M_2, with a total of N atoms, then, if $c_1 = N_1/N$, $c_2 = N_2/N$, we can write almost immediately

$$p = \frac{h^2}{5m}\left(\frac{3}{8\pi}\right)^{2/3} \left[\frac{c_1 Z_1 + c_2 Z_2}{c_1 M_1 + c_2 M_2}\right]^{5/3} d^{5/3} \tag{I 8}$$

where Z_1 and Z_2 are the valencies.

Using eqn (I 6) we find

$$v_s^2 = \tfrac{2}{3} E_f^a \left[\frac{c_1 Z_1 + c_2 Z_2}{c_1 M_1 + c_2 M_2}\right] \tag{I 9}$$

where we have eliminated the number of electrons per unit volume n in favour of the Fermi energy in the alloy E_f^a using

$$n = \frac{8\pi}{3h^3} (2m)^{3/2} (E_f^a)^{3/2}. \tag{I 10}$$

Clearly eqn (I 5) is regained from eqn (I 9) in the limit of a one-component system, and obviously eqn (I 9) has, at best, all the limitations of the Bohm–Staver formula.

Finally, we point out that a formula which is a little less restrictive than eqn (I 9) may be obtained by eliminating the valencies Z_1 and Z_2 in favour of the velocities of sound v_{s1} and v_{s2} of the pure components, using eqn (I 5). Then, from eqn (I 9)

$$v_s^2 = E_f^a \frac{\left(\dfrac{c_1 M_1 v_{s1}^2}{E_{f1}} + \dfrac{c_2 M_2 v_{s2}^2}{E_{f2}}\right)}{(c_1 M_1 + c_2 M_2)}. \tag{I 11}$$

The Bohm–Staver theory ought to become more realistic at high pressures, since the linear approximation involved is better as the kinetic energy of the electrons becomes more important relative to the potential energy. Again, however, core electrons (cf Fig. I 1) will play an essential role over the pressure range at present attainable.

Appendix II

Analogy Between Excitonic Phase and Superconducting State

In eqn (41) a Bardeen–Cooper–Schrieffer description of the excitonic wave function was recorded. Minimizing the total energy with respect to the quantities $u_{\mathbf{k}}$ and $v_{\mathbf{k}}$ leads to a gap function $\Delta_{\mathbf{k}}$ satisfying

$$\Delta_{\mathbf{k}} = \sum_{\mathbf{p}} V(\mathbf{k}-\mathbf{p})(\Delta_{\mathbf{p}}/2[\xi_p^2 + |\Delta_{\mathbf{p}}|^2]^{1/2}) \quad \text{(II 1)}$$

where

$$\xi_p = \tfrac{1}{2}[\epsilon_b(\mathbf{k}) - \epsilon_a(\mathbf{k})] \quad \text{(II 2)}$$

and

$$V(\mathbf{q}) = \frac{4\pi e^2}{\epsilon(q)q^2} \quad \text{(II 3)}$$

is the interaction term in the Hamiltonian, $\epsilon(q)$ being an appropriate dielectric function for the material under discussion. As Jérome and co-workers emphasize, the phase of $\Delta_{\mathbf{k}}$ is arbitrary and the energy is independent of it.

In terms of a quantity defined by

$$\phi_p = \Delta_p/2[\xi_p^2 + |\Delta_p|^2]^{1/2} \quad \text{(II 4)}$$

the gap equation can be written

$$\left[\left(E_g + \frac{k^2}{2\mu}\right)^2 + 4|\Delta_{\mathbf{k}}|^2\right]^{1/2} \phi_{\mathbf{k}} = \sum_{\mathbf{p}} V(\mathbf{k}-\mathbf{p})\phi_{\mathbf{p}}. \quad \text{(II 5)}$$

For comparison we notice that the equation for the exciton wave function is

$$\left[\frac{k^2}{2\mu} + |E_x|\right] \chi_{\mathbf{k}} = \sum_{\mathbf{p}} V(\mathbf{k}-\mathbf{p})\chi_{\mathbf{p}}, \quad \text{(II 6)}$$

where μ is the reduced mass ($\mu^{-1} = m_a^{-1} + m_b^{-1}$). The eqn (II 5) and (II 6) together can be readily used to verify that $\Delta = 0$ when $E_g \geqslant |E_x|$. However, a non-trivial solution corresponding to $|\Delta| > 0$ exists for all values of E_g, either positive or negative, which are $\leqslant |E_x|$.

There is only formal similarity with the theory of superconductivity. For, as is well known, in this case the two-particle density matrix $\langle \mathbf{r}_1'\mathbf{r}_2' | \rho_2 | \mathbf{r}_1\mathbf{r}_2 \rangle$ remains finite in the limit $|\mathbf{r}_1 - \mathbf{r}_1'| \to \infty$, $\mathbf{r}_1 \sim \mathbf{r}_2$, $\mathbf{r}_1' \sim \mathbf{r}_2'$. This circumstance is referred to as off-diagonal long-range order and is a basic characteristic of the superconducting state. In the case of the excitonic insulator at absolute zero, Jérome and co-workers show that the two-particle density matrix has no such off-diagonal long-range order. However, a new periodicity in this two-particle density matrix characterized by $\mathbf{w}$ does appear, and this is a basic feature of this excitonic phase. The d.c. conductivity at zero temperature can then be shown to be zero and we are dealing with an insulating phase.

Turning to a discussion of the finite temperature resistivity of the excitonic insulator, a calculation paralleling the superconducting gap leads to the result

$$\Delta_p = \sum_{\mathbf{k}} V(\mathbf{p}-\mathbf{k})(\Delta_k/2E_k) \tanh(\tfrac{1}{2}\beta E_{\mathbf{k}}) \tag{II 7}$$

where $E_k^2 = \xi_k^2 + |\Delta_k|^2$ and $\beta = 1/k_B T$. This should be compared with the Bardeen–Cooper–Schrieffer result at elevated temperatures (see for example the discussion of March *et al.*, 1967).

The calculation of the resistivity is then most simply carried out in the semimetal limit and in the case when the masses are equal. The result for the frequency dependent conductivity may then be shown to take the form, for low frequencies

$$\sigma(\omega) = -\frac{ie^2}{m\omega} \sum_p \left\{\frac{p^2\Delta^2}{3mE_p^3}(1-2n_p) - \left(1-\frac{\xi_p}{E_p}\right)\right\} \tag{II 8}$$

where

$$n_p = [\exp(\beta E_p) + 1]^{-1}. \tag{II 9}$$

If we denote the d.c. conductivity by $\sigma_1(0)$ and write this in the usual form

$$\sigma_1(0) = \frac{n_{\text{effective}} e^2 \tau}{m} \tag{II 10}$$

where τ is the relaxation time, then using the Kramers–Kronig relationship the effective number of carriers $n_{\text{effective}}$ may be shown to take the form

$$n_{\text{effective}} = 2n_c \int_0^\infty \frac{dt}{(t^2+1)^{3/2}} \{\exp[\beta\Delta(t^2+1)^{1/2}] + 1\}^{-1}, \tag{II 11}$$

where n_c is the number of carriers which in the semimetal limit is equal to $k_f^3/3\pi^2$. The temperature dependence of the conductivity can now be written formally as

$$\frac{d\sigma}{dT} = \frac{\tau e^2}{m}\left\{\frac{\partial n_{\text{effective}}}{\partial T} + \frac{\partial n_{\text{effective}}}{\partial \Delta}\frac{d\Delta}{dT}\right\}. \tag{II 12}$$

As $T \to T_c$ the second term on the right-hand side diverges and $d\sigma/dT \to \infty$. On the other hand, as $T \to 0$, $d\sigma/dT$ tends to 0. The qualitative behaviour sketched in Fig. 24 arises in this way. An entirely similar calculation leads to the behaviour of σ at a fixed temperature as the energy gap E_g is changed by varying the external pressure.

CHAPTER 5

Adsorption of Gases at High Pressures

P. G. MENON†

Department of Chemical Engineering, Technological University Twente, Enschede, The Netherlands

I. INTRODUCTION

The adsorption of gases on solids at pressures up to 1 atm has been extensively investigated and hundreds of papers are being published

† Present address: Ketjen N.V. Amsterdam, The Netherlands.

TABLE I. Investigations in the Field of High-pressure Adsorption since 1930

Adsorbent	Adsorbate	Temp. (°C)	Maximum pressure (atm)	Exptl method[a]	Reference
Sugar charcoal	N_2O, C_2H_4 and N_2		50	1	McBain and Britton (1930)
Charcoal	H_2, CH_4	25 to 100	150	3	Frolich and White (1930)
Charcoal	CO_2, N_2O and SiF_4		100	1	Coolidge and Fornwalt (1934)
Alumina	C_3H_6			1	Morris and Maass (1933)
Alumina	dimethyl ether			1	Edwards and Maass (1935)
Coal (French)	CH_4	20	150		Audibert (1935)
Coal (Belgian)	CH_4	20	400		Coppens (1936, 1937)
Coal (Russian)	CH_4	20	1000	3	Palvelev (1948)
Coal (Russian)	CH_4	20	1000	3	Khodot (1948)
Coal (German)	CH_4	20	110		Beckmann (1954)
Coal (Dutch)	CH_4	20	500	5	van der Sommen *et al.* (1955)
Coal (British)	CH_4	20	1000	3	Moffat and Weale (1955)
Coal (Polish)	CO_2	20	60	3	Czaplinski and Lason (1965)
Coconut shell charcoal	N_2, CO, CO_2, CH_4, C_2H_2, C_2H_4, C_2H_6, C_3H_6, C_3H_8 and n-C_4H_{10}	100 to 450	14	3	Ray and Box (1950)
Silica gel and activated carbon	C_1 to C_4 hydrocarbons		20	3	Lewis *et al.* (1950)
Silica	CO_2	-85 to $+40$	80	4	Vasil'ev (1957)
Supported Ni catalyst	H_2	20	140	4	Vaska and Selwood (1958)
Active carbon	Ar and N_2	-76 to $+20$	400	3	von Antropoff (1952, 1954, 1955)
Active carbon	He, Ne and H_2	-196	30	3	Czaplinski and Zielinssi (1958)
Silica gel	He, Ne and N_2	-196	30	3	Czaplinski and Zielinski (1959)
Alumina	He, Ne and H_2	-196	30	3	Czaplinski and Zielinski (1959)
Active carbon	(CH_4, CO, or N_2) $+H_2$	20	100		Zhukova and Kel'tsev (1959)

TABLE I (*continued*)

Adsorbent	Adsorbate	Temp. (°C)	Maximum pressure (atm)	Exptl method[a]	Reference
Porous plug of lampblack	CO_2	19, 30 and 32	80	1	Jones *et al.* (1959)
Fe catalyst for NH_3 synthesis	N_2, H_2, $1N_2+3H_2$	97 to 400	50	3	Sastri and Srikant (1961)
Alumina	N_2	$-7\cdot6$ to $+100$	3000	5	Michels *et al.* (1961)
Alumina	CO	0 to 50	3000	5	Menon (1965)
Alumina	CO_2	25 to 35	100	6	Jones and Evans (1966)
Active carbon	CH_4, C_2H_6		20	2	Boehlen *et al.* (1964)
Silica gel	CH_4	-40 to $+20$	140	2	Gilmer and Kobayashi (1964)
Silica gel	C_2H_6, C_3H_8 or C_4H_{10} with CH_4	-40 to $+40$	135	7	Gilmer and Kobayashi (1964)
Silica gel	$CH_4+C_3H_8$	20	65	7	Gilmer and Kobayashi (1965)
Silica gel	$CH_4+C_3H_8$	0 to 40	65	7	Haydel and Kobayashi (1967)
Silica gel	$CH_4+C_2H_6$	5 to 35	95	7	Masukawa and Kobayashi (1968)
Molecular sieves	N_2, CH_4, N_2+CH_4	-150 to $+20$	80	3	Lederman and Williams (1964)
Charcoal	C_2–C_5 hydrocarbons	20	50		Rasulov and Velikovskii (1965)
Silica gel	Binary and ternary hydrocarbon mixtures		120	7	Mason and Cooke (1966)
Carbon black (Spheron 6)	CH_4	37 to 121	650	3	Stacy *et al.* (1968)
Charcoal	CH_4, C_3H_8, C_4H_{10} and their binary mixtures	10 to 60	130	7	Payne *et al.* (1968)

[a] Experimental methods: 1 sorption balance; 2 other gravimetric methods; 3 measuring gas volumes at atmospheric pressure; 4 Vasil'ev's method; 5 glass-piezometer technique; 6 from dielectric constant measurements; 7 gas chromatography.

every year, but the data at high pressures are very scarce. During the last twenty years there has been an enormous increase in the volume of high-pressure research. Many types of apparatus for generation and measurement of pressures up to 10,000–20,000 atm have become commercially available in this period. This, in turn, has enabled many studies of chemical reactions and equilibria at pressures up to 20,000 atm. The progress in this field has been reviewed by Gonikberg (1960) and quite recently in an excellent treatise by Weale (1967). In spite of all these developments adsorption studies at high pressures are only very few in number and often limited to pressures below 100 atm.

A brief survey of the early work on adsorption at high pressures has been given by McBain and Britton (1930). Most of the papers published since 1930 have been covered in a recent review (Menon, 1968b), in which the results obtained from high-pressure adsorption measurements and their interpretation, accuracy of the results, especially after the numerous corrections which have to be applied to them, and the physico-chemical significance of these results have been discussed in detail. Hence these aspects of high-pressure adsorption will be only briefly referred to, if at all, in the present paper. Greater emphasis is laid here on the different experimental methods and techniques used in high-pressure adsorption measurements. Some theoretical considerations and empirical correlations with potentialities for industrial applications of high-pressure adsorption are included in the present paper along with a survey of the recent work on adsorption from multi-component gas mixtures—these have not been covered at all in the earlier review. Finally, some interesting studies and problems are grouped in the Section headed " Miscellaneous "; they cannot be called high-pressure adsorption as such, but their importance in a wider context is obvious.

A list of high-pressure adsorption work done since 1930 is given in Table I. The experimental techniques employed, adsorbents and adsorbates used and the temperature and pressure ranges covered are also given in the Table.

II. Experimental Methods

The experimental techniques used for measurement of adsorption at high pressures may be broadly divided into the following six groups: A Gravimetric methods, B Method of measuring gas volumes at atmospheric pressure, C Vasil'ev's method, D The glass-piezometer technique, E Adsorption from multi-component gas mixtures, F Gas chromatographic methods. The techniques employed in the various investigations in this field have been indicated in Table I. Some details of the above six methods are given below.

A. GRAVIMETRIC METHODS

1. *The McBain Sorption Balance*

The earliest outstanding work in the field of high-pressure adsorption is the investigation of McBain and Britton (1930) on the sorption of N_2O, C_2H_4 and N_2 on activated sugar charcoal up to 50 atm pressure. The adsorbent, taken in a basket, was suspended from a quartz spiral, which was mounted in a thick-walled glass tube. The amount of gas adsorbed was determined gravimetrically from elongation of the calibrated spiral as measured with a precision cathetometer.

Jones *et al.* (1959) have developed an all-metal sorption balance (Fig. 1) for adsorption measurements up to 80 atm pressure. The quartz spiral is mounted in a brass tube which is surrounded by a steel bomb. Two perspex windows, fitted on diametrically opposite sides of the steel bomb, enable a beam of light to pass through the bomb. The elongation of the spiral is measured with a cathetometer; the pressure of the gas is measured by a Bourdon gauge; and the whole apparatus is thermostatted. The reproducibility of the extension is within 0·1 mm, corresponding to a weight of 1 mg; the bucket for the adsorbent is made of copper foil 0·0025 cm thick and it has a weight of 0·588 g and a metal volume of 0·064 cm^3. About 500 mg of the adsorbent is taken for the study. Details of buoyancy corrections are given in the original paper.

The use of the McBain sorption balance to measure adsorption at relatively high pressures has two disadvantages both related to the buoyancy correction: (1) the buoyancy depends on the quantity being measured, that is the amount of adsorbate in the pores of the adsorbent, and (2) at and near the critical point, gradients of density can exist in the gaseous phase and consequently there is some uncertainty as to the value of the density in the vicinity of the bucket containing the adsorbent. To overcome the above difficulties a dielectric method of measuring the adsorption has been used by Jones and Evans (1966) for CO_2 on alumina up to 100 atm pressure. A cell consisting of two concentric brass cylinders contains the adsorbent powder in the annular space between the cylinders. A formula has been derived relating the apparent dielectric constant of the powder to the dielectric constant of the material of the powder and that of the fluid in the intergranular space, which enables the adsorption of CO_2 to be calculated.

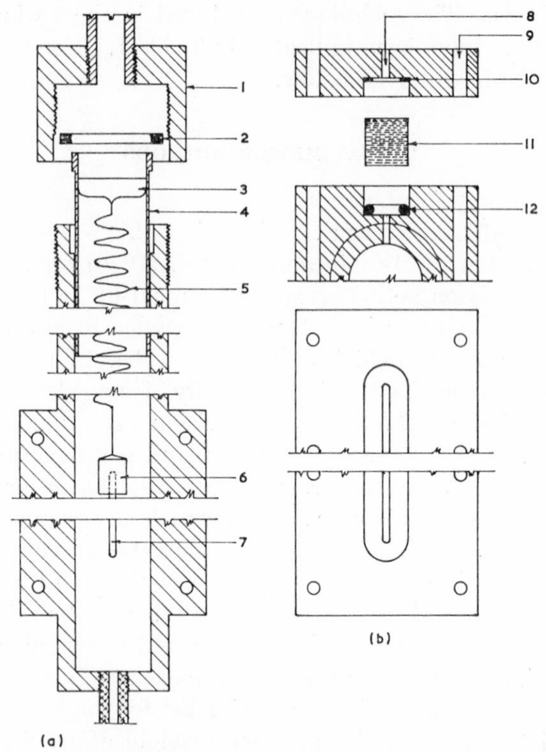

FIG. 1. All-metal sorption balance of Jones *et al.* (1959). 1 cap, 2 Dowty bonded seal, 3 moveable support for quartz spring, 4 brass tube, 5 quartz spring, 6 sample bucket, 7 window, 8 slot for view, 9 bolt hole, 10 rubber cushion, 11 perspex window, 12 rubber ring. (Unpublished figure, kindly supplied by W. M. Jones.) \\\\ Steel, × × copper, + + aluminium, /// brass, ● rubber, – – perspex.

Jones and Evans, however, admit that the general applicability of the dielectric constant method is severely limited by the lack of a complete theory to explain the results. Although the method is partly chosen for work at high pressures to avoid the large buoyancy of the gravimetric method, the same type of problem arises in the dielectric method, that is it fails to handle the unadsorbed gas in the pores any better than the gravimetric method. But there are some experimental advantages: (1) the cell is smaller and its design easier, it can also stand higher pressures than in the gravimetric method, and (2) fluctuations in density near the critical point do not affect the results.

2. *Direct Weighing*

Boehlen and Guyer (1964) have converted a Mettler analytical balance for adsorption measurements at pressures up to 20 atm. In this the analytical balance simply replaces the calibrated quartz spiral of the McBain sorption balance. Larger adsorbent samples can be taken on this, but it has all the disadvantages and uncertainties of the sorption balance, with the additional limitation of the rather too low pressure range obtained.

Gilmer and Kobayashi (1964) have used a direct gravimetric method to determine the adsorption isotherms of methane on silica gel up to 140 atm pressure. They equilibrated the chromatograph column using a steady flow of methane through it at the desired temperature and pressure, then closing off the column with valves at both ends, removing the assembly from the rest of the system and weighing it on an analytical balance. The mass of total gas (gas adsorbed plus that in equilibrium in the gaseous phase) in the column is the difference between the weight of the column assembly and that when it was evacuated. The mass adsorbed was obtained by subtracting from this difference the product of the free volume (as determined by helium) in the column and the gas density at the equilibrium conditions of temperature and pressure.

B. METHOD OF MEASURING GAS VOLUMES AT ATMOSPHERIC PRESSURE

This is the easiest and most commonly used type of apparatus for measurement of adsorption at high pressures. In general it consists of an all-steel side to handle the gas under pressure and a glass side to measure the volume of gas on expansion to 1 atm pressure. The adsorbent is first charged with gas at the highest pressure possible; then known amounts of gas (measured in a gas burette at 1 atm) are withdrawn from the high-pressure side and the resulting equilibrium pressures are noted. Strictly speaking only the desorption isotherms are measured in this way. A recent version of this type of apparatus, used by Kini (1964), is shown in Fig. 2. The adsorbent is taken in the sample tube A (internal volume 2–3 cm^3), which is connected through an Aminco valve V_1 to a Bourdon pressure gauge by means of capillary tubing (0·055 in i.d.). The valves V_2 and V_3 connect the system to high vacuum and to a condensing coil C. The other end of the coil is closed by the valve V_4. A known quantity of the gas is transferred from the glass bulb to the condensing coil and the required pressure is generated by thermal compression. The quantity of gas contained at different pressures in the manometric space and that in the manometric space plus the empty specimen tube are determined by prior calibration by bleeding it out into the burette B. The amount of gas in the specimen tube at different pressures is then given by the difference between the two calibration curves. A correction is applied for the volume of the

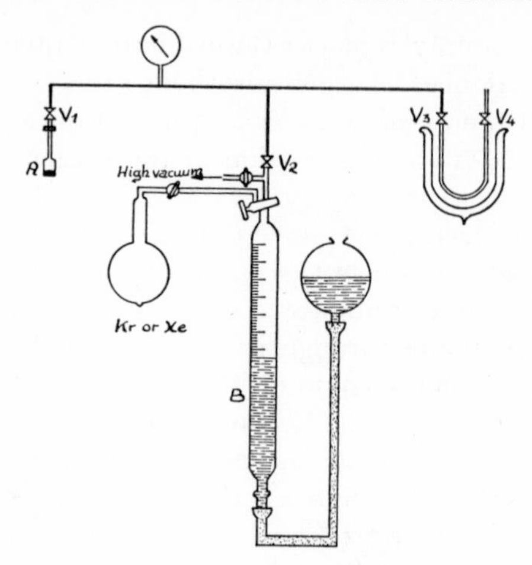

FIG. 2. Typical example of apparatus based on measuring gas volumes at atmospheric pressure (after Kini, 1964).

sample using the specific volume of the material determined by helium displacement at 110°C.

To measure adsorption isotherms directly (instead of desorption isotherms) Czaplinski and Zielinski have developed the apparatus shown in Fig. 3. The quantity of gas was measured in the gas burette A. The gas was then introduced through the stopcock Z_1 into the mercury-

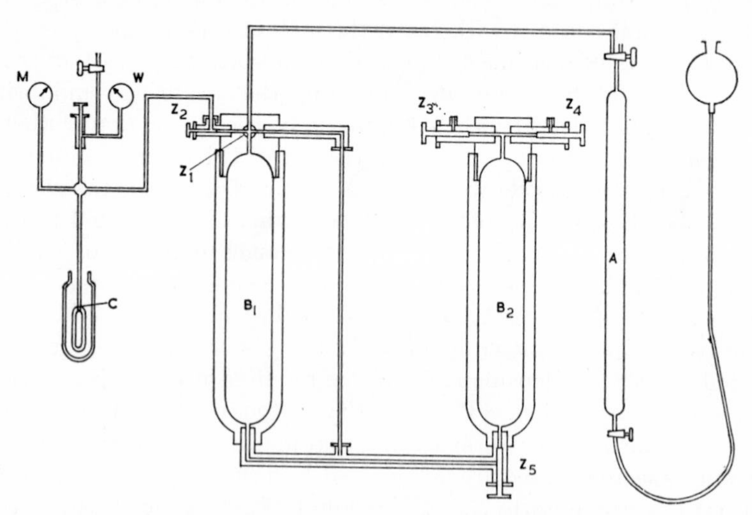

FIG. 3. Apparatus of Czaplinski and Zielinski (1958). A gas burette, B_1, B_2 mercury-piston gas compressor, C adsorbent tube, M pressure gauge, W vacuum gauge, Z_1–Z_5 needle valves. (Reproduced by permission of E. Zielinski.)

piston gas compressor, B_1, B_2, which consisted of two steel cylinders connected at the bottom and provided with valves Z_1 to Z_5. The level of mercury in B_1 could be observed in a glass capillary tube serving as a level gauge. The gas in B_1 was compressed by letting in compressed air into B_2 and forcing the mercury to flow from B_2 to B_1. The compressed gas was introduced into the adsorption tube C, which is connected to a pressure gauge and a vacuum line. The apparatus was first calibrated with helium. Then glass beads of the same volume as the adsorbent were taken in C, known amounts of gas were compressed into it and the pressure change occurring on cooling C with liquid nitrogen was observed. Replacing the glass beads by the adsorbent and repeating the measurements would yield lower values of pressure. From these data the volume of gas adsorbed could be calculated. The adsorption of He, Ne and H_2 on active carbon, alumina and silica gel have been measured in this way at liquid nitrogen temperature and up to 30 atm pressure. One serious source of error here can be from the unequal thermal contraction of the adsorbent and the dummy glass beads causing appreciable alteration of the dead space volume.

C. VASIL'EV'S METHOD

An elegant method, which requires no compressor or oil press to raise the pressure, nor a gauge for measuring the high pressure, was used by Vasil'ev (1957) for the adsorption of CO_2 on silica gel up to 80 atm pressure. Thermal compression was employed to obtain the high pressure. A simplified version of the apparatus is shown in Fig. 4. A

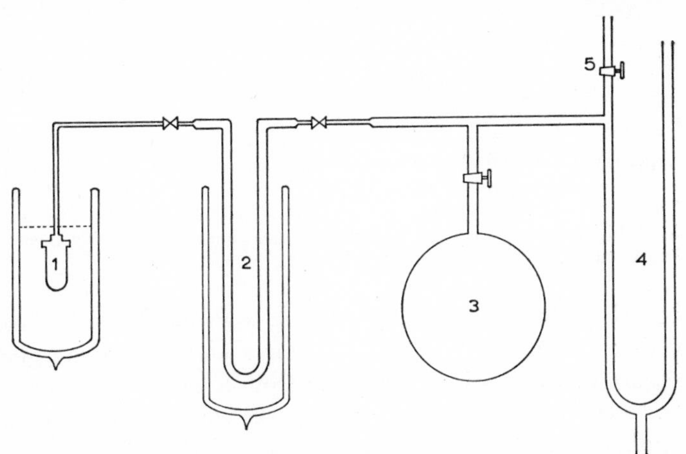

FIG. 4. Vasil'ev's apparatus. 1 Adsorbent tube, 2 thermal compressor, 3 flask serving as a gas burette, 4 mercury manometer.

known amount of gas measured at 1 atm pressure was transferred into a calibrated steel U-tube cooled in liquid nitrogen. The coolant was removed and the U-tube was opened to the evacuated adsorption vessel, the helium dead space of which was already known. At equilibrium the U-tube was isolated from the vessel and the gas in it was measured after expansion to 1 atm pressure. From these data, the amount of CO_2 adsorbed and also the equilibrium pressure were calculated using the accurate compressibility data of Michels *et al.* (1935). Further details of this method, given in English by Dubinin *et al.* (1958), have been reproduced in a recent monograph (Ross and Olivier, 1964).

Vaska and Selwood (1958) used a similar method for simultaneous measurement of hydrogen adsorption and of specific magnetization up to 140 atm pressure on a Kieselguhr-supported nickel catalyst. Their apparatus is shown in Fig. 5. A small stainless steel autoclave A was

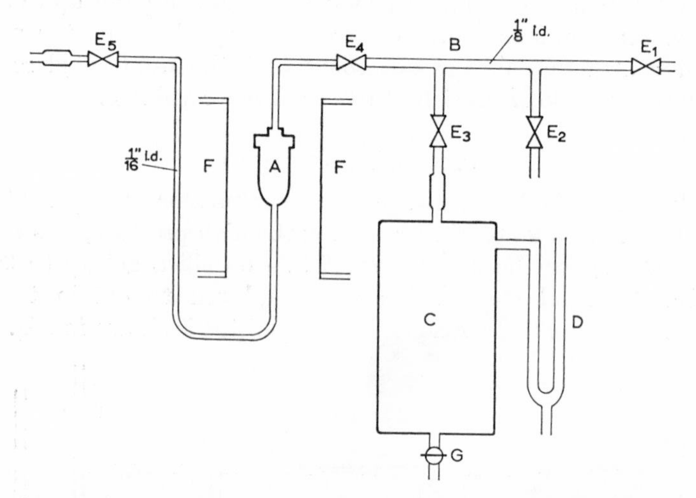

FIG. 5. Apparatus for simultaneous measurement of hydrogen adsorption and magnetization (after Vaska and Selwood, 1958). A adsorbent chamber, B gas pipette, C glass exhaust chamber serving as burette, D constant-volume manometer, E_1–E_5 high-pressure valves, F primary solenoid for magnetization measurements, G to vacuum.

connected to a steel tube B serving as a gas pipette. The latter was connected to a constant pressure gas source and to a glass reservoir C. The pressure of the gas in C was read from a constant-volume manometer D. Hydrogen was admitted from the constant pressure source to B by opening valve E_2. Then E_2 was closed and the gas was expanded into C through valve E_3. From the pressure of hydrogen in C and B and from the volumes of C and B, the original pressure of the gas as well as its original amount in B could be calculated. After evacuation of chambers B and C, B was filled with the same amount of gas as before (from the constant pressure source). Then the gas was admitted to A by opening the valve E_4. After attaining equilibrium E_4 was closed and the pressure of the gas and its amount remaining in B were determined by expansion into C as before. From this adsorption pressure and the known dead-space in A, the amount of unadsorbed gas in A was calculated. The difference between the original amount of hydrogen in B and the sum of the amounts in B and A after adsorption gave the amount adsorbed by the adsorbent.

D. THE GLASS-PIEZOMETER TECHNIQUE

A significant advance in technique and much higher accuracy were achieved by van der Sommen *et al.* (1955) by resorting to the glass-piezometer technique developed two decades earlier by Michels and co-workers (1935) for gas compressibility measurements of gases. The experimental technique was further improved and for the first time the pressure range in gas adsorption measurements was extended to 3000 atm by Michels *et al.* (1961). Two distinctive features of this method over almost all the earlier measurements in the field of high pressure adsorption are: (1) the adsorbent and the adsorbate gas are enclosed in glass and not in metal even at a pressure of 3000 atm, and (2) the use of free-piston-type pressure balance of Michels (1923, 1924) not only ensures an accuracy of 1 in 10,000 or better in pressure measurements, but also maintains the pressure constant even if the amount of gas adsorbed varies with time during any particular measurement.

1. *The Piezometer*

The glass-piezometer technique for determining the compressibility isotherms of gases has been described previously (Michels and Michels, 1935; Michels *et al.*, 1935). For purposes of adsorption measurements some details of the piezometer were altered. It consisted of two bulbs with volumes of about 32 and 38 ml at the lower end, one bulb (10 cm long and 1·3 cm outside diameter) at the top and 5 or 6 bulbs varying in volume from 0·9 to 0·35 ml in between, interconnected by short sections of capillary tubing (Fig. 6). A glass capsule containing the dehydrated and evacuated adsorbent (alumina) and provided with a thin spherical diaphragm, was enclosed in the top bulb of the piezometer. The diaphragm was made to break at an external pressure of 20–30 atm. For the adsorption measurement it was necessary to know the total volume outside the capsule and the free volume inside it. These two

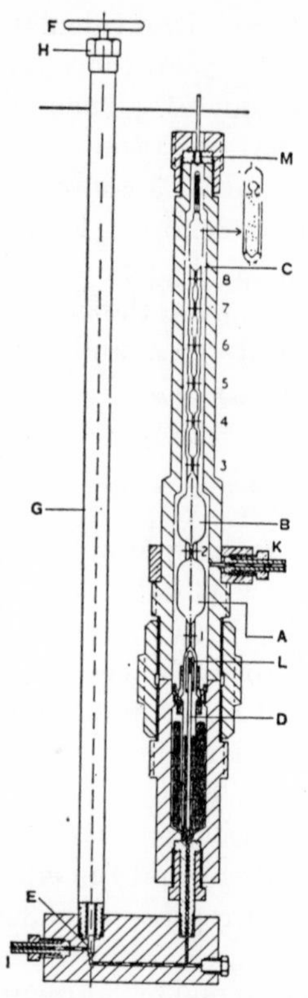

FIG. 6. Glass piezometer, mounted in the high-pressure vessel, for adsorption measurements up to 3000 atm pressure (Michels *et al.*, 1961). The glass capsule containing the adsorbent is enclosed in the top bulb. (Reproduced by permission of Royal Netherlands Chemical Society.)

volumes were determined separately. The volumes between the platinum contacts 1, 2, 3 . . . 8 could be calibrated with mercury in the usual way (Michels *et al.*, 1935), but this was not possible for the top bulb. To overcome this difficulty the piezometer was first filled with nitrogen (adsorbate) to obtain a pressure of about 5 atm at contact 1 at 25°C. The pressures were then measured when the mercury reached contacts 1 and 2. From these data and the known volumes below contact 8 and the known P–V–T relation of the gas, the volume of the top bulb (outside the capsule) as well as the amount of gas taken in the piezometer could be readily calculated. The pressure was then raised slowly till the diaphragm of the capsule broke and the adsorbent became accessible to the gas. The pressures were measured when the mercury reached successive platinum contacts while increasing the pressure and later while decreasing it.

2. *The Glass Capsule*

The capsule was made to conform to the following requirements. (1) The dehydration of the adsorbent had to be carried out in the capsule at 300 and 600°C under vacuum (soft glass for 300° and supermax glass for 600°C). (2) To allow the pressure measurements at contacts 1 and 2 for determining the top-bulb volume, the glass diaphragm should not break below 20 atm pressure. (3) To restrict the pressure drop on breaking, the diaphragm should break at a pressure not higher than 30 atm and the volume of the capsule should not be large. By a few trials a thickness of 0·04 mm at the thinnest part of the spherical diaphragm was found to be the optimum.†

The capsule consisted initially of two parts A and B, separated by the diaphragm (Fig. 7). The section A, later to be filled with alumina, was evacuated and sealed off. The capsule was then pressure tested (in a steel bomb filled with oil) at an external pressure of 20 atm for 1 hr. To

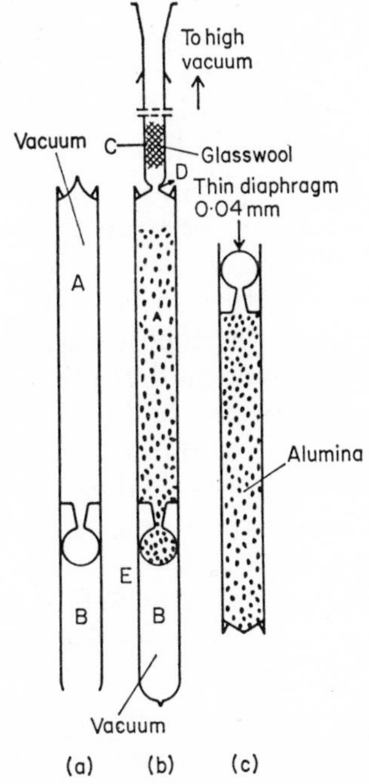

FIG. 7. The glass capsule in its three stages: (a) during pressure testing, (b) during the dehydration of alumina, (c) in the final form. From Michels *et al.* (1961). (Reproduced by permission of Royal Netherlands Chemical Society.)

† To determine the thickness of the diaphragm, the capsule was immersed in a slightly coloured liquid of the same refractive index as the glass. The thickness could then be measured with a microscope.

determine the free volume inside the capsule it was necessary to know: (1) the internal volume of the empty capsule, and (2) the volume of alumina in it. The former was obtained from the external volume of the capsule and the weight and density of the glass, while the latter was calculated from the weight and helium density of the dehydrated adsorbent (alumina, in this case). The loss of weight on dehydration was also determined. The sequence of nine steps involved in the determination of these quantities are given in the paper of Michels *et al.* (1961), the three stages of the capsule, however, are shown in Fig. 7.

3. *The High-pressure Experiment*

The experimental set-up for normal P–V–T measurements and also for adsorption measurements up to 3000 atm pressure is shown in Fig. 8. The piezometer was mounted in a high-pressure steel vessel, evacuated and filled with the adsorbate gas (N_2 or CO) as described by Michels *et al.* (1935). The amount of the gas was enough to give a pressure of about 5 atm at contact 1 of the piezometer (see Fig. 6). The oil thermostat in which the steel vessel was immersed was regulated at $25 \pm 0{\cdot}005$°C. Oil was pumped in into the steel vessel from the press until mercury inside the piezometer just reached contact 1—as seen by a Wheatstone bridge to which the platinum contacts are all connected in series with known resistances in between the contacts. The pressure was measured with the Michels pressure balance (Michels, 1923, 1924) and the temperature of the thermostat with a platinum resistance thermometer. The pressure was then raised till the mercury reached contact 2 at about 9 atm and one more reading was taken. From these two P–V–T measurements the top-bulb volume (outside the capsule) could be calculated. The procedure was repeated with different fillings of the gas. Each set gave an independent value for the volume of the bulb; the deviation from the average amounted to 1 : 5000. For the last filling (about 20 atm at contact 2) the amount of gas taken in the piezometer was also calculated from the known P–V–T data for the gas. During all these measurements the diaphragm of the capsule was still unbroken.

On compressing the gas from contact 2 to contact 3 the pressure would increase fourfold. In this pressure range the diaphragm of the capsule collapsed and the adsorbent became accessible to the gas. Pressures were then measured at contacts 1–8 and in the reverse order at 0°, 25° and 50°C. The pressures during the adsorption and desorption series agreed in general within 1 : 10,000 (cf Menon, 1968b, p. 292) and hence the adsorption was fully reversible and there was no hysteresis in the adsorption–desorption cycle. More gas was then filled in to reach

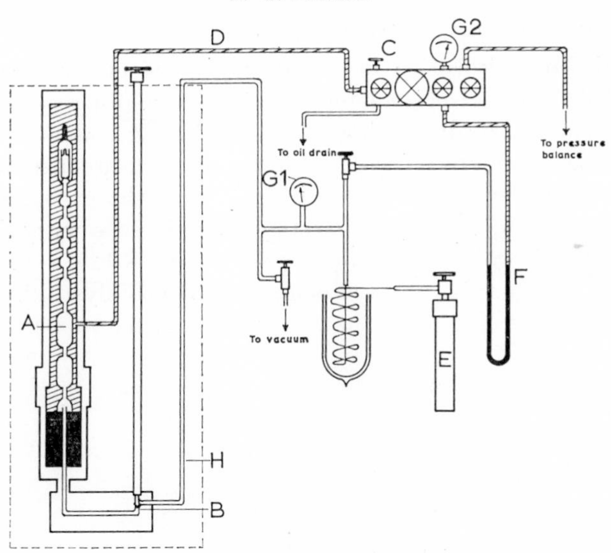

Fig. 8. The experimental set-up for compressibility and also adsorption measurements of gases up to 3000 atm pressure. *A* glass piezometer mounted in the high-pressure vessel (for details see Fig. 6), *B* valve for isolating the piezometer from the gas-filling side, *C* oil press, *D* steel capillary, *E* gas cylinder (pure gas), *F* mercury-in-glass differential manometer for gas fillings up to 10 atm, *G*1 pressure gauge (gas, 0–100 atm), G_2 pressure guage (oil, 0–4000 atm), *H* connection to gas-filling side. (After Michels *et al.*, 1935.)

higher pressures. For every new gas filling, pressures at contacts 1 and 2 should be in the range already covered by the higher contacts for the earlier filling; this overlap enabled the calculation of the gas added, taking into account the gas already adsorbed at those pressures. In this way the pressure range was extended to 3000 atm. After finishing the work at lower temperatures, measurements at 75° and 100°C were made. This sequence of measurements was followed to avoid endangering the lower temperature results by the possible poisoning of the adsorbent due to the enhanced vapour pressure of mercury and also the solvent-effect of compressed gas on mercury at higher temperatures. However, a redetermination of the adsorption at 25°C at the end of all other measurements did not indicate any alteration in the activity of the alumina in the capsule due to exposure to mercury vapours at 100°C for a few days. Further details of the experimental technique along with the exact values for the volumes of the piezometer bulbs, the total gas in the piezometer at each gas filling, the corrections applied for the thermal expansion and compressibility of the piezometer glass, and so on, in the case of N_2 on alumina have been given by Michels *et al.* (1961) and for CO on alumina by Menon (1965).

4. *Adsorption at Intermediate Pressures*

Low-pressure adsorption units can cover the pressure range below 1 atm, while the glass-piezometer technique for high-pressure adsorption measurements starts from about 6 atm pressure. As an independent check for the accuracy of the high-pressure measurements it was felt necessary to set up an apparatus just to cover the pressure range 1–6 atm (Menon, 1961). This apparatus also served very well for low-temperature adsorption measurements at −100, −125, −150°C, and so on. In this, the gas burette was connected to a catalyst tube which was kept in a cryostat (Fig. 9). The temperature of the copper block in the

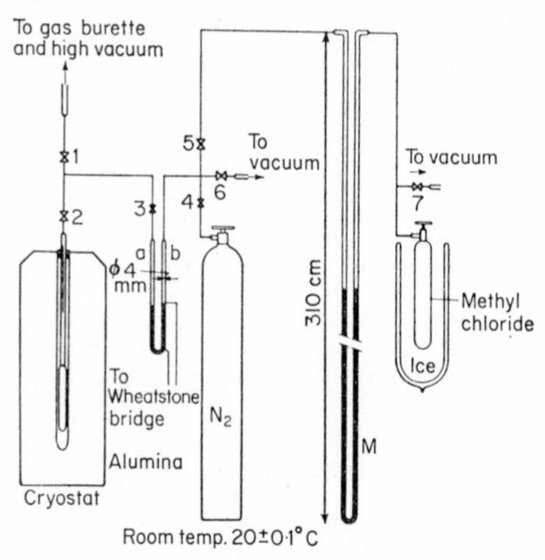

FIG. 9. Apparatus for adsorption measurements at intermediate pressures (1–6 atm). (From Menon, 1961.)

middle of the cryostat could be kept constant anywhere between −188° and +50°C. The glass catalyst tube was connected through copper capillaries (2 mm i.d.) to a short mercury manometer which was provided with a sharp platinum contact in one limb, *b*. The mercury level in this tube was kept exactly at the platinum contact (as checked with a Wheatstone bridge circuit) by manipulating the connecting valves to vacuum or to a nitrogen gas cylinder. In this way the volume of the manometric space in limb *a* was always kept constant. The pressure in limb *b* (same as in *a* since the mercury levels in both limbs were equal) was measured on a mercury manometer, *M*, 310 cm long. This manometer was used for measuring pressure directly up to 4 atm. Counter-pressure applied by means of the vapour pressure of methyl chloride at 0°C (191·5 cm Hg) enabled measurements up to 6·5 atm.

The whole apparatus was kept in a constant-temperature room at 20 ± 0·1°C. A requisite amount of nitrogen was transferred from the gas burette when the adsorbent is at low temperature so that on closing valve 1 and raising the temperature of the cryostat to 0°C, the desorbed gas in the catalyst tube could exert a pressure of about 6·5 atm. Thereafter known amounts of gas from the catalyst tube were withdrawn into the gas burette through valve 1, noting the pressure of the remaining gas. This set of measurements actually constituted a desorption series. In practice, the low-pressure adsorption at −150°C was first measured and, on raising the temperature to 0°C, the desorption series at medium pressures (6·5–1 atm) was followed in the same cycle.

The adsorption of nitrogen on alumina at 0°C was measured using the medium pressure apparatus. The results are plotted in Fig. 10 along

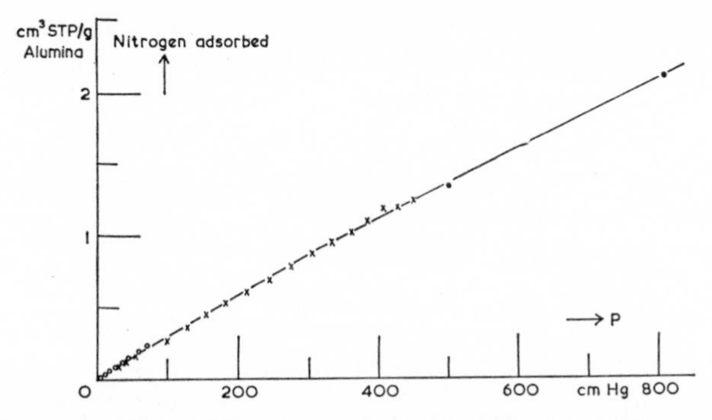

FIG. 10. Adsorption of N_2 on alumina at 0°C, measured in three different experimental units. (From Menon, 1961.) ○ from low-pressure adsorption apparatus, × from medium-pressure adsorption apparatus, ● from high-pressure adsorption apparatus.

with the low-pressure data and the first two points of the high pressure data at the same temperature. The continuity of the adsorption isotherm right from 0 to 10 atm pressure, covering experimental points measured with three entirely different apparatus, indicates that there cannot be any serious inherent flaw or fault in the glass-piezometer technique as applied to high pressure adsorption measurements.

E. ADSORPTION FROM MULTI-COMPONENT GAS MIXTURES

Lewis *et al.* (1950) used a reverse-pass apparatus to study pure gas and also multi-component gas adsorption equilibria on active carbons and silica gel up to 20 atm pressure. Known quantities (volumetric) of each of the components of the mixture were introduced to the adsorbent and were permitted to equilibrate under isobaric and isothermal conditions. The unadsorbed gas was analysed. From the previous calibration of the dead space with helium, and by material balance, both x–y diagrams (where x and y are the mole fractions in the adsorbate and vapour respectively) and the quantity of the mixture adsorbed were calculated. As an experimental check on the calculated balances, the adsorbate was pumped off and its volume and composition determined. To ascertain the attainment of equilibrium, either of the gases of the mixture was permitted to contact the adsorbent first, thus approaching the equilibrium from both sides. Gas mixing was achieved by passing the gas back and forth over the adsorbent by means of a motor-driven crank arm and siphon and by heating the steel mercury reservoir.

For studying the chemisorption of nitrogen and hydrogen from a $1N_2 : 3H_2$ mixture up to 50 atm pressure on an iron catalyst Sastri and Srikant (1961) used a volumetric method, measuring the gas volumes on expansion to 1 atm pressure. The composition of the gas mixture was followed with a katharometer.

The adsorption isotherms of N_2 and CH_4 and their binary mixtures on molecular sieves were measured (Lederman and Williams, 1964) at -150, -98, -78 and $+20$°C up to 80 atm pressure. A mass spectrometer was used to analyse the samples of the mixture drawn from the apparatus at equilibrium.

More recent studies of multi-component adsorption equilibria are made by gas chromatographic methods (see next Section).

F. GAS CHROMATOGRAPHIC METHODS

1. *Kobayashi's Method*

The first gas chromatographic apparatus modified for high-pressure operation was developed by Stalkup and Kobayashi (1963) and further refined by Koonce *et al.* (1965). Using it, the measurement of vapour–solid distribution coefficients or K values of a solute distributed between a gas phase and an adsorbed phase at essentially infinite dilutions was carried out (Gilmer and Kobayashi, 1964) for C_2H_6, C_3H_8 and n-C_4H_{10} in a CH_4–silica gel system up to 2000 p.s.i. The method was extended to a study of multi-component gas-solid equilibrium at high pressures with CH_4–C_3H_8–silica gel system (Gilmer and Kobayashi, 1965) and CH_4–C_2H_6–silica gel system (Masukawa and Kobayashi, 1968). The total adsorption, component adsorption and the K value for each component are related to the retention volume for the components as measured by a technique which distinguishes the molecules in the elution gas from those in the perturbing sample. Radioactively traced hydrocarbons were used to obtain the appropriate retention volumes.

A schematic diagram of the chromatographic apparatus used in the above work by Kobayashi and co-workers is given in Fig. 11. The tubings, fittings and valves were $\frac{1}{8}$ in nominal tubing size, the columns were $\frac{1}{4}$ in and $\frac{3}{16}$ in for the adsorption and retention determinations,

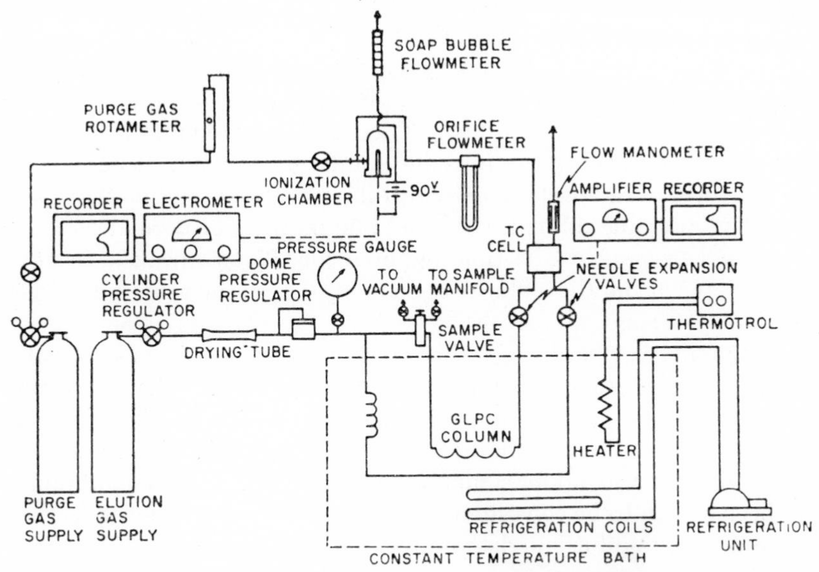

Fig. 11. Kobayashi's high-pressure gas chromatographic apparatus (cf Koonce *et al.*, 1965.) (Reproduced by permission of R. Kobayashi and Am. Inst. Chem. Engrs.)

respectively. The tubing assembly was designed to minimize dead-space volume in the system. Because radioactive materials were used, the entire apparatus was enclosed by a large hood which was connected to an exhaust duct. The reference gas stream passed through a coil in the thermostat and was then expanded through a needle valve before flowing through the reference side of the thermal conductivity cell. The carrier gas stream was passed through a six-way valve through which the flow could be diverted to sweep out a small sample tube containing the solute sample. The carrier gas, now containing solute, was passed through the chromatographic column (packed with firebrick impregnated with n-decane or n-heptane), expanded through a needle valve to 1 atm pressure and passed through the thermal conductivity cell. Then it was mixed with a large volume of nitrogen purge gas and passed through an ionization chamber and finally vented out into the exhaust hood.

The signal from the thermal conductivity cell was fed through a d.c. linear amplifier to a chart recorder (as in normal gas chromatographs) to give the elution history of the solute. The current from the ionization chamber passed through a preamplifier equipped with a resistance selector where the signal was converted to a voltage. The signal was then monitored by a vibrating read electrometer and fed to a chart recorder to produce the elution diagram of a radioactive solute sample.

The experimental procedure was to adjust the flow of elution gas (CH_4 or a mixture of CH_4 and C_3H_6, for instance) to approximately 70 cc/min at 760 mm Hg and 300°K after the desired column pressure and temperature were established. Then a 0·25 cc (760 mm Hg, 300°K) sample of gaseous solute was introduced into the column. The samples were mixtures of radioactive methane diluted with normal methane and radioactive propane diluted with a 50 : 50 mixture of normal methane and normal propane. The time required for the two peak concentrations to appear in the effluent stream was then measured for both detection systems.

The simplicity, speed, versatility and sensitivity of gas chromatography as an analytical technique are well recognized. The application of this technique with radioactive samples to the study of multicomponent adsorption eliminates the necessity for calibrating the system volume with helium and for sampling the gas phase. The data obtained have been shown to be consistent with those from a gravimetric method and from infinite-dilution adsorption K values.

2. *The Break-out Curve Method*

Mason and Cooke (1966) used a simpler and more straightforward gas-chromatographic flow method based on a material balance as the components are transported by a carrier gas through an adsorbent-packed column. The inlet composition was kept constant during a measurement: carrier gas and components to be adsorbed for an adsorption measurement, or pure carrier gas for a desorption measurement. Chromatographic analysis of the effluent gas from the column and knowledge of the gas flow rate permit a material balance to be made.

The experimental apparatus consisted of a gas supply system which delivered gas mixture of constant composition at a steady rate. Mixtures were prepared by injecting hydrocarbon liquids into the methane carrier upstream from the column. The column was packed with a weighed quantity of adsorbent heated previously to 260°C for 3 hr. After the column was purged with the carrier gas and the pressure regulated, flow was diverted to a bypass line while composition of the inlet stream was adjusted. When the desired composition was reached the flow was turned on into the adsorption column. Samples of the outlet gas from the column were withdrawn periodically and analysed by gas chromatographical methods until their composition became steady and equal to that of the inlet gas. An adsorption break-out curve was obtained from a plot of composition *versus* time. For desorption, pure carrier gas was passed through the column until the exit gas analysis showed no trace of the component gas being studied. A desorption break-out curve was thus obtained.

The area behind the break-out curve is proportional to the quantity of component adsorption at the concentration of the inlet gas and the temperature and pressure of the column. In this way the adsorption of all components other than the carrier can be determined in any gas mixture. However, because of re-equilibrium between gaseous and adsorbed phases as successive adsorption fronts move along the adsorbent column, the column must be saturated with the most highly adsorbed component before the break-out curve for any component is complete.

For binary mixtures, Eberly (1961) has shown that the values of adsorption for the heavier component at all levels of concentration up to that of the mixture can be calculated from a single desorption break-out curve. This implies instantaneous equilibrium between gaseous and adsorbed phases and absence of longitudinal diffusion in the adsorbent column. Typical break-out curves for adsorption and desorption are shown in Fig. 12. For a concentration shown as point *B*, the area *ABCD* under the desorption curve is integrated. The quantity of material represented by this area equals that which would

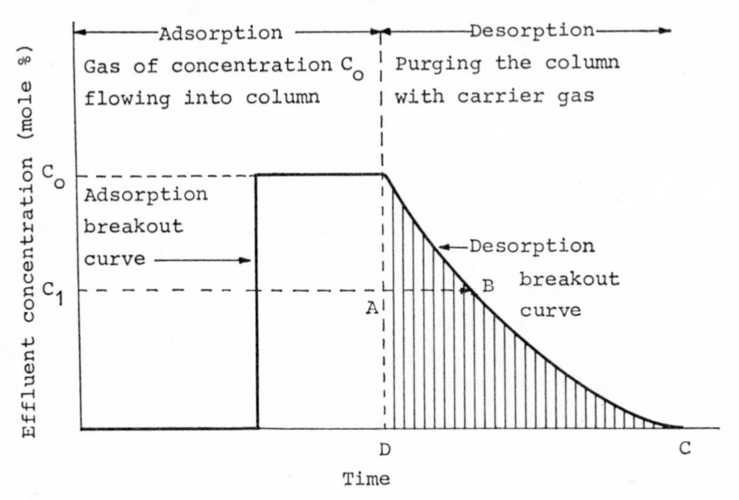

FIG. 12. Adsorption and desorption break-out curves (after Eberly, 1961).

be adsorbed from a gas of the composition represented by B at the conditions of temperature and pressure in the column. Mason and Cooke (1966) used desorption experiments for measuring adsorption of the heavier components in all binary mixtures with methane. The adsorption break-out curve was used for all components heavier than methane from other gas mixtures.

3. *Gas-circulating Method*

Recently Payne *et al.* (1968) have used a gas-circulating apparatus for measuring the adsorption of methane, propane and n-butane, and methane–propane and methane–n-butane mixtures on activated charcoal in the temperature range 0–70°C and up to 2000 p.s.i. pressure. Their apparatus (Fig. 13) consists of two main sections: an injection system

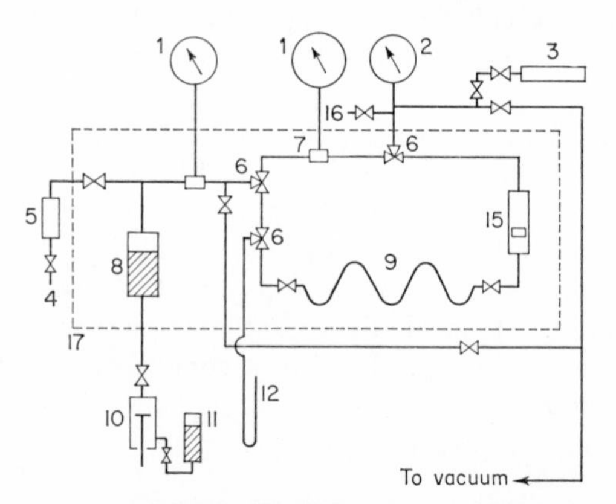

FIG. 13. Circulation-type adsorption apparatus of Payne *et al.* (1968). 1 0–200 p.s.i. Heise gauge, 2 vacuum-pressure gauge, 3 sample bomb, 4 gas supply, 5 dryer, 6 3-way valve, 7 diaphragm gauge separator, 8 Hg vessel, 9 adsorption column, 10 volumetric Hg pump, 11 reservoir, 12 manometer, 15 magnetic pump, 16 vent, 17 thermostat bath. 13, 14, details of vacuum system, not shown. (Reproduced by permission of T. W. Leland and American Chemical Society.)

and an adsorption loop. The latter consists of a closed coil packed with a weighed amount of adsorbent. The dead space in the loop is determined by expanding helium into it. Measured quantities of gas are admitted to the loop from the injection system and circulated in the closed loop with a magnetic pump. All parts of the apparatus containing gas are enclosed in a thermostat. The excess or differential adsorption (Gibbs adsorption) is the difference between the total moles injected and the number of moles filling the dead-space volume at equilibrium conditions. For the mixture isotherms, a measured quantity of the heavier hydrocarbon is first injected and the methane added subsequently in measured increments. A small sample of the non-adsorbed gas is removed for analysis by a gas chromatograph. The gas composition and total equilibrium pressure are measured when equilibrium is established after each methane injection. From these measurements and the compressibility data for the mixture, the total adsorption with the equilibrium composition of adsorbate and gas phases can be determined by a material balance on each component.

III. Results

A. Adsorption in the Critical Region of the Adsorbate

The anomalous properties of gases in the critical region are well known and have also been reviewed recently (Green and Sengers, 1966; Sengers and Levelt-Sengers, 1968). In the field of adsorption, one unsolved problem has been whether multi-molecular adsorption above the critical temperature of the adsorbate gas and necessarily under high pressures is possible or not. In theory, this seems to be not impossible, since molecular clusters are known to exist in compressed gases even above the critical temperature. Morris and Maass (1933) and Edwards and Maass (1935) investigated the adsorption of propylene and dimethyl ether on alumina near the critical temperature and up to saturation pressures. They obtained S-shaped isotherms at even 9° above the critical temperature. There is an element of uncertainty in their results on propylene adsorption since propylene can undergo polymerization on alumina as catalyst.

The investigation of adsorption of CO_2 on porous plugs of lampblack by Jones *et al.* (1959) is a reliable recent work on adsorption in the critical region. Their results are shown in Fig. 14. The shape of the adsorption isotherms at 19° and 30°C are not greatly altered by taking the buoyancy into account, since most of the adsorption occurs sharply near the saturation vapour pressure. But the shape of the isotherm at 32°C is much altered and the effect of filling the plug with sorbate is apparent. Earlier Jones and Isaac (1959) found that the flow of CO_2 through plugs of lampblack was much greater than that anticipated from established equations for streamline flow; this excess flow was assumed to occur in the sorbed phase. Comparison of the pattern of flow with that of the adsorption isotherms in Fig. 14 justifies this assumption (see also Section III.G).

The results of Jones and co-workers, as also the earlier work of Morris and Maass (1933), Edwards and Maass (1935) and Coolidge and Fornwalt (1934), all show that there is no change in the type of adsorption as the critical temperature is exceeded. A denser fluid forms in the porous

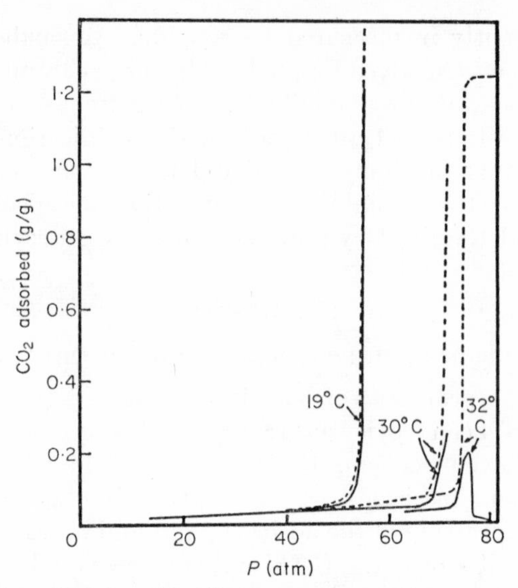

FIG. 14. Adsorption of CO_2 by a lampblack plug (data of Jones *et al.*, 1959). Isotherms determined neglecting the buoyancy of the adsorbate (full curves) and after correcting for it (broken curves). (Reproduced by permission of W. M. Jones and the Faraday Society.)

material. Flow can occur in this denser fluid in the same way as it occurs in a condensate below the critical temperature. This conclusion has been further confirmed by Jones and Evans (1966) from adsorption of CO_2 on alumina up to 100 atm pressure, determined from measurement of dielectric constant; typical S-shaped adsorption isotherms were obtained at 32° and 35°C. An analogy between the pressure (P_{max}) at the maximum in adsorption isotherms at high pressures and the critical pressure P_c, so that $P_{max} = P_c$, when $T = T_c$, pointed out recently (Menon, 1968a) is discussed in Section III. E.

B. PARTIAL MOLAR VOLUME OF ADSORBATE (GONIKBERG, 1960)

Krichevskii and Kal'varskaya (1940) have shown that the dependence of the chemical potential of an adsorbed substance on pressure can be expressed by the equation:

$$\left(\frac{\partial \mu_{is}}{\partial P}\right)_T = \bar{v}_{is}, \tag{1}$$

where $\bar{v}_{is}$ is the partial molar volume of the adsorbed species. (A derivation of this equation, given by Temkin (1950), has been reproduced in Gonikberg's monograph (1960).)

Taking the simplest case of adsorption on a uniform or homogeneous surface in the absence of lateral interactions among adsorbed molecules (Langmuir's concepts), the expression for the chemical potential of the adsorbed substance can be written as

$$\mu_s = \mu_s^0 + RT \ln \frac{\theta}{1-\theta} \tag{2}$$

where μ_s^0 is the standard value of the chemical potential at $\theta = 0{\cdot}5$ and θ is the degree of coverage of the surface by the adsorbed molecules.

At adsorption equilibrium

$$\mu_s = \mu_{\text{gas}} \tag{3}$$

where μ_{gas} is the chemical potential of the adsorbate in the gas phase.

At low pressures,

$$\mu_{\text{gas}} = \mu_{\text{gas}}^0 + RT \ln P. \tag{4}$$

Langmuir's adsorption isotherm assumes the form:

$$\theta = \frac{P}{P+b}. \tag{5}$$

From eqns (2), (4) and (5) it can be derived that

$$b = \exp\,[(\mu_s^0 - \mu_{\text{gas}}^0)/RT]. \tag{6}$$

At high pressures

$$\mu_{\text{gas}} = \mu_{\text{gas}}^0 + RT \ln f \tag{7}$$

where f is the fugacity of the gas.

In view of eqn (1), one may write:

$$\left(\frac{\partial \mu_s^0}{\partial P}\right)_T = \bar{v}_s \tag{8}$$

$$\left(\frac{\partial \ln b}{\partial P}\right)_T = \frac{\bar{v}_s}{RT}. \tag{9}$$

Neglecting the compressibility of the adsorbed substance, eqn (9) leads to

$$b = b^0 \exp\,(\bar{v}_s P/RT) \tag{10}$$

where b^0 is the limiting value of b at $P = 0$.

Substituting eqn (10) in eqn (5) yields:

$$\theta = \frac{f}{f + b^0 \exp\,(\bar{v}_s P/RT)}. \tag{11}$$

Differentiating with respect to pressure,

$$\frac{\partial}{\partial P}\left[\ln\left(\frac{1}{\theta}-1\right)\right]=\frac{\bar{v}_s}{RT}-\frac{\partial \ln f}{\partial P}. \tag{12}$$

But

$$\left(\frac{\partial \ln f}{\partial P}\right)_T=\frac{v_{\text{gas}}}{RT} \tag{13}$$

where v_{gas} is the molar volume of gas at pressure P.

Hence

$$\frac{\partial}{\partial P}\left[\ln\left(\frac{1}{\theta}-1\right)\right]=\frac{\bar{v}_s-v_{\text{gas}}}{RT} \tag{14}$$

or

$$\frac{\partial \theta}{\partial P}=\theta(1-\theta)\,\frac{v_{\text{gas}}-\bar{v}_s}{RT}. \tag{14a}$$

Thus $\partial\theta/\partial P$ is positive (adsorption increases with pressure) as long as $v_{\text{gas}}>\bar{v}_s$, and it becomes negative when $\bar{v}_s>v_{\text{gas}}$. At some intermediate value of pressure $v_{\text{gas}}=\bar{v}_s$, this state corresponds to maximum adsorption. A further increase in pressure leads to a decrease in adsorption. This behaviour of adsorption is analogous to that of the solubility of gases in liquids under pressure, where also a maximum point of solubility is observed at high pressures (Gonikberg, 1960). Krichevskii and Kal'varskaya (1940) studied the adsorption of benzene and carbon tetrachloride vapours on charcoal under pressures of N_2, H_2 and N_2–H_2 mixture up to 600 atm. They found that the maximum adsorption of the vapours was somewhat below 100 atm. Their experimental data show that the partial molar volume of the adsorbed liquid considerably exceeds that of the pure liquid.

C. ADSORPTION ISOTHERMS AT HIGH PRESSURES

A maximum in adsorption isotherms at high pressures and above the critical temperature of the adsorbate has been observed in all cases where the measurements have been carried out to sufficiently high pressures. As typical examples, the adsorption of argon on active carbon and nitrogen on alumina are shown in Figs 15 and 16. According to the ordinary definition of adsorption, the amount adsorbed is the *excess* material present in the pores and on the surface of the adsorbent over and above that corresponding to the density of the gas in the gas

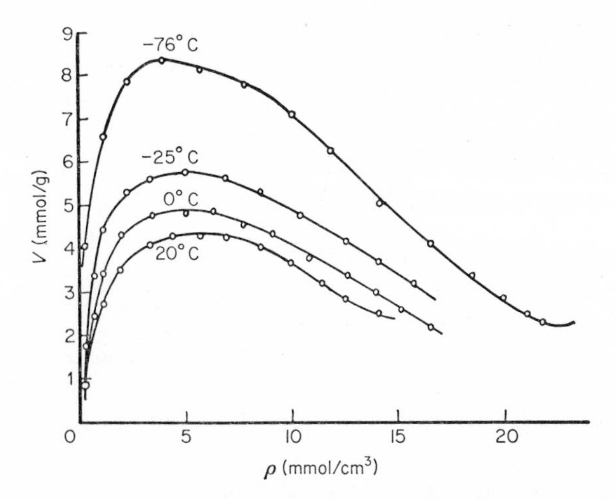

Fig. 15. Adsorption of argon on active carbon at high gas densities (pressures up to 400 atm) (data of von Antropoff, 1954a).

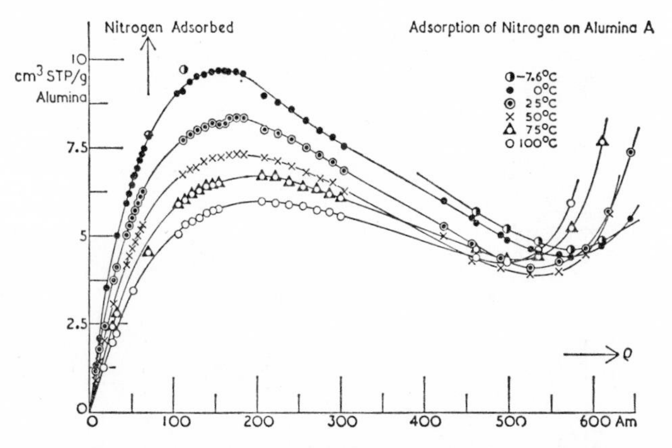

Fig. 16. Adsorption of N_2 on alumina at high gas densities (pressures up to 3000 atm). Densities are in amagat units. (Data of Michels *et al.*, 1961. Reproduced by permission of Royal Netherlands Chemical Society.)

phase at that temperature and pressure. All the experimental methods for measurement of adsorption yield only this excess gas or vapour on the surface, termed differential adsorption, Δg. The total gas on the surface (absolute adsorption, g_a) will be the sum of this excess amount and the amount of gas which will be present on the surface in any case due to the applied gas density only, even in the absence of adsorption forces. As von Antropoff (1952) has pointed out, the original derivation of the Langmuir adsorption isotherm is for absolute adsorption, not for differential adsorption, although only the latter can be experimentally measured. In all low-pressure adsorption isotherms, however, the difference between the above two definitions of adsorption is too little to be of any consequence.

As a convenient and practical way, absolute adsorption may be calculated as (Menon, 1961, 1968b)

$$g_a = \Delta g + \rho S d \tag{15}$$

where ρ is the gas density corresponding to the pressure applied at the experimental temperature, S is the total (BET) surface area of the adsorbent and d is the " thickness " or height of the adsorption layer on the surface. The term $\rho S d$ gives the amount of gas in the adsorption layer (of volume Sd) owing to the applied gas density only. The value of d has to be chosen on an empirical and arbitrary basis, for example the diameter of the adsorbate atom or molecule from the van der Waals constant b, from the Lennard–Jones model, or from the second virial coefficient for the gas.

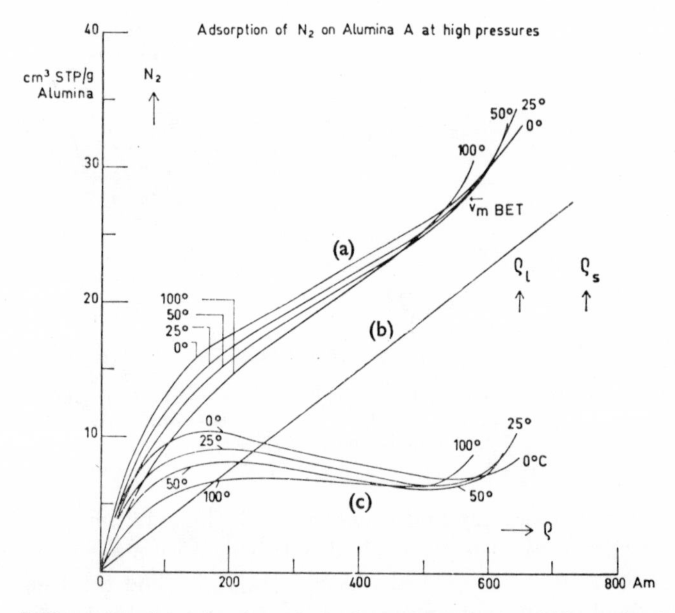

FIG. 17. Differential adsorption Δg and absolute adsorption g_a of N_2 on alumina at high gas densities (pressures up to 3000 atm). (a) Total gas g_t, in the adsorption layer. (b) Gas in a 3·15 A layer on the entire surface of the adsorbent due to gas density only if there were no adsorption. (c) Excess gas Δg in the surface layer due to adsorption. (Data of Menon, 1968b; reproduced by permission of American Chemical Society.)

A more detailed discussion on these two definitions of adsorption and on the difficulties and uncertainties in calculating absolute adsorption has been given elsewhere (Menon, 1968b). The course of the differential and absolute adsorption isotherms for nitrogen on alumina at gas densities up to 650 amagats (pressures up to 3000 atm) is shown in Fig. 17. Especially noteworthy is the fact that the absolute adsorption isotherms intersect and cross over in a region corresponding to the v_m value for the alumina from a BET plot of the nitrogen isotherms at liquid nitrogen temperature. This represents some sort of a " complete surface coverage ". Beyond this the isotherms exhibit a second ascending and apparently endothermic part, suspected to be caused by a reorientation of the adsorbed molecules into a more compact configuration on the adsorbent surface. Further details are given in the reference mentioned above.

D. SORPTION BY COAL

1. *Sorption of Methane*

The earliest attempts in the field of high-pressure adsorption were to measure the sorption of methane by coal. The purpose was to get a better insight into the mechanism of methane hold-up in coal and the release of this methane when seams of coal were exposed during mining, sometimes leading to disastrous fires and explosions in coal mines. In recent years, Palvelev (1948) and Khodot (1948, 1949, 1951) have obtained methane sorption isotherms for Russian coals up to 1000 atm pressure. The apparent sorption they found either increases only slightly with pressure above 200 atm, or declines from a maximum near this pressure to a limiting value which is well above zero at 1000 atm. Another recent study is by van der Sommen *et al.* (1955) up to 500 atm, where for the first time accurate techniques were employed in volume and pressure measurements in high-pressure adsorption work.

The most extensive study on methane sorption is by Moffat and Weale (1955) on a series of 10 coals ranging in C-content from 93·7 to 80·7% and at pressures up to 1000 atm. For calculation of sorbed volumes they used both the lump density and the helium density of coal. The lump density is obtained by coating suitably shaped coal lumps with vaseline and weighing them in air and in water. This gives the density of coal including the pore space. The total sorption is the methane sorbed by the lump volume of the coal and includes the gas which is simply compressed in the pore space. The apparent sorption is calculated from the volume of the coal calculated from its helium density; here the pores in the coal are regarded as part of the dead space

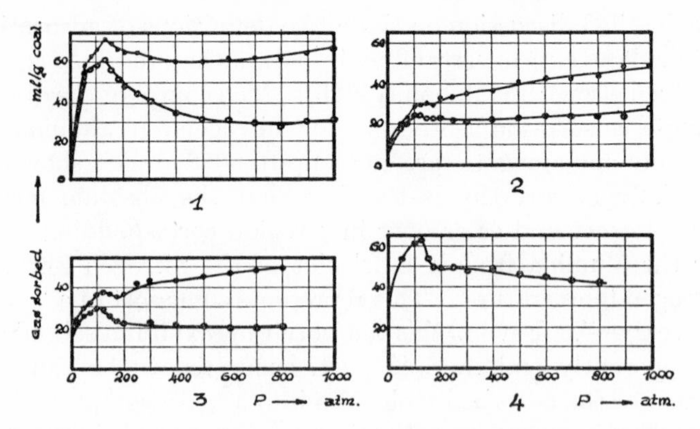

FIG. 18. Sorption of methane by coals up to 1000 atm pressure. 1 coal *A*, 2 coal *E*, 3 coal *H*, 4 beechwood charcoal. Carbon content of coals: *A* 93·7, *E* 85·7, *H* 83·6%. The closed points are for total sorption and the open points for apparent sorption. (Reproduced by permission from Moffat and Weale (1955).)

or free volume in the sorption apparatus. Typical isotherms at 25°C of total sorption and apparent sorption for three coals and beechwood charcoal are shown in Fig. 18. The total sorption usually increases rapidly with pressure at first and then more slowly above 100 or 150 atm, sometimes showing a maximum in this pressure range. The highest value of the total sorption varies between coals from about 28–70 ml/g (1000–2500 cu ft/ton). The apparent adsorption also increases rapidly with pressure to 100–150 atm where it reaches a maximum of 20–60 cm^3/g and of varying sharpness according to the coal, but it does not come down to zero even at 1000 atm. The general conclusion is that methane is physically adsorbed on the large internal surface of coal, and is not chemisorbed or held in the type of solid solution formed by compressed gases in rubber and linear high polymers. The adsorbed amount at 1000 atm is higher than that to be expected. This may be due to the compressibility of coal increasing the pore volume at high pressure and hence changing the dead space in the apparatus. A more probable cause, according to Moffat and Weale, is the penetration of methane molecules at high pressures into spaces between the coal lamellae, which are not included in the dead-space determination with helium at 1 atm pressure. This interstitial adsorption is accompanied by a slight expansion of the coal structure which has also been measured experimentally (see next Section).

Moffat and Weale (1955) found that the total sorption at 1 atm does not vary much with the ranks of coal, but at any particular high pressure there is an ill-defined minimum at a C-content of about 89% in the coal

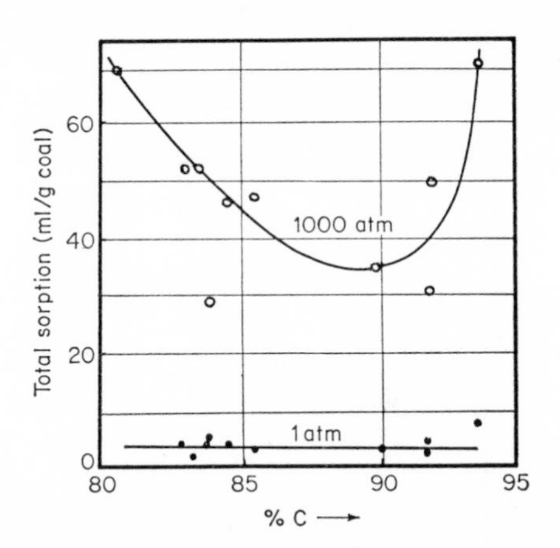

FIG. 19. Total sorption at 1 atm and at 1000 atm of methane by coal as a function of coal rank. Data of Moffat and Weale (1955). (Reproduced by permission of K. E. Weale.)

(Fig. 19). Such minima have been noted for properties of coals such as porosity and heat of wetting in methanol. A minimum in CO_2 sorption as a function of coal rank at any fixed pressure has also been reported recently (see Section D.3).

2. *Swelling of Coal During Methane Sorption*

The swelling of coal during methane sorption up to 150 atm was first observed by Audibert (1942) and later confirmed by de Braaf *et al.* (1952). Moffat and Weale (1955) measured it for their coal samples by attaching electrical resistance strain gauges to suitably cut blocks of coal, with connections from the pressure vessel to a Wheatstone bridge. A reference strain gauge attached to a piece of steel inside the vessel was found not to alter during the experiments. Measurements were made parallel and perpendicular to the bedding plane, and were combined to give bulk expansions. The change parallel to the plane was assumed to be uniform in every direction. The bulk expansion measured during sorption–desorption cycle for four representative coals are shown in Fig. 20. In methane the greatest expansion is shown by the lowest rank coal *J*, but the middle coals like coal *H* show more anisotropy and greater hysteresis than coal *J*. The anthracite *A* expanded only very slightly on sorption, but an appreciable expansion of about 0·2% in each direction occurred on desorption and the original dimensions were regained only after evacuation. The association of bulk expansion with methane sorption is supported by the absence of expansion when (a) coal *J* was measured in compressed helium which is not adsorbed, and

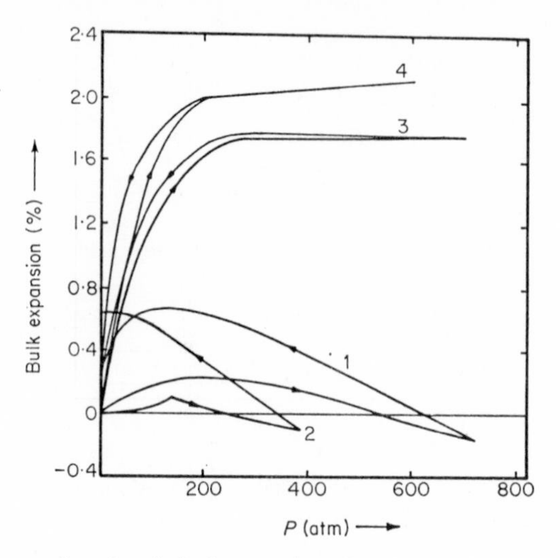

FIG. 20. Bulk expansion of coals during sorption of methane at high pressures. 1 coal *H*, 2 coal *A*, 3 coal *J*, 4 beechwood charcoal. Carbon content: coal *A* 93·7, *H* 83·6, *J* 80·7. Data of Moffat and Weale (1955). (Reproduced by permission of K. E. Weale.)

(b) coal *H* was encased in metal which mechanically prevented expansion.

3. *Adsorption of CO_2 on Coals*

The adsorption of CO_2 on Polish coals at different stages of metamorphosis has been studied by Czaplinski and Lason (1965) at 20°C and pressures up to 60 atm. The sorbed amount at about 35 atm pressure is 25–35 cm^3 STP/g coal. At a fixed pressure the amount sorbed slightly varies with the rank or carbon content of coal (Fig. 21), showing a minimum of sorption for coals with 88–89% C. This behaviour is quite similar to the results for methane sorption on coal discussed earlier. In another series of measurements Czaplinski (1965) has reported that on coals about 65% of the CO_2 sorption already takes place at 5·6 atm, while in the 30–45 atm range the sorption isotherms exhibit a broad maximum. Indications are that the major part of the CO_2 taken up by coals remains in the adsorbed state.

E. PREDICTION OF THE PRESSURE AT THE MAXIMUM IN HIGH-PRESSURE ADSORPTION

As mentioned in Section III.C, experimentally one can measure only the *excess* material present on the surface of the adsorbent over and

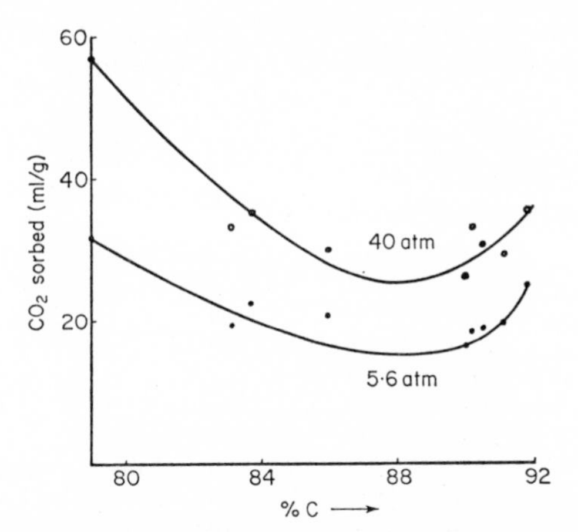

FIG. 21. Adsorption of CO_2 on Polish coals as a function of coal rank (from the data of Czaplinski and Lason, 1965).

above that corresponding to the density of the gas in the gas phase at that temperature and pressure. With increasing pressure this excess or differential adsorption soon reaches a maximum. As the pressure is increased further the density of the gas phase (ρ) gradually approaches that of the adsorbed phase (ρ_a). This can continue till $\rho = \rho_a$ and the amount adsorbed, measured experimentally and calculated according to the above definition, must become zero.† Hence the high-pressure adsorption isotherms must exhibit a maximum even by elementary considerations.

Quite recently Menon (1968a) has found that the pressure $P_{\max}$ at which the high-pressure adsorption isotherms exhibit a maximum can be calculated from the relation

$$P_{\max} = (T/T_c)^2 P_c \tag{16}$$

where T_c and P_c are the critical constants for the adsorbate gas and T is the temperature of the adsorption isotherm. Thus $P_{\max}$ can be directly calculated from the properties of the gas; it does not seem to depend on the type or nature of the adsorbent surface. In Table II the values of $(T/T_c)^2 P_c$ are compared with literature data on pressure $P_{\max}$ at which experimental adsorption isotherms exhibit maximum. The agreement between the two is quite satisfactory, especially in view of the numerous inaccuracies and uncertainties in usual high-pressure adsorption measurements, and the added difficulty of locating $P_{\max}$ in cases

† In practice this has not been observed for any rigid non-compressible adsorbent. The behaviour of the adsorption isotherms at very high pressures is quite different (see Section III.C, also Menon (1968b)).

TABLE II. Comparison of $(T/T_c)^2P_c$ with P_{max} from High Pressure Adsorption Data for Adsorbents of Second Structural Type (from Menon, 1968a)

Gas	Adsorbent	Temp. (°C)	$(T/T_c)^2P_c$ (atm)	P_{max} (atm)	Reference
CH_4	Coal	25	112	100–115	See Table I
N_2	Alumina	−0·76	156	154	Michels *et al.* (1961)
		25·1	188	200	
		50·0	220	227	
		74·9	256	260	
		99·7	293	300	
CO	Alumina	−0·77	142	146	Menon (1965)
		25·0	170	180	
		50·0	200	216	
CH_4	Silica gel	0	94	96	Gilmer and Kobayashi (1964)
		−20	81	82	
		−40	68	63	
C_2H_4	Silica gel	25	56	52	Gilmer and Kobayashi (1964)
CO_2	Porous plug of lampblack	32	73·6	74	Jones *et al.* (1959)

where the maximum in the isotherm is not sharp but is rather flattened over a pressure range of 10–30 atm.

This relation $(T/T_c)^2P_c = P_{max}$ is found to be valid only for adsorbents with relatively large micropores. Molecular sieves, charcoal and active carbons of large surface area are typical examples of adsorbents with extremely small micropores of molecular dimensions, for which the effect of increasing adsorption potentials as a result of the overlapping of the fields of opposite walls of the pore is prominent. For these adsorbents (of the first structural type, according to Dubinin, 1960), P_{max} is always much lower than $(T/T_c)^2P_c$, the ratio of the two often being 0·6–0·8.

A consequence of the relation $(T/T_c)^2P_c = P_{max}$ will be that $P_c = P_{max}$ when $T = T_c$. Thus P_{max} seems to have the role of a critical pressure for the gas adsorbed above its critical temperature, as if above this pressure the adsorbed gas forms a continuous film or " condensed " layer. Also, an adsorption isotherm measured at the critical temperature of a gas should show a maximum at the critical pressure. The closest approach to critical temperature can be seen in the measurement of adsorption of CO_2 on a porous plug of lampblack (Jones *et al.*, 1959) at 32°C ($T_c = 31$°C) where a maximum in the isotherm is observed at 74 atm, which agrees very well with the calculated value $(T/T_c)^2P_c = 73·6$ atm and the critical pressure of CO_2 of 73 atm.

The theoretical significance of the above relation is still not clear. But, in industrial applications of adsorption the pressure to be applied to obtain maximum adsorption on an adsorbent can be approximately calculated from the following rules of the thumb:

(1) Adsorbents of first structural type $P_{max} = 60\text{–}80\%$ of $(T/T_c)^2 P_c$. (Average pore diameter $D < 20$ Å.)
(2) Adsorbents of second structural type ($D > 20$ Å), $P_{max} = (T/T_c)^2 P_c$.

F. SURFACE AREAS FROM HIGH-PRESSURE ADSORPTION

The importance of a knowledge of surface area values for adsorbents and catalysts in use in research and industry has been generally recognized. The surface areas are almost always calculated by application of the Brunauer–Emmett–Teller (BET) equation to adsorption data. Often the adsorption itself may be measured by simple and rapid methods (where only approximate and comparative surface area values are required, or when areas of a large number of samples are required on a routine basis). Sometimes it may also be necessary to resort to special procedures to get reliable adsorption data (in cases where the nature of the adsorbent brings in additional complications). High-pressure adsorption has been adapted to serve both these roles as follows:

(1) Haley (1963) has extended the Nelson–Eggertsen (1958) continuous flow method of adsorption measurements to determination of surface areas by the BET method. A mixture of 10% N_2 in He was passed over the sample cooled in liquid nitrogen at a fixed rate at various pressures up to 10 atm, causing the N_2 partial pressure to reach its liquefaction point. Nitrogen adsorbed or desorbed by increasing or decreasing pressure in the sample tube was measured continuously (at 1 atm pressure) by thermal conductivity with a reference stream of 10% N_2 in He. Isotherms determined in this way show good agreement with those determined by conventional volumetric adsorption measurements. The distinct advantages claimed for the former method are that it does not involve vacuum techniques, it gives a permanent record automatically, is faster and simpler for routine application and requires less skill.
(2) Measurement of the surface areas of coals and cokes offer unusual difficulties (Kini, 1963). Methods based on heat of wetting or on adsorption of methanol are subject to doubt, since dipole interaction or hydrogen bonding may occur between the adsorbate and the oxygenated groups in the coal. The standard BET method using argon or nitrogen at liquid nitrogen temperature is considered to give generally low values because of contraction of pores and slow diffusion of adsorbate into the

pores at the low temperature, and, rarely, high values caused by capillary condensation at rather low relative pressures. Surface area values obtained from adsorption measurements of hydrocarbons near ambient temperature may be unreliable owing to the possibility of solution of the adsorbate in coal, while adsorption of CO_2 may be influenced by the quadrupole moment of the molecule. The method of low angle X-ray scattering is not applicable either, since it cannot distinguish the open pores of coal from the closed pores inaccessible to gases.

To avoid the above difficulties and uncertainties Kini (1964) has resorted to adsorption of xenon at 0°C and of krypton at −78°C at pressures up to 40 atm. Specific surface areas of a number of coals, cokes and partly gasified cokes obtained in this way using Xe and Kr are compared with those obtained with the same materials by low-pressure adsorption of argon and methanol.

G. SURFACE DIFFUSION AT HIGH PRESSURES

Adsorption at high pressures seems to make an important contribution to diffusion of gases through porous catalysts. Satterfield and Cadle (1968) have recently studied gaseous diffusion and flow in commercial methanol synthesis catalysts and in supported nickel and palladium catalysts up to 65 atm pressure. Surface diffusion of nitrogen made an increasing contribution to the total nitrogen flux with increase in pressure and at the highest pressure was of comparable magnitude to the volume diffusion flux. The surface diffusion fluxes for the different catalysts increase linearly with pressure. The linear relationship suggests that the surface diffusion can be well correlated by Fick's law with Henry's law relating the surface and gas phase concentrations:

$$N_{\mathrm{N},S} = S.\left(\frac{D_{SS}k_S}{\tau}\right)\frac{P}{RTL}(y_{\mathrm{N}0} - y_{\mathrm{N}L}) \quad (17)$$

where $N_{\mathrm{N},S}$ = flux of nitrogen caused by surface diffusion, S = surface area per unit pellet volume of the catalyst, D_{SS}/τ = surface diffusion coefficient, k_S = Henry's law constant, L = pellet thickness, $y_{\mathrm{N}0}$ and $y_{\mathrm{N}L}$ are the mole fractions of nitrogen at distance through pellet $x = 0$ and at $x = L$.

A comparison of the surface diffusion flux to the theoretical nitrogen flux shows (Satterfield and Cadle, 1968) that at 65 atm pressure the ratio varies from 0·5 to 0·9, depending upon the particular make of the commercial catalysts. This suggests that there may be cases where the surface diffusion flux may be negligible at atmospheric pressure but significant at high pressures.

H. ADVANTAGE OF HIGH PRESSURE IN INDUSTRIAL ADIABATIC ADSORPTION UNITS

In industrial practice gas-phase adsorption techniques may be broadly divided into two groups: (1) the adsorption of organic gases and vapours on activated carbons, with possibility to recover them if required, and (2) the drying and purification of air or gases using activated alumina or silica gel. A distinct advantage of elevated pressure in industrial adiabatic adsorption units has been pointed out by Worthington (1958). By increasing the pressure of a gas or airstream, the pressure of water vapour in it is increased in most cases to the saturation pressure or above and hence some water condenses out. Thus, an increase in pressure from 1 to 5 atm for a gas originally at or above 20% relative humidity (r.h.) would, for a given constant temperature, result in the gas becoming saturated. Condensed water in the form of mist would be removed by a physical separation process which, incidentally, should also remove the oil mist (from the lubrication of the compressor).

The problem of drying a water-saturated gas or air under pressure is similar to drying at atmospheric pressure, but there is an important difference in the heat balance. By increasing the gas pressure so that condensation occurs, the ratio of the weight of gas to the weight of water vapour in the gas stream is increased. In an adiabatic drier, heat balance considerations show that the heat of adsorption released is primarily removed by the carrier gas which gets heated up. As the amount of heat released during adsorption is directly proportioned to the amount of water adsorbed, this increase in gas–water ratio results in a smaller temperature rise during adsorption. If an air drier operating at atmospheric pressure on 100% r.h. inlet air is compared with one operating at 5 atm pressure and 100% r.h., then at the same inlet air temperatures, the temperature rise in the high-pressure drier will be one-fifth of that in the atmospheric pressure unit. Thus, with increasing pressure, adiabatic adsorption units tend to isothermal conditions. The heat balance equation for operation at pressure p atm is (Worthington, 1958):

$$\frac{H}{P} = \Delta\theta \left(\frac{c_g}{h} + \frac{c_a}{a}\right) \tag{18}$$

where H = heat of adsorption, $\Delta\theta$ = temperature rise, c_g = specific heat of the gas or air, h = absolute humidity of gas, c_a = specific heat of adsorbent and a = water vapour adsorption capacity of the adsorbent.

The assumption made here is that the increase in pressure does not affect the adsorption capacity of the adsorbent for water vapour.

Since the physical adsorption of ordinary gases or air is not increased very much by pressures of 5 or 10 atm, this assumption is quite reasonable.

IV. Adsorption Equilibria in Multicomponent Gas Mixtures

The increasing use of adsorption separation techniques in the natural gas industry demands basic adsorption data and also useful correlations of these data for efficient design of high-pressure adsorption processing equipment. Gas streams processed in adsorption plants usually contain over 90% methane, the remainder consisting of higher paraffin hydrocarbons. Pressures commonly range from 300 to 1200 p.s.i. and temperatures from 25 to 50°C. Equilibrium adsorption capacities of commonly used adsorbents for the individual components of these complex, high-pressure, natural gas systems are therefore needed for a wide range of conditions. As pointed out by Lewis *et al.* (1950), the main advantages of adsorption as a separation technique, as compared to absorption, distillation or extractive distillation, are that in many cases much higher selectivity can be obtained by adsorption than by any of the other techniques, and adsorbents have a relatively high capacity for volatile materials even at low partial pressure. Higher selectivity promises more effective separations, while high adsorbent capacity allows the use of higher temperatures (eliminating refrigeration) and lower pressures than would usually be required with conventional separation methods.

A. Correlations of Lewis *et al.*

Based on experimental study of pure gas and binary gas adsorption isotherms of C_1 to C_4 hydrocarbons on silica gel and active carbons, Lewis *et al.* (1950) have tried to develop methods to predict adsorption equilibria from Dubinin's modification of Polanyi's potential theory of adsorption. Their procedure is as follows:

1. *Pure Gas Isotherms*

Dubinin and Radushkevich (1947) suggested that similar types of compounds on a given adsorbent would have equal adsorption potentials when equal amounts were adsorbed. Equal amounts adsorbed were defined as the product of moles adsorbed (N) and molal volume (V) of adsorbate measured as a saturated liquid. Thus,

$$N_1 RT \ln (P_s/P)_1 = N_2 RT \ln (P_s/P)_2 \tag{19}$$

for $N_1 V_1' = N_2 V_2'$.

Dividing the two equalities,

$$\left[\frac{RT}{V'}\ln\,(P_s/P)\right]_1 = \left[\frac{RT}{V'}\ln\,(P_s/P)\right]_2 \tag{20}$$

where V' is the molal volume of saturated liquid at a temperature corresponding to the adsorption isotherm, P_s is the saturated liquid vapour pressure of the material being adsorbed at the temperature in question, and P is the adsorption pressure. For better fit with their experimental data (up to 20 atm pressure), Lewis and co-workers modified eqn (2) to eqn (3):

$$\left[\frac{RT}{V}\ln\,(f_s/f)\right]_1 = \left[\frac{RT}{V}\ln\,(f_s/f)\right]_2 \tag{21}$$

for equal values of NV, where f_s is the fugacity of the vapour corresponding to vapour pressure, f is the fugacity of the vapour at adsorption pressure, and V is the molal volume of saturated liquid at vapour pressure equal to adsorption pressure. If this relation is to correlate the data, the value of $(RT/V)\ln\,(f_s/f)$ should be the same for all gases at equal volumes adsorbed.

Data for eight different gases adsorbed on silica gel and active carbon at 25°C in the pressure range 0·2–20 atm are shown in Fig. 22. Paraffins,

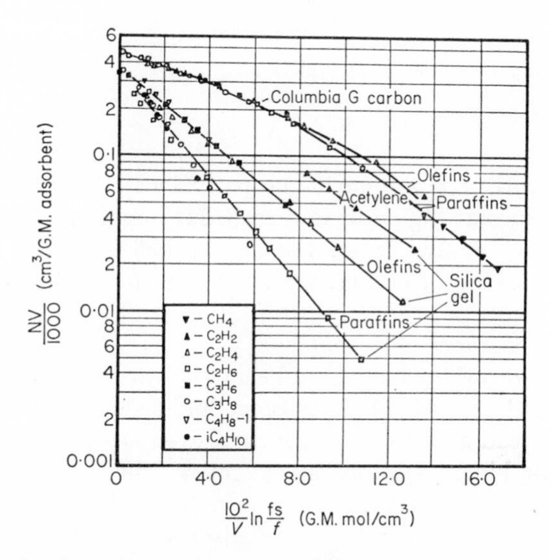

FIG. 22. Generalized correlation of Lewis *et al.* (1950) for adsorption isotherms on silica gel and active carbon at 25°C and 0·2–20 atm pressure. (Reproduced by permission of E. R. Gilliland.)

olefines and acetylene form three distinct straight lines on silica gel, while on active carbon a single line can represent practically all the data. The temperature dependence of adsorption can also be successfully correlated by eqn (3); as a typical example the data for ethylene and propane on silica gel are shown in Fig. 23. When this type of correlation

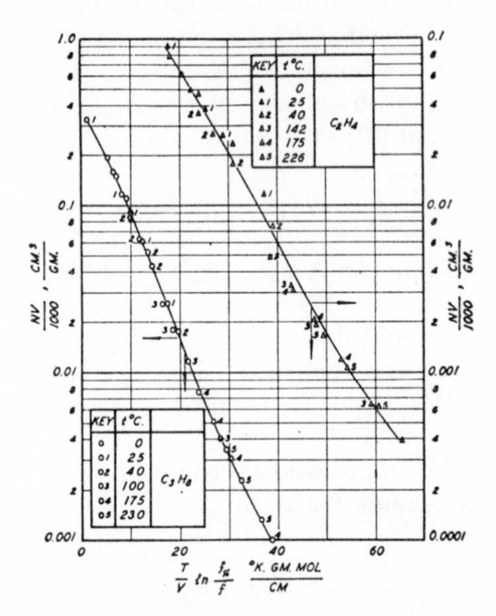

FIG. 23. Correlation of Lewis *et al.* (1950) for isotherms of ethylene and propane on silica gel at various temperatures. (Reproduced by permission of E. R. Gilliland and American Chemical Society.)

is employed with a minimum of properly located experimental data, adsorption values at higher or lower pressures and at various temperatures can be estimated; also, the adsorption isotherms of different hydrocarbon gases on the same adsorbent can be predicted.

2. *Binary and Ternary Gas Mixtures*

Lewis and co-workers have found that the variation of total quantity adsorbed is not linear with composition expressed as mole fraction in the adsorbate, instead a plot of N_1 against N_2 yields a satisfactory straight line for all the binary systems investigated, where N_1 and N_2 are the moles of each component adsorbed from the mixture. The terminal values of the line are N_1' and N_2', the adsorption capacities for the pure components. The equation of the straight line is

$$(N_1/N_1') + (N_2/N_2') = 1 \tag{22}$$

and in general $\sum_{1}^{n} N_1/N_1' = 1$ for a system of n component gases in the mixture. A plot of (N_1/N_1') against (N_2/N_2') is shown in Fig. 24 for the

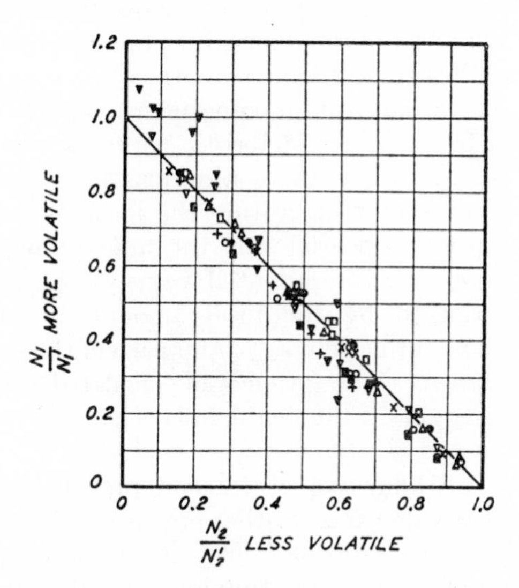

FIG. 24. Total adsorption correlation of Lewis *et al.* (1950) for silica gel and active carbons. (Reproduced by permission of E. R. Gilliland and American Chemical Society.)

data of Lewis and co-workers. This correlation yields mixture adsorption values which are within 6% of the experimental value and hence is quite satisfactory.

In all, thirty-five mixtures were investigated by Lewis and his co-workers. They observed that the pure gas for which the adsorbent exhibits the higher capacity is the gas preferentially adsorbed from a mixture. At constant partial pressure there is less of a given hydrocarbon adsorbed from a mixture than from the pure gas, thus each component interferes with and decreases the adsorption of the other components.

For binary mixtures, the adsorption relative volatility (α) is defined as

$$\alpha = (y_1 x_2 / y_2 x_1) \tag{23}$$

where y_1 and y_2 are mole fractions in the vapour for the two components and x_1 and x_2 are those in the adsorbate. The α of any two components of a ternary mixture is approximately the same as that of the two components in a binary mixture. Besides, for most cases the α for a binary mixture can be calculated from the two known binary systems in which one component is common, the other components constituting the new binary.

The method of Lewis and co-workers for using only the adsorption isotherms of the pure gases to predict adsorption relative volatilities, α, was based on the Polanyi–Dubinin potential theory discussed earlier. It was assumed that the adsorption potential for each gas in the mixture could be obtained from the characteristic curve for pure gases of the type shown in Fig. 22. Since the correlations for the pure gases had indicated that at constant temperature the most important factor in determining adsorption potential was the amount adsorbed, it was assumed that in mixture, the potential value of the individual components was determined by the total amount of adsorbate. The potential value was used to calculate f_s/f for each of the components and the value of the vapour composition was calculated using the Lewis–Randall type fugacity rule. Further details of the calculation are given in the paper of Lewis *et al.* (1950).

If the values of α are known at two or more pressures, interpolation is permissible to get the value at any other pressure. Alternatively, since α varied inversely as the ratio of N_1'/N_2', the value of α at any other pressure could be obtained by multiplying the value at 1 atm by the ratio of N_1'/N_2' for the two pressures.

B. APPLICATION OF BET EQUATION TO HIGH-PRESSURE ADSORPTION DATA

Mason and Cooke (1966) found the Brunauer–Emmett–Teller (BET) equation, as extended to gas mixtures by Hill (1946), was the most satisfactory one for reproducing experimental data obtained by them. This is not surprising, since the pressures of interest are sufficiently high to cause multilayer adsorption. The Langmuir theory and the Polanyi potential theory, as applied by Lewis and co-workers (see Section IV.A), were found to be inadequate. One difficulty in using the BET equation here is that the temperatures of interest are above the critical temperatures for methane and ethane. For these two gases pseudo values of the saturation vapour pressure p_0 were obtained by extrapolation of their vapour pressure *versus* temperature ($\log p$ *versus* $1/T$) plots. To account for non-ideal behaviour of gases over the pressure ranges used, fugacities were used instead of pressures. With these two modifications the BET equation could be satisfactorily applied to the heavier components in natural gas mixtures. The details of the calculation procedure have been given in the paper of Mason and Cooke.

C. A TWO-DIMENSIONAL EYRING EQUATION OF STATE TO PREDICT ADSORPTION

Payne *et al.* (1968) have used a two-dimensional (2D) version of the equation derived by Eyring from a simplified cell model for liquids. The 3D Eyring equation (Hirshfelder *et al.*, 1954) is

$$P = \frac{RT}{\overline{V} - cb^{1/3}\overline{V}^{2/3}} - \frac{a}{\overline{V}^2} \tag{24}$$

where the constants a and b have roles similar to the van der Waals constants, and the constant c depends on the shape of the cell assumed for the fluid model. The 2D analogue of this equation is

$$\pi = \frac{RT}{\alpha - c_2 b_2{}^{1/2}\alpha^{1/3}} - \frac{a_2}{\alpha^2} \tag{25}$$

where π = 2D pressure exerted parallel to the gas–solid interface and α = molal adsorption area (reciprocal of moles adsorbed per unit area). The a_2 and b_2 constants are analogous to the corresponding 2D van der Waals constants, while c_2 depends ideally on the shape of an average 2D cell structure assumed for the fluid model and should vary with the adsorbent used. The adsorption isotherm for high pressures, derived from the above is

$$f^g = \frac{RT[K_f^* \exp(\overline{V}^s(P^g - P^0)/RT]}{(\alpha^{1/2} - c_2 b_2{}^{1/2})^2} . \exp\left(\frac{c_2 b_2{}^{1/2}}{\alpha^2 - c_2 b_2{}^{1/2}}\right) - \frac{2a_2}{RT\alpha} \tag{26}$$

where f^g = fugacity in gas phase, $\overline{V}^s$ = molal volume of surface phase (averaged over the various homogeneous sub-regions making up the surface), P^g = pressure of gas phase, P^0 = pressure of reference state ($P = 1$ atm), and K_f = equilibrium constant expressing ratio of fugacities in gas phase and in 2D surface phase parallel to gas–solid interface, the asterisk indicating that it is for ideal gas conditions.

Payne *et al.* (1968) have found that the mobile fluid model using the 2D Eyring equation is effective in developing high-pressure adsorption isotherms on heterogeneous surfaces for both pure components and mixtures. Their results for the absolute adsorption of methane on charcoal are shown in Fig. 25. Constants in the above equation can be obtained from low-pressure data; for simple molecules these can be calculated from Van der Waals-type interaction parameters. The analysis of these constants by the de Boer–Kruyer (1958) method suggests a more localized adsorption of propane than of either methane or butane. The constants serve as simple characterizing parameters which indicate the ordering properties of the surface (through the c term)

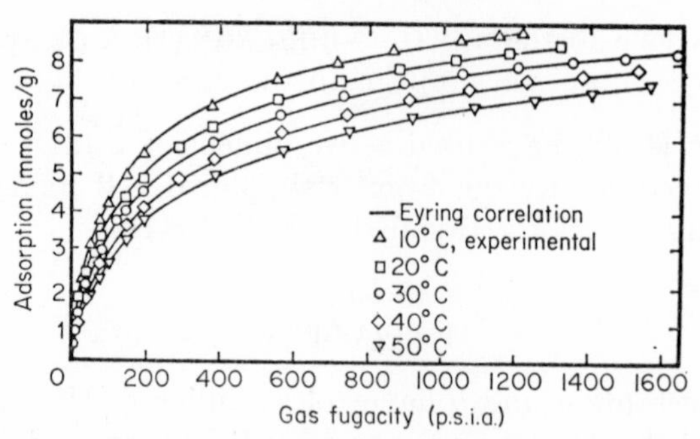

FIG. 25. The absolute adsorption of methane on charcoal and the 2D Eyring correlation of Payne *et al.* (1968). (Reproduced by permission of T. W. Leland and the American Chemical Society.)

and the distortion of the intermolecular potential on the surface (through the a_2 term). Furthermore, adsorption isotherms for mixtures can be calculated from the isotherms for pure components using relatively simple combining rules for the mixture equation of state constants. This allows direct computation of the total moles adsorbed and the surface-phase composition from pure-component adsorption data.

D. A 2D VIRIAL EQUATION OF STATE

As mentioned in Section II.F, the adsorption of methane, ethane and propane and their mixtures on silica gel has been extensively studied by Kobayashi and co-workers by gas chromatography combined with the tracer perturbation technique.

As a typical example of their results Fig. 26 shows the adsorption of

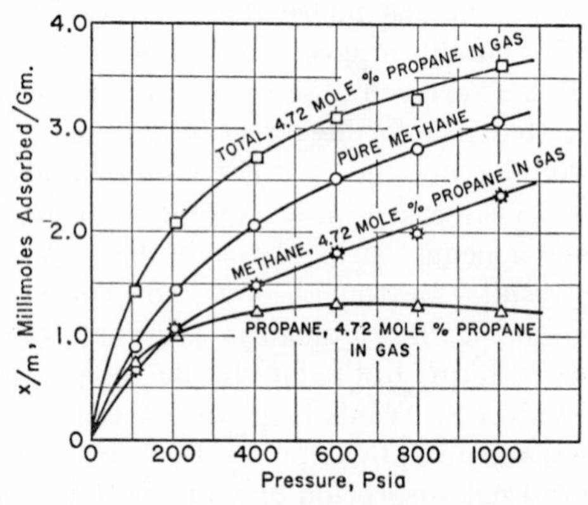

FIG. 26. Comparison of various adsorption isotherms on silica gel at 20°C, obtained by Gilmer and Kobayashi (1965). (Reproduced by permission of R. Kobayashi and Am. Inst. Chem. Engrs.)

methane and propane at 20°C from a mixture containing 4·72 mole percentage of propane (Gilmer and Kobayashi, 1965).

Haydel and Kobayashi (1967) fitted their mixed adsorption data for the methane–propane–silica gel system to a 2D virial equation of state for the adsorbed phase with the series truncated after the third virial coefficient. The 2D virial equation of state is expressed as:

$$\phi A/RT = w_1 + w_2 + B_{11}w_1^2 + B_{12}w_1w_2 + B_{22}w_2^2 + C_{111}w_1^3 + C_{112}w_1^2w_2 + C_{122}w_1w_2^2 + C_{222}w_2^3 + \ldots \tag{27}$$

where ϕ is the 2D spreading pressure, A the specific area of adsorbent, w_1 and w_2 the millimoles adsorbed per g adsorbent, of components 1 and 2, and the B and C terms the second and third virial coefficients. A 2D equation of state cannot be applied until it is converted into a form utilizing measurable physical quantities. This is done through the use of the Gibbs–Duhem equation for the adsorbed phase

$$[(Am)d\phi = \sum n_i d\mu_i]_{T,P} \tag{28}$$

which for a binary mixture can be written as

$$\left[\frac{A}{RT}\, d\phi = w_1 d \ln f_1 + w_2 d \ln f_2\right]_{T,P} \tag{29}$$

where m is the mass of adsorbent, n number of moles, and μ the chemical potential which can be related to the fugacity f of the gas. An isotherm for each component in a binary mixture can now be written as follows:

$$f_1 = w_1 \exp\left(A_1 + 2B_{11}w_1 + B_{12}w_2 + \tfrac{3}{2}C_{111}w_1^2 + C_{112}w_1w_2 + \tfrac{1}{2}c_{122}w_2^2 + \ldots\right) \tag{30}$$

$$f_2 = w_2 \exp\left(A_2 + 2B_{22}w_2 + B_{12}w_1 + \tfrac{3}{2}c_{222}w_2^2 + c_{122}w_1w_2 + \tfrac{1}{2}c_{112}w_1^2 + \ldots\right). \tag{31}$$

This introduces two more constants A_1 and A_2, which are related to the initial slope of each pure component isotherm for components 1 and 2 respectively and are characteristic of the adsorbate–adsorbent interaction while the virial coefficients are characteristic of the absorbate–absorbent interactions. Thus, there are a total of nine parameters evaluated from the experimental data. Each isotherm has six, three of which are repeated. The mixed virial is related to the pure-component virials by the so-called mixing rules, for example

$$\sqrt{B_{12}} = \tfrac{1}{2}(\sqrt{2B_{11}} + \sqrt{2B_{22}}). \tag{32}$$

As a typical example of the success of the above correlation, Fig. 27

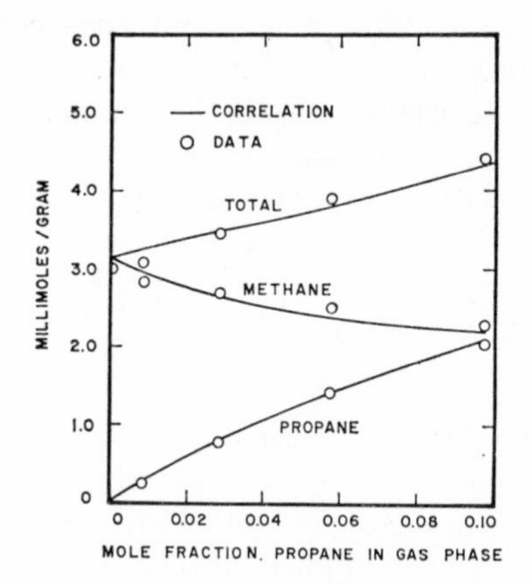

FIG. 27. The adsorption of methane and propane and their mixtures on silica gel at 20°C and 1000 p.s.i., both experimentally measured by Haydel and Kobayashi (1967) and calculated from a 2D virial equation. (Figure kindly supplied by R. Kobayashi.)

shows the adsorption of methane and propane and their mixtures on silica gel at 20°C and 1000 p.s.i., both calculated as well as experimentally measured by Haydel and Kobayashi (1967). The results can also be expressed in terms of K values (=gas-phase concentration/adsorbate concentration) or of selectivity coefficients (=ratio of methane and propane K values). The large value of the selectivity coefficient indicates that propane has a much greater affinity for silica gel than methane and hence a mixture of these two hydrocarbons can be readily separated by adsorption on silica gel.

E. SEPARATION AND PURIFICATION OF GASES

The potentialities of fractional separation of hydrocarbons by selective adsorption at high pressures seem to be very high in the processing of petroleum and natural gases. This has led to several new investigations in this field during the last five years. The work of Kobayashi and co-workers, Mason and Cooke, and Payne and his co-workers have already been mentioned earlier (Sections II.F, III.B–D). Martin (1964) has patented a cyclic adsorption process for fractionating mixtures by selective adsorption. Based on studies up to 50 atm pressure, Rasulov and Velikovskii (1965) found charcoal and zeolite as the most effective adsorbents for short-cycle extracting columns for C_2–C_5 hydrocarbons. Marks *et al.* (1963) studied adsorption rates in an experimental unit designed to duplicate conditions existing in the treatment of high-pressure natural gas mixtures; a simplified mathematical model, together with a correlation for a parameter describing the absolute rate of adsorption, was shown to be useful for describing the fixed-bed adsorption process.

From a study of the isotherms of CH_4, CO and N_2 and binary mixtures of these with H_2 on active carbon at 20° and pressures up to 100 atm, Zhukova and Kel'tsev (1959) found that the presence of H_2 decreased the adsorption capacity of the other gases. This study led to a purification method (Kel'tsev and Zhukova, 1961) for H_2 in which a moving bed of activated carbon yields H_2 of purity 99·9% or better. Simpson and Cummings (1964) developed an equation for predicting the performance of a solid adsorbent in a dynamic system. As a test case they studied the drying of air on silica gel at 30–120°F, gas velocity 10–90 ft/min, pressure 40–280 p.s.i. and concentration of moisture in air 10–90% of the saturation value. Based on the results of high-pressure adsorption studies (Czaplinski and Zielinski, 1958) of He, Ne and H_2 on active carbon at liquid nitrogen temperature, Zielinski (1959) developed a process for separation of He and Ne from He–Ne concentrates containing N_2 and H_2.

V. Miscellaneous

A. Solubility of Gases in Polymers

Rubber and polymers are capable of absorbing large quantities of gases at high pressures. This is a bulk effect (as distinct from the purely surface phenomena encountered in adsorption), comparable to the

Table III. Some Investigations on Gas–Polymer Solubility

Gas	Solid	Temp. (°C)	Pressure range (atm)	Reference
H_2	rubber		550 and 1150	Tammann and Bochow (1928)
N_2, H_2	polystyrene	120	250	Newitt and Weale (1948)
CH_4, N_2	polyethylene	125–228	650	Lundberg *et al.* (1962)
CH_4	polystyrene	100–188	350	Lundberg *et al.* (1962)
C_2H_6, C_3H_8, C_4H_{10}, C_5H_{12}	polyethylene	90–160	2000	Ehrlich and Kurpen (1963)
C_2H_4	polyethylene	200	2000	Ehrlich (1965)
C_2H_4	polyethylene	100–140	2000	Diepen *et al.* (1965)
C_2H_4	polyethylene	theoretical work		Cernia and Mancini (1965)

solubility of gases in liquids. A complimentary equilibrium of great theoretical and technical interest, the solubility of polymers in compressed gases, has also been studied in some cases. A list of some important investigations in these two fields is given in Table III.

B. SOLUBILITY OF GASES IN METALS

At high pressures hydrogen can diffuse into and dissolve in metals and the hydrogen-embrittlement of steel is well known. Metal–hydrogen equilibria at high pressures have not been much investigated. It is usually difficult to obtain reproducible results because of variations in the micro-structure of the metal and the effects of surface contamination.

The only system which has been investigated in detail is the palladium–hydrogen equilibrium in three distinct ranges by three groups of workers:

(1) Two-phase region (160–313°C, up to 40 atm) Gillespie and Hall (1926) and Gillespie and Gelstaun (1936).
(2) Low-temperature region (– 78 to 100°C, up to 1000 atm) Perminov *et al.* (1952).
(3) High-temperature region (326–477°C, up to 1000 atm) Levine and Weale (1960).

The isotherms of the palladium–hydrogen system obtained in the above three investigations are shown in Fig. 28. The atomic ratio H/Pd at 1000 atm varies from 0·61 at 477° and 0·69 at 326° to 0·92 at – 78°C.

Extrapolation of the data indicates that pressures of the order of 10^5 to 10^6 atm will be required to attain the composition H/Pd = 1.

The sorption of hydrogen by palladium is an exothermic process. Levine and Weale (1960) have calculated the isosteric heats of sorption ($-\Delta H$) as a function of the H/Pd ratio. The decrease in $-\Delta H$ is suspected to be caused by an increasing expansion of the metal lattice.

The electrical resistance of the solid phase has also been measured by Levine and Weale (1960) at 366, 396 and 456°C and at hydrogen pressures up to 1000 atm. The resistance increases with H content until a maximum value of the relative resistance $R/R_0 = 1{\cdot}49$ is reached at H/Pd value of about 0·4. As H/Pd is increased further, R/R_0 decreases and then levels off between 1·368 and 1·395. There is a remarkable coincidence of the maxima on the isotherms at the three temperatures. This behaviour is in contrast to the continuous increase of R/R_0 up to H/Pd = 0·8 at 25°C, found by Flanagan and Lewis (1959). The existence of the maximum has been attributed to the onset of co-conduction by the sorbed hydrogen which at higher temperatures begins well before the *d*-band is filled.

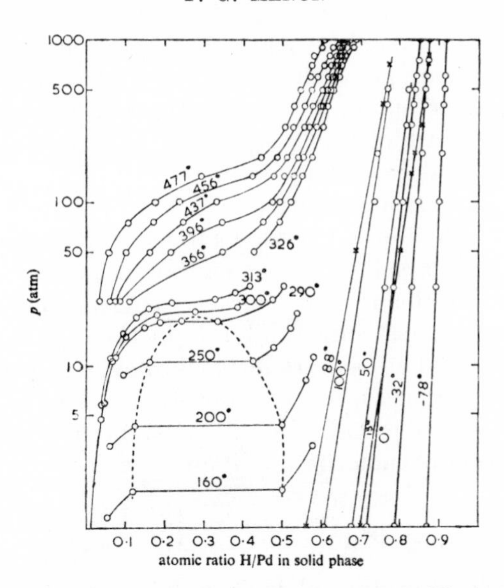

FIG. 28. The hydrogen–palladium system at high pressures. The isotherms in the linear low-temperature region are of Perminov and co-workers, those in the two-phase region are of Gillespie and his co-workers, and those at 326–477°C are of Levine and Weale. (Reproduced by permission of K. E. Weale and the Faraday Society.)

C. ADSORPTION AT WATER-GAS INTERFACE

Using a pendant drop method Hough *et al.* (1952) determined interfacial tensions in the water–He, water–CH_4 and water–N_2 interfaces. The adsorption of He, CH_4 and N_2 at the interface was estimated from the isothermal change of interfacial tension with pressure, by the application of the Gibbs adsorption isotherm. The pressure range covered was 1–1000 atm at temperatures of 27–138°C. The existence of at least monolayer concentration at high pressures was indicated for the water–N_2 and water–CH_4 systems. The presence of plastic films in these two systems was observed at high pressures (presumably due to the formation of hydrates), and this was taken as an indication that the actual values of surface concentration might be considerably larger than the lower limit values calculated from the experimental data.

D. ADSORPTION IN EVERYDAY LIFE

Adsorption does appear to be involved in numerous everyday problems and processes: respiration (transport of gases from the atmosphere *via* lungs to the blood and *vice versa*), air pollution (effect on organism and ways to reduce it), cancer research and the effect of smoking, fumigation procedures involving destruction of micro-organisms and their spores by sterilization with gases, and so on. Scientists working in radiation research are looking for specific adsorbents to be used as decontaminants. In food science and technology, the techniques for food dehydration have been very much improved in recent years and dehydrated foods are gaining increased importance not only as foods for special purposes (military services, tourism, space flight) but also in the regular market. Adsorption, together with diffusion, plays the most important role in the stability and quality (shelf-life) of dehydrated food products (Boskovic, private communication). These and other aspects of adsorption phenomena in everyday life have been reviewed recently by Giles (1964). The effect of pressure on these adsorption phenomena is still an unexplored field.

E. ADSORPTION IN CELLULAR PROCESSES

The role which adsorption plays in cellular processes has not been extensively studied, although it is very likely that living cells and their various components are involved in adsorption somewhere along the line. It is a well-known fact that in living cells certain chemical reactions occur which *in vitro* occur at only high pressures. If this means that cells are capable of developing high pressures even in very minute regions, then adsorption at high pressures also becomes worth considering. Since it is now known that cells do develop localized chemical gradients, the idea of localized high pressures in cells also seems to be a possibility (Norton, private communication).

F. HYPERBARIC OXYGEN IN MEDICINE AND SURGERY

The advantages and also the difficulties of using oxygen at elevated pressure (1–3 atm) in medicine and surgery have been recognized in recent years (Boerema, 1964). In hyperbaric oxygen exposure, oxygen after fully saturating haemoglobin goes in solution in the plasma of the blood. At a sufficiently high arterial partial pressure of oxygen, the oxygen requirements can be met entirely by the dissolved oxygen. Boerema (1959) demonstrated this possibility with pigs. The circulating volume of blood was exchanged with macrodex. The animals survived the removal of haemoglobin when breathing oxygen at ambient pressure of 3 atm. An operating room with high atmospheric pressure has also been described by Boerema (1961).

The two main toxic effects of breathing oxygen in high concentrations upon animals and human beings are convulsions and symptoms of pulmonary edema. In oxygen breathing at pressures above 3 atm convulsions dominate the picture, while exposure at moderate pressures (1–3 atm) results in pulmonary pathology without convulsions. The pulmonary pathology, called the Lorrain Smith effect, is usually the fatal complication. The pathologenesis of oxygen poisoning is still not clear. However, the importance and irreversibility of the Lorrain Smith effect have been widely recognized (cf Boerema 1964). The sequence of events leading to this pulmonary damage are: first vascular congestion with alveolar exudation and haemorrhages, and thereafter bronchial obstruction and atelectasis, resulting in the typical appearance of Lorrain Smith lungs.

The biochemical basis of oxygen poisoning is also not fully elucidated although many protectives, some of them anti-oxidants like Vitamin E, seem to delay the onset of symptoms. Vitamin C is also considered as an antioxidant and is moreover required to maintain capillary permeability.

The sorption (adsorption and absorption) by the lung tissues of oxygen at elevated pressures must be playing a very important part in the above crucial problems of hyperbaric oxygen therapy and surgery. Recently Dattatreya (1967) has found that both Vitamin C and the humidification of oxygen served as protectives in hyperbaric oxygen poisoning in guinea pigs. This protective action of water vapour may perhaps be due to its preferential sorption by the lung tissues and consequently a suppression of their capacity for excessive oxygen sorption under hyperbaric conditions.

References

Audibert, E. (1935). *Annls Mines Carbltr., Paris* **13**(8), 225.
Audibert, E. (1942). *Annls Mines Belg.* **14**(1), 71.
Beckmann, F. (1954). *Brennst.-Chem.* **35,** 6.
Boehlen, B. and Guyer, A. (1964). *Helv. chim. Acta* **47,** 1815.
Boehlen, B., Hausmann, W. and Guyer, A. (1964). *Helv. chim. Acta* **47,** 1821.
Boerema, I. (1959). *Arch. Chir. Neerl.* **11,** 70.
Boerema, I. (1961). *Surgery* **49,** 291.
Boerema, I. (1964). (ed.) "Clinical Application of Hyperbaric Oxygen", Elsevier, Amsterdam.
Boskovic, M. Private communication.
Cernia, E. M. and Mancini, C. (1965). *J. Polym. Sci.* B**3**, 1093.
Coolidge, A. S. and Fornwalt, H. J. (1934). *J. Am. chem. Soc.* **56,** 561.
Coppens, L. (1936). *Annls Mines Belg.* **37,** 173.
Coppens, L. (1937). *Annls Mines Belg.* **38,** 137.
Czaplinski, A. (1965). *Archwm Gorn.* **10,** 239.

Czaplinski, A. and Lason, M. (1965). *Archwm Gorn.* **7,** 53.
Czaplinski, A. and Zielinski, E. (1958). *Przem. chem.* **37,** 640.
Czaplinski, A. and Zielinski, E. (1959). *Przem. chem.* **38,** 87.
Dattatreya, R. M. (1967). " Vitamin C and Humidification of Oxygen as Protectives in Oxygen Poisoning (Lorrain Smith Effect) ", Doctorate Thesis in Medicine. University of Amsterdam.
de Boer, J. H. and Kruyer, S. (1958). *Trans. Faraday Soc.* **54,** 540.
de Braaf, W., Itz, G. N. and Maas, W. (1952). *C.R. 3rd Congr. Stratigr. Geol. Carb.* (*Heerlen*) **1,** 51.
Diepen, G. A. M., Swelheim, T. and des Arons, J. (1965). *Recl Trav. chim. Pays-Bas* **84,** 261.
Dubinin, M. M. (1960). *Chem. Rev.* **60,** 235.
Dubinin, M. M. and Radushkevich, L. V. (1947). *Dokl. Akad. Nauk SSSR* **55,** 327; also Dubinin, M. M. (1960). *Chem. Rev.* **60,** 235.
Dubinin, M. M., Bering, S. P., Serpinski, V. V. and Vasil'ev, B. N. (1958). *In* " Surface Phenomena in Chemistry and Biology ", p. 172–188. Pergamon Press, New York.
Eberly, P. E. (1961). *J. phys. Chem.* **65,** 1261.
Edwards, J. and Maass, O. (1935). *Can. J. Res.* **13**B, 133.
Ehrlich, P. (1965). *J. Polym. Sci.* **A3,** 131.
Ehrlich, P. and Kurpen, J. J. (1963). *J. Polym. Sci.* **A1,** 3217.
Flanagen, T. B. and Lewis, F. A. (1959). *Trans Faraday Soc.* **55,** 1400.
Frohlich, P. K. and White, A. (1930). *Ind. Engng Chem.* **22,** 1058.
Giles, C. H. (1964). *Chem. Inds*, Lond. **724,** 760.
Gillespie, L. J. and Gelstaun, L. S. (1936). *J. Am. chem. Soc.* **58,** 2656.
Gillespie, L. J. and Hall, F. P. (1926). *J. Am. chem. Soc.* **48,** 1207.
Gilmer, H. B. and Kobayashi, R. (1964). *A. I. Ch. E. Jl* **10,** 797.
Gilmer, H. B. and Kobayashi, R. (1965). *A. I. Ch. E. Jl* **11,** 702.
Gonikberg, M. G. (1960). " Chemical Equilibria and Reaction Rates at High Pressures." English Translation (1963). Israel Programme for Scientific Translations, Jerusalem.
Green, M. S. and Sengers, J. V. (1966). (ed.) " Critical Phenomena—Proceedings of a Conference Held in Washington, D.C., April 1965 ". National Bureau of Standards, Washington, Miscellaneous Publication 273.
Haley, A. J. (1963). *J. appl. Chem.*, Lond. **13,** 392.
Haydel, J. J. and Kobayashi, R. (1967). *Ind. Engng Chem. Fundam.* **6,** 546.
Hill, T. L. (1946). *J. chem. Phys.* **14,** 268.
Hirschfelder, J. O., Curtiss, C. F. and Bird, R. B. (1954). " The Molecular Theory of Gases and Liquids ", p. 5. John Wiley, New York.
Hough, E. W., Wood, B. B. and Rzasa, M. J. (1952). *J. phys. Chem.* **56,** 996.
Jones, W. M. and Evans, R. E. (1966). *Trans. Faraday Soc.* **62,** 1596.
Jones, W. M. and Isaac, P. J. (1959). *Trans. Faraday Soc.* **55,** 1947.
Jones, W. M., Isaac, P. J. and Phillips, D. (1959). *Trans. Faraday Soc.* **55,** 1953.
Kel'tsev, N. V. and Zhukova, Z. A. (1961). *Trudy vses. nauchn.-issled. Inst. prir. Gazov.* **12,** 143; *Chem. Abstr.* (1961) **57,** 12287.
Khodot, V. V. (1948). *Izv. Akad. Nauk SSSR, Otd. Tekh. Nauk* 733.
Khodot, V. V. (1949). *Izv. Akad. Nauk SSSR, Otd. Tekh. Nauk* 991.
Khodot, V. V. (1951). *Izv. Akad. Nauk SSSR, Otd. Tekh. Nauk* 1085.
Kini, K. A. (1963). *Fuel*, Lond. **42,** 103.
Kini, K. A. (1964). *Fuel*, Lond. **43,** 173.

Koonce, K. T., Deans, H. A. and Kobayashi, R. (1965). *A. I. Ch. E. Jl* **11,** 259.
Krichevskii, I. R. and Kal'varskaya, R. S. (1940). *Acta Phys-chim. URSS* **13,** 49.
Lederman, P. B. and Williams, B. (1964). *A. I. Ch. E. Jl* **10,** 30.
Levine, P. L. and Weale, K. E. (1960). *Trans. Faraday Soc.* **56,** 357.
Lewis, W. K., Gilliland, E. R., Chertow, B. and Cadogan, W. P. (1950). *Ind. Engng Chem.* **42,** 1319, 1326.
Lundberg, J. L., Wilk, M. B. and Huyett, M. J. (1962). *J. Polym. Sci.* **57,** 275.
McBain, J. W. and Britton, G. T. (1930). *J. Am. chem. Soc.* **52,** 2198.
Marks, D. E., Robinson, R. J., Arnold, C. W. and Hoffmann, A. E. (1963). *J. Petrol. Technol.* **15,** 443.
Martin, H. Z. (1964). U.S. Patent 3,149,934; (1964). *Chem. Abstr.* **61,** 12969.
Mason, J. P. and Cooke, C. E. (1966). *A. I. Ch. E. Jl* **12,** 1097.
Masukawa, S. and Kobayashi, R. (1968). *J. chem. Engng Data* **13,** 197.
Menon, P. G. (1961). " Adsorption of Nitrogen and Carbon Monoxide on Alumina up to 3000 Atmospheres Pressure ", Doctorate Thesis, Technological University, Delft.
Menon, P. G. (1965). *J. Am. chem. Soc.* **87,** 3057.
Menon, P. G. (1968a). *J. phys. Chem.* **72,** 2695.
Menon, P. G. (1968b). *Chem. Rev.* **68,** 277.
Michels, A. (1923). *Ann. Phys.* **72,** 285; *Proc. K. ned. Akad. Wet.* **26,** 805.
Michels, A. (1924). *Ann. Phys.* **73,** 577; *Proc. K. ned. Akad. Wet.* **27,** 930.
Michels, A. and Michels, C. (1935). *Proc. R. Soc.* Lond. A**153,** 201.
Michels, A., Menon, P. G. and Ten Seldam, C. A. (1961). *Recl Trav. chim.* **80,** 483.
Michels, A., Michels, C. and Wouters, H. (1935). *Proc. R. Soc.* Lond. A**153,** 214.
Moffat, D. H. and Weale, K. E. (1955). *Fuel,* Lond. **34,** 349.
Morris, H. E. and Maass, O. (1933). *Can. J. Res.* **9,** 240.
Nelsen, F. M. and Eggertsen' P. T. (1958). *Anal. Chem.* **30,** 1387.
Newitt, D. M. and Weale, K. E. (1948). *J. chem. Soc.* 1541.
Norton, D. A. Private communication.
Palvelev, V. T. (1948). *Dokl. Akad. Nauk SSSR* **62,** 779; (1949). *Chem. Abstr.* **43,** 1237.
Payne, H. K., Sturdevant, G. A. and Leland, T. W. (1968). *Ind. Engng Chem. Fundam.* **7,** 363.
Perminov, P. S., Orlov, A. A. and Frumkin, A. N. (1952). *Dokl. Akad. Nauk SSSR,* **84,** 749.
Rasulov, A. M. and Velikovskii, V. S. (1965). *Gaz. Prom.* **10,** 45; (1965). *Chem. Abstr.* **62,** 9831.
Ray, G. C. and Box, E. O. (1950). *Ind. Engng Chem.* **42,** 1315.
Ross, S. and Olivier, J. P. (1964). " On Physical Adsorption ", pp. 72–75. Interscience Publishing, New York.
Sastri, M. V. C. and Srikant, H. (1961). *J. scient. ind. Res.* (India) **20**D, 321.
Satterfield, C. N. and Cadle, P. J. (1968). *Ind. Engng Chem. Fundam.* **7,** 202.
Sengers, J. V. and Levelt-Sengers, A. (1968). *Chem. Engng News,* June 10, 104.
Simpson, E. A. and Cummings, W. P. (1964). *Chem. Engng Progr.* **60,** 57.
Stacy, T. D.. Hough, E. W. and McCain, W. D. (1968). *J. Chem. Engng Data,* **13,** 74
Stalkup, F. I. and Kobayashi, R. (1963). *A. I. Ch. E. Jl* **9,** 121.
Tammann, G. and Bochow, K. (1928). *Z. anorg. allg. Chem.* **168,** 322.
Temkin, M. I. (1950). *Zh. fiz. Khim.* **24,** 1312.
van der Sommen, J., Zwietering, P., Eillebrecht, B. J. M. and van Krevelen, D. W. (1955). *Fuel,* Lond. **34,** 444.

Vasil'ev, B. N. (1957). *Zh. fiz. Khim.* **31,** 498.
Vaska, L. and Selwood, P. W. (1958). *J. Am. chem. Soc.* **80,** 1331.
von Antropoff, A. (1952). *Kolloid.-Z.* **129,** 11.
von Antropoff, A. (1954a). *Kolloid.-Z.* **137,** 105.
von Antropoff, A. (1954b). *Kolloid.-Z.* **137,** 108.
von Antropoff, A. (1955). *Kolloid.-Z.* **143,** 98.
Weale, K. E. (1967). " Chemical Reactions at High Pressures." E. & F. N. Spon, London.
Worthington, R. (1958). *In* " Chemical Engineering Practice ". Ed. by H. W. Cremer, and T. Davies, **6,** 302. Butterworths, London.
Zhukova, Z. A. and Kel'tsev, N. V. (1959). *Trudy vses. nauchn.-issled. Inst. prir. Gazov.* **6,** 154; (1961). *Chem. Abstr.* **54,** 14633.
Zielinski, E. (1959). Polish Patent 42,225; (1961). *Chem. Abstr.* **55,** 6803.

Author Index

Numbers in italics are the pages on which the references are listed

H

I

L

M

T

U

V

W

Subject Index

G

N

O

P

T